Diagnosis and Management of Skin Diseases, Related Disorders and Their Comorbidities

Diagnosis and Management of Skin Diseases, Related Disorders and Their Comorbidities

Guest Editors

Alin Laurentiu Tatu
Lawrence Chukwudi Nwabudike

MDPI

Basel • Beijing • Wuhan • Barcelona • Belgrade • Novi Sad • Cluj • Manchester

Guest Editors
Alin Laurentiu Tatu
Clinical Department
Dunarea de Jos University
Galati
Romania

Lawrence Chukwudi
Nwabudike
National Institute of Diabetes
Nutrition and Metabolic
Diseases, N.C. Paulescu
Bucharest
Romania

Editorial Office
MDPI AG
Grosspeteranlage 5
4052 Basel, Switzerland

This is a reprint of the Special Issue, published open access by the journal *Diagnostics* (ISSN 2075-4418), freely accessible at: https://www.mdpi.com/journal/diagnostics/special_issues/skin_diseases.

For citation purposes, cite each article independently as indicated on the article page online and as indicated below:

Lastname, A.A.; Lastname, B.B. Article Title. *Journal Name* **Year**, *Volume Number*, Page Range.

ISBN 978-3-7258-4845-4 (Hbk)
ISBN 978-3-7258-4846-1 (PDF)
https://doi.org/10.3390/books978-3-7258-4846-1

Contents

About the Editors

Alin Laurentiu Tatu

Alin Laurentiu Tatu is a Full Professor of Dermatology and Venerology Clinical Department, Faculty of Medicine and Pharmacy, Dunarea de Jos University, Galati, Romania. He also works as a Chief of the Dermatology Department, Saint Parascheva Clinical Hospital of Infectious Diseases, Galati, Romania, and also is the Director of the Multidisciplinary Integrative Center for Dermatological Interface Research, Galati, Romania. His interests include demodex; endosymbionts; cutaneous microbiome; integrative medicine; optical coherence tomography; dermoscopy; confocal microscopy; cutaneous comorbidities; skin cancer; psoriasis; rosacea; atopic dermatitis; acne; hidradenitis suppurativa. He has expertise in a multitude of skin-imaging techniques and in various fields of specialist dermatology.

Lawrence Chukwudi Nwabudike

Lawrence Chukwudi Nwabudike is a senior dermatologist at the N. Paulescu National Institute of Diabetes in Bucharest, Romania. He also owns his own private practice, which is mainly focused on complementary and alternative therapies. He holds a Doctorate in Dermatology (Ph.D.) and is a Fellow of the Royal College of Physicians (FRCP). His areas of special interest are complementary and alternative therapies in dermatology, diabetic foot ulcer care, and general wound care, as well as the history of dermatology. He has published extensively in these areas and his works include journal articles and book chapters with internationally recognised publishers. He has acted as Guest Editor for several international dermatology journals.

diagnostics

MDPI

Editorial

Special Issue "Diagnosis and Management of Skin Diseases, Related Disorders and Their Comorbidities"

Alin Laurentiu Tatu [1,2,3,*] and Lawrence Chukwudi Nwabudike [4]

1 Clinical Department, Faculty of Medicine and Pharmacy, Dunarea de Jos University, 800008 Galati, Romania
2 Dermatology Department, Saint Parascheva Clinical Hospital of Infectious Diseases, 800179 Galati, Romania
3 Multidisciplinary Integrated Center of Dermatological Interface Research (MICDIR), 800010 Galati, Romania
4 N. Paulescu National Institute of Diabetes, 030167 Bucharest, Romania; chukwudi.nwabudike@live.com
* Correspondence: alin.tatu@ugal.ro

Dermatology is a continuously evolving specialty and touches on every part of the field of medicine. Our Special Issue is aimed at the examination of this continuous evolution and the vastness of our discipline. This Issue will be a collection and presentation of data on the latest developments in the field of dermatology and is intended to be a continuation of articles previously published by the Guest Editors [1,2].

Patients with lichen planus may suffer from mucosal disease, including oral lichen planus (OLP). In district hospitals, lichen planus in all its forms affects about 4% of patients. A single-centered Spanish study with a cohort of 275 OLP patients and 275 controls set out to determine if there was any link between OLP and prediabetes. The diagnosis of prediabetes was based on the American Diabetes Association's criterion of blood sugar levels of 100–125 mg/dL. The authors found that OLP was significantly associated with prediabetes and with oral antidiabetic medication (i.e., overt type 2 diabetes, as inferred from oral medication use). The study authors concluded that it would be reasonable to check glycemia levels in patients with OLP.

Non-melanoma skin cancers continue to be challenging to diagnose and treat, partly due to their heterogeneity as a group and partly due to their protean manifestations, which sometimes mimic those of other cutaneous disorders. A Romanian study set out to address these difficulties with the use of ultrasound. Their cohort comprised 31 patients with head and neck tumors, who were assessed using 13, 20, and 40 MHz ultrasound transducers, as well as histopathology. The authors found that the 13 MHz transducers offered larger, but less detailed images and may be more useful for large tumors. More detailed images were produced by the 20 and 40 MHz transducers, which were able to discern hyperechoic spots; this could be useful in the diagnosis of basal cell carcinomas, according to the study authors.

Pseudoxanthoma elasticum (PXE) is a rare autosomal recessive disorder, which often begins in the skin. Genetic testing may aid in diagnosis. A Hungarian cohort ($n = 5$) was assessed using autofluorescence (AF) to determine potential diagnostic features of PXE. The authors found that high-intensity AF signaling using a 450 nm LED light device was suitable for patients with PXE and concluded that this process might represent a useful non-invasive method for the early diagnosis of PXE.

Artificial intelligence is increasingly being used in diagnostics. Some authors trained a self-developed neural network to diagnose actinic keratosis. They found that their artificial intelligence program performed better than traditional convoluted neural networks, requiring only 123 s of training.

Psoriatic arthritis can be a challenging disorder to diagnose and to treat. Magnetic resonance imaging (MRI) is increasingly used in the diagnosis of this disorder, as opposed to traditional radiographs. MRI findings of psoriatic arthropathy were reviewed. Bone erosion, the inflammation of vertebral bodies, and spondylodiscitis were amongst the vertebral issues identified in the literature. In addition, ankylosis, joint erosion, enthesitis, capsular inflammation, and fat metaplasia were notable MRI findings at the sacroiliac joint level. These findings may aid dermatologists in the diagnosis of psoriatic arthritis without using overt skin lesions when there is a suspicion of disease.

A group of authors carried out an update on the clinical presentation, differential diagnosis, and diagnostic characteristics of atypical dermatofibroma. They were able to provide data to enable the differentiation of this disorder from other malignant disorders, especially amelanotic melanoma.

Correlations between atopic dermatitis and other skin disorders were identified in a review by another group of authors [3]. They also reviewed the clinical aspects of atopic dermatitis itself, including the ocular manifestations of the disorder, such as blepharitis, keratoconus, and retinal detachment. Correlations were also found between atopic dermatitis and cutaneous infections, as well as non-melanoma skin cancer and respiratory disorders.

In another paper, a case of metastatic melanoma treated with various therapy modes, including immune-checkpoint inhibitors, was presented. The authors reviewed the potential benefits to patients and the adverse effects of multimodal therapy on this complicated disorder, as well as potential approaches to these side effects [4].

A rare association of cutaneous melanoma and glioblastoma was presented as a case report, with a review of the literature. Potential genetic and familial causes were explored. Melanoma may have protean manifestations and multiple associations, as this case report and review shows. Its regressing phase may also resemble collision tumors, thereby intensifying the challenge.

A rare case of peripheral T-cell lymphoma not otherwise specified (PTCL-NOS) was reported. The diagnosis, therapy, and—happily—its subsequent remission were documented. The authors noted the potential for this form of lymphoma to be misdiagnosed as mycosis fungoides or Sézary syndrome. The importance of arriving at the right diagnosis cannot be understated, as the authors remind us, since PTCL-NOS can become fatal very rapidly, while mycosis fungoides is generally an indolent disorder.

The combination of a rare tumor with a more commonly encountered disorder was reported in another paper, namely a case of somatostatinoma and neurofibromatosis type 1. The patient in this case was a 50-year-old, the usual age for the manifestation of somatostatinoma. Metastases had occurred at the time of diagnosis. Neurofibromatosis has protean manifestations, including orthopedic ones. Thus, such combinations may prove challenging to diagnose.

This Special Edition offers 11 diverse and stimulating manuscripts. The significance of general internal medicine in dermatology is illustrated in a significant proportion of these articles. Lichen planus continues to be important in this regard and there are continuous updates to its diagnosis and increasing therapeutic possibilities at our disposal. These broadening therapeutic possibilities may also be applied in the therapy of psoriasis.

Funding: This research received no external funding.

Data Availability Statement: Not applicable.

Acknowledgments: The paper was academically supported by Dunarea de Jos University of Galati through the Multidisciplinary Integrated Center of Dermatological Interface Research MIC-DIR (Centrul Integrat Multidisciplinar de Cercetare de Interfata Dermatologica—CIM-CID), "Dunărea de Jos" University, Galați, Romania.

Conflicts of Interest: The authors declare no conflicts of interest.

References

1. Nwabudike, L.C.; Tatu, A.L. Reply to Gambichler T et al.: Altered epigenetic pathways and cell cycle dysregulation in healthy appearing skin of patients with koebnerized squamous cell carcinomas following skin surgery. *J. Eur. Acad. Dermatol. Venereol.* **2019**, *33*, e3–e4. [CrossRef] [PubMed]
2. Nicolescu, A.C.; Strilciuc, S.; Lăpădatu, R.; Grad, D.A.; Vlădescu, C.; Olteanu, R. The economic burden of atopic dermatitis in Romania: A broad perspective. *Front. Public Health* **2025**, *12*, 1531042. [CrossRef] [PubMed]
3. Munteanu, A.C.; Toader, M.P.; Zemba, M.; Porumb, V.; Cozmin, M.; Moraru, A.D.; Nicolescu, A.C.; Brănișteanu, D.C. Differences and Similarities in Epidemiology and Risk Factors for Cutaneous and Uveal Melanoma. *Medicina* **2023**, *59*, 943. [CrossRef] [PubMed]
4. Li, R.; Mu, X.; Liu, Z.; Huang, R.; Peng, X. Association of type 2 diabetes, hypertension, and hyperlipidemia with immune-related adverse events in patients undergoing immune checkpoint inhibitors therapy. *Front. Immunol.* **2025**, *16*, 1472197. [CrossRef] [PubMed]

diagnostics

MDPI

Article

Risk of Prediabetes and Diabetes in Oral Lichen Planus: A Case–Control Study according to Current Diagnostic Criteria

Lucía Rodríguez-Fonseca [1], Santiago Llorente-Pendás [2] and María García-Pola [1,*]

[1] Department of Surgery and Medical-Surgical Specialties, Faculty of Medicine and Sciences of the Health, Oviedo University, 33006 Oviedo, Spain
[2] Private Practice, 33004 Oviedo, Spain
* Correspondence: mjgarcia@uniovi.es; Tel.: +34-985-103635

Abstract: Objective: To estimate the prevalence of prediabetes and diabetes in patients with oral lichen planus (OLP). Methods: Prospective cohort, including consecutive patients diagnosed clinically and histologically with OLP from 2018 to 2022. Patients and controls were matched by age and gender. Fasting plasma glucose value collection from all patients. Multivariate regression analysis evaluated the relationship between prediabetes and diabetes variables according to current diagnostic criteria. Results: The sample comprised 275 patients (207 women; 75.3%), mean age 59.60 ± 12.18 years for both groups. Prediabetes was diagnosed according to the American Diabetes Association (ADA, 100–125 mg/dL), in 21.45% of OLP patients (59/275) and 14.55% (40/275) of control patients ($p = 0.035$). Patients with the atrophic-erosive form exhibited stronger association with taking oral antidiabetics ($p = 0.011$). Multivariate analysis showed that being over >60 years and having a cutaneous location was associated with ≥3 sites (OR 1.81 and OR 2.43). ADA prediabetes and oral antidiabetics drugs increased the probability of OLP (OR 1.60 (1.04–2.51), $p = 0.03$ and OR 2.20 (1.18–4.69), $p = 0.017$) after adjustment for sex and age. Conclusions: Because glycemia 100–125 mg/dL was associated with OLP, testing serum fasting plasma glucose seems reasonable in order to prevent development of diabetes and deal with possible complications until new studies are complete.

Keywords: oral lichen planus; lichen planus; prediabetes; diabetes; oral antidiabetic; comorbidity

1. Introduction

Lichen planus (LP) is an inflammatory, mucocutaneous, chronic disease with periods of remission and exacerbation. LP affects the skin, nails, scalp, and mucous membranes, especially oral and genital [1]. Oral lichen planus (OLP) has a worldwide prevalence of 1.01%, is more common in women, and is increasingly common after the age of 40 [2].

Diagnosis is made by the presence of papular and reticular lesions with a more or less symmetrical location whether accompanied by any of the other clinical forms or not. It must be corroborated by microscopic study confirming the sub-basal inflammatory infiltrate, the liquefactive degeneration of the basal layer, and the absence of dysplasia [3]. OLP can present in association with other clinical forms, such as atrophic, erosive, or bullous forms. These non-keratotic forms appear more frequently with symptoms of burning or pain than keratotic forms such as papular, reticular or plaque-like [4,5].

Pathogenically, there is an immunological basis that is affected by various factors, to which oral commensal flora has recently been added [6]. In the oral inflammatory process, keratinocytes from OLP lesions produce proinflammatory cytokines. Additionally, in the pathogenic mechanism, the interaction of an antigen-presenting dendritic cell would initiate activation of cytotoxic CD8+ cells and CD4+ cells that differentiate into Th1 and Th17 cells. The complex protein cascade of cytokines, chemokines and adhesion molecules results in the promotion of apoptosis of keratinocytes and mucosal basement membrane destruction [7].

OLP is associated with viral infections and systemic comorbidities. In a general sense, Liu et al. highlighted the association with hepatitis C, diabetes mellitus (DM), thyroid diseases, dyslipidemias, cardiovascular diseases, and anxiety [8].

Diabetes is a heterogeneous group of diseases defined by chronically elevated blood glucose levels. The typical subclassification of diabetes includes type 1 diabetes (previously known as insulin-dependent, juvenile or childhood-onset diabetes), characterized by deficient insulin production in the body, and type 2 diabetes (non-insulin-dependent or adult-onset diabetes), resulting from the body's ineffective use of insulin. Type 1 diabetes accounts for approximately 5–10% of all diabetes, whereas type 2 diabetes accounts for 90–95% [9]. The number of new cases of type 2 diabetes increases in the over 55s [10].

The pooled prevalence of DM amongst OLP patients has been indicated as 2.72%, with the hypothesis of being associated with autoimmune diseases such as DM type −1 [11]. A recent meta-analysis highlighted that the prevalence of diabetes in patients with OLP ranged from 1.62 for type 1 to 9.77 for non-specified diabetes [12]. That meta-analysis included a large number of varied study designs, including retrospective, prospective, cross-sectional, and longitudinal methodologies. Some studies collected data on diabetes through medical records, questionnaires, and interviews, while others had varying interpretations of diabetes based on the classic World Health Organization (WHO) criteria or the authors' own criteria, among others.

Despite many studies having examined the prevalence of diabetes in OLP, only some reported statistically significant differences in glycemia levels compared to the normal population [13–17], while others found no differences [5,18–23]. Even fewer studies have indicated the link between OLP and prediabetes [24,25]. Therefore, our objective is to determine the prevalence and risk of prediabetes and diabetes in patients with OLP and control subjects according to the most current diagnostic criteria for prediabetes and diabetes based on serum parameters.

2. Materials and Methods

The present case–control study was designed according to the Strengthening Reporting of Observational Studies in Epidemiology (STROBE) guidelines (Supplementary Materials Table S1) [26]. The study was approved by the Ethics Committee of the Principality of Asturias as conforming to data protection regulation and the Declaration of Helsinki.

2.1. Patient Selection

The case group consisted of 275 OLP patients attending the Oral Medicine Unit at Oviedo University between January 2018 and November 2022. The sample size was calculated on the basis of a single proportion formula by the Epidat program with a 95% confidence interval (CI). Inclusion criteria for diagnosis of OLP were considered clinically and histologically. The clinical criteria included the presence of bilateral, mostly symmetrical lesions; the presence of a lace-like network of slightly raised gray-white lines in a reticular pattern; and by erosive, atrophic, bullous and plaque-type lesions [27]. Histopathological diagnosis by modified WHO criteria included: (1) well-defined band-like zones of cellular infiltration confined to the superficial part of the connective tissue; (2) "liquefactive degeneration" in the basal layer; and (3) the absence of epithelial dysplasia [3].

The protocol for clinical findings included: sex, age, daily tobacco and alcohol consumption, ex-smoker or ex-drinker, medical history, and prescription medication focusing on treatment of diabetes. The characteristics of OLP included a record of sites of involvement (cutaneous, genital, and scalp), predominant type of OLP (non-atrophic-erosive and atrophic-erosive), and number of sites (two sites or three or more sites).

The criteria for exclusion from the case group included: oral lichenoid lesion [28,29]; being under 18 years old; pregnant or breastfeeding women; and patients treated with topical or systemic corticosteroids, radiotherapy, or cancer chemotherapy.

The control group consisted of 275 consecutive patients who were examined in the same Oral Medicine Unit during the same period for other types of benign oral pathology-such as traumatic or benign tumors without any sign or history of OLP (n = 185) or for general oral health check-ups (n = 90). The controls were matched for age and sex to the patients with OLP.

Criteria for exclusion from the control group included: being under 18, pregnant or breastfeeding patients, and patients treated with radiotherapy or cancer chemotherapy.

2.2. Determination of Serum Parameters and Drugs

Glucose levels were examined in samples drawn before 10 a.m. after a 12 h fasting period. Serum parameters were collected after a first-day exploration of the mouth. The reference values for the parameters were based on: (1) the American Diabetes Association (ADA) prediabetes definition (100–125 mg/dL) [11]; (2) the WHO prediabetes definition (110–125 mg/dL) [12]; and (3) the WHO or ADA diabetes definition (≥126 mg/dL) [30,31].

Data was collected about drugs for treating diabetes, including insulin and oral antidiabetic drugs.

2.3. Statistical Analysis

Statistical analysis was done using R (R Development Core Team, version 4.1.3), and the information was recorded in an Excel file. Initially, we conducted a descriptive study of all of the variables. First, mean age was considered and subsequently split into two categories, selecting 60 years and under or over 60 years as break points. The predominant clinical form was classified as non-atrophic-erosive or atrophic-erosive (Figures 1 and 2). The location of OLP was recorded as two locations or three or more locations. The ADA prediabetes, WHO prediabetes, WHO diabetes variables and medication were recorded as either present or absent. We added the variable type of treatment and recorded whether treatment was insulin-dependent or used antidiabetic drugs. Student's t-test was used to compare the numerical variables between the two groups. The relationships between qualitative variables were assessed by Pearson's chi-square test. The relationships were quantified by crude ratios and a multivariate logistic regression model produced using stepwise selection with variables whose p value was <0.20 in the univariate analysis and by age and sex. Results were considered statistically significant when $p < 0.05$.

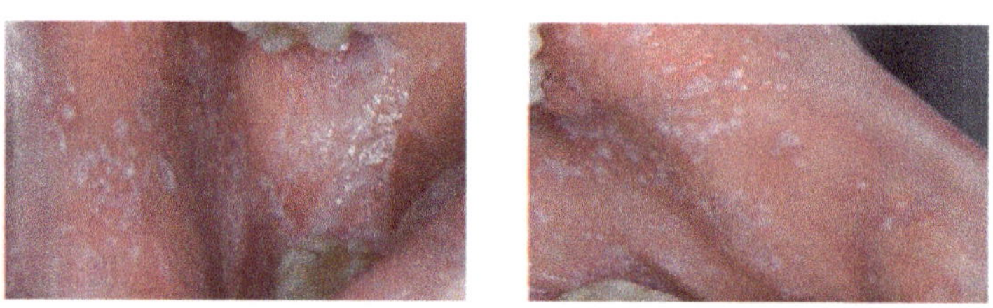

Figure 1. Clinical image of oral lichen planus of type non-atrophic-erosive located on both buccal mucosa.

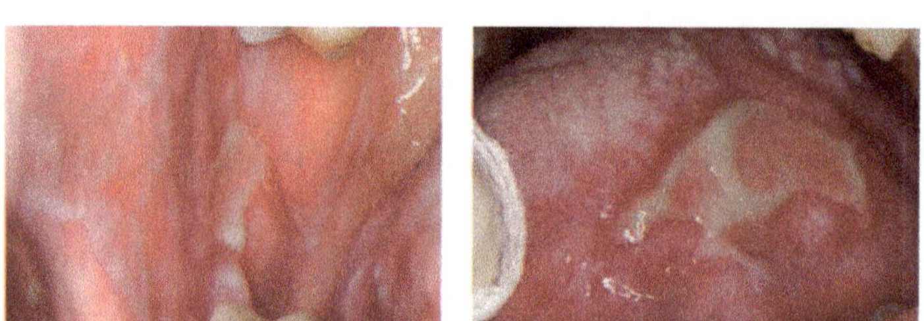

Figure 2. Clinical image of oral lichen planus of type atrophic-erosive located on buccal mucosa and lateral border of tongue.

3. Results

A total of 550 patients were analyzed (275 with OLP and 275 control patients). Both sample groups consisted of 207 women (75.27%) and 68 men (24.73%), and the average age was 59.60 ± 12.18 years (18–86). Table 1 shows the general characteristics of the case and control groups (Table 1).

Table 1. Characteristics of patients from control group and oral lichen planus group (OLP). Numbers in parentheses represent percentages. [1]: In addition to oral location. ADA (American Diabetes Association) (100–125 mg/dL). WHO prediabetes: World Health Organization (110–125 mg/dL). *: Statistically significant.

Variable		OLP	Control	*p* Value
Sex	Female	207 (75.27)	207 (75.27)	1
	Male	68 (24.73)	68 (24.73)	
Age	Mean (SD)	59.61 (12.19)	59.61 (12.19)	1
	≤60	143 (52.0)	143 (52.0)	1
	>60	132 (48.0)	132 (48.0)	
Tobacco user	No	203 (73.82)	193 (70.18)	0.09
	Yes	28 (10.18)	45 (16.36)	
	Ex-smoker	44 (16.0)	37 (13.45)	
Alcohol drinker	No	221 (80.36)	223 (81.09)	0.956
	Yes	52 (18.91)	52 (18.91)	
	Ex-drinker	2 (0.73)		
Clinical-form OLP				
	Non-atrophic-erosive	98 (35.64)		
	atrophic-erosive	177 (64.36)		
Oral location	2	246 (89.45)		
	≥3	29 (10.55)		
Cutaneous [1]	No	246 (89.45)		
	Yes	29 (10.55)		
Genital [1]	No	253 (92.0)		
	Yes	22 (8.0)		
Scalp [1]	No	268 (97.45)		
	Yes	7 (2.55)		
Glycemia	(mean, mg/dL (SD)	95.45 ± 17.62	98.15 ± 21.32	0.106
ADA prediabetes	No	235 (85.45)	216 (78.55)	0.035 *
	Yes	40 (14.55)	59 (21.45)	
WHO prediabetes	No	253 (92.0)	259 (94.18)	0.313
	Yes	22 (8.0)	16 (5.82)	
Diabetes (≥126 mg/dL)	No	262 (95.27)	251 (91.27)	0.061
	Yes	13 (4.73)	24 (8.73)	
Treatment of diabetes	No	257 (93.45)	243 (88.36)	0.038 *
	Yes	18 (6.55)	32 (11.64)	
Type of treatment	Insulin	5 (1.82)	4 (1.45)	0.015 *
	Oral antidiabetic	13 (4.73)	28 (10.18)	

Non-atrophic-erosive OLP was recorded in 98 patients (35.64%), and atrophic-erosive recorded in 177 (64.36%). Two intraoral sites were recorded in 139 patients (50.5%) and three or more sites were recorded in 136 (49.45%) patients. Forty-five patients (16.36%) exhibited extraoral manifestations.

The mean glucose level in the patients with OLP was 98.15 ± 21.32 and 95.45 ± 17.62 in the control group without statistically significant differences ($p = 0.106$). The prevalence of prediabetes according to ADA criteria was 21.45% (59/275) in the OLP group and 14.55% (40/275) in the control group [$p = 0.035$]. The prevalence of glycemic levels (≥126 mg/dL) was 8.73% (24/275) in the OLP group and 4.73% in the control group (13/275) [$p = 0.061$]. The prevalence of patients with treatment for diabetes was 11.64% (32/275) in the OLP group and 6.55% in the control group (18/275) [$p = 0.038$]. Therefore, despite receiving treatment, 15 (46.87%) of the 32 patients with diabetes still had a high glycemic level, as did 9 (50%) of the 18 subjects in the control group. A total of 9 (3.27%) patients with OLP and 4 (1.45%) from the control group did not know they were diabetics. The prevalence of patients with oral antidiabetic treatments was 10.18% (28/275) in the OLP group and 4.73% in the control group (13/275) [$p = 0.015$] (Table 1).

Table 2 shows the logistic regression analysis according to the clinical form (Table 2). The mean age of patients with atrophic-erosive forms of OLP was significantly higher (mean 60.97 yr; $p = 0.014$) than those with non-atrophic-erosive types (mean 57.15 yr), and in those over 60 years, the atrophic-erosive forms of OLP were more frequent ($p = 0.006$). There were no statistically significant differences in relation to clinical form or sex ($p = 0.946$) or location in genitals ($p = 0.396$), skin ($p = 0.891$) or scalp ($p = 0.694$).

Table 2. Statistically significant associations ($p < 0.05$ *) according to the logistic regression by clinical form. OR: Odd ratio. CI: confidence interval. Percentages in parentheses. ADA prediabetes: American Diabetes Association (100–125 mg/dL). WHO prediabetes: World Health Organization (110–125 mg/dL). [1]: In addition to oral location. SD: standard deviation.

Variable	Non-Atrophic-Erosive	Atrophic-Erosive	OR Univariate (CI, p Value)	OR Multivariate (CI, p Value)
Sex				
Female	74 (35.7)	133 (64.3)		
Male	24 (35.3)	44 (64.7)	0.98 (0.55–1.73, $p = 0.946$)	
Mean age (SD)	57.15 (11.68)	60.97 (12.39)	0.97 (0.95–0.99, $p = 0.014$) *	-
Age ≤ 60 years				
≤60 years	62 (43.4)	81 (56.6)		
>60 years	36 (27.3)	96 (72.7)	0.49 (0.29–0.81, $p = 0.006$) *	0.54 (0.32–0.91, $p = 0.021$) *
Genitals [1]				
No	92 (36.4)	161 (63.6)		
Yes	6 (27.3)	16 (72,7)	0.66 (0.23–1.66, $p = 0.396$)	-
Cutaneous [1]				
No	88 (35.8)	158 (64.2)		
Yes	10 (34.5)	19 (65.5)	0.94 (0.41–2.08, $p = 0.891$)	-
Scalp [1]				
No	96 (35.8)	172 (64.2)		
Yes	2 (28.6)	5 (71.4)	0.72 (0.10–3,39, $p = 0.694$)	-
Glycemia (mean)	95.2 (19.9)	99.8 (22.0)	0.99 (0.97–1.00, $p = 0.097$)	1.02 (1.00–1.05, $p = 0.053$)
ADA prediabetes				
No	81 (37.5)	135 (62.5)		
Yes	17 (28.8)	42 (71.2)	0.67 (0.35–1.24, $p = 0.219$)	
WHO prediabetes				

Table 2. *Cont.*

Variable	Non-Atrophic-Erosive	Atrophic-Erosive	OR Univariate (CI, *p* Value)	OR Multivariate (CI, *p* Value)
No	93 (35.9)	166 (64.1)		
Yes	5 (31.2)	11 (68.8)	0.81 (0.25–2.30, $p = 0.706$)	-
Diabetes (≥126 mg/dL)				
No	91 (36.3)	160 (63.7)		
Yes	7 (29.2)	17 (70.8)	0.72 (0.27–1.75, $p = 0.490$)	2.51 (0.76–8.64, $p = 0.131$)
Type of treatment				
No	95 (39.1)	148 (60.9)		
Oral antidiabetics	2 (7.1)	26 (92.9)	0.12 (0.02–0.42, $p = 0.004$) *	0.17 (0.03–0.62, $p = 0.011$) *
Insulin	1 (25.0)	3 (75.0)	0.52 (0.03–4.12, $p = 0.573$)	0.64 (0.03–5.44, $p = 0.704$)

There were statistically significant differences in relation to clinical form and oral antidiabetic treatment ($p = 0.005$), which was also higher in atrophic-erosive forms of OLP (26; 92.9%) than in non-atrophic-erosive types (2; 7.1%).

The logistic regression confirmed the statistically significant association of age >60 years ($p = 0.007$) and oral antidiabetic treatments ($p = 0.021$) with the atrophic-erosive clinical form. In patients who were over 60, there was also a statistically significant difference in terms of sites ($p = 0.036$), with three or more sites of OLP (74; 56.1%) being more common than two (58; 43.9%).

The logistic regression confirmed the statistically significant association of WHO prediabetes ($p = 0.015$) with two sites (Table 3). Nevertheless, the multivariate study determined that there was a higher probability of ≥3 sites being affected at >60 years (OR 1.81; $p = 0.018$) and in patients with cutaneous involvement (OR 2.43; $p = 0.036$).

Table 3. Statistically significant associations ($p < 0.05$ *) according to the logistic regression by number of locations. OR: Odd ratio. CI: confidence interval. Percentages in parentheses. ADA: American Diabetes Association criteria. WHO: World Health Organization. [1]: In addition to oral location. SD: standard deviation.

Variable	2 Sites	≥3 Sites	OR Univariate (CI, *p* Value)	OR Multivariate (CI, *p* Value)
Sex				-
Female	102 (49.3)	105 (50.7)		
Male	37 (54.4)	31 (45.6)	0.81 (0.47–1.41, $p = 0.463$)	
Mean age (SD)	58.5 (12.68)	60.8 (12.0)	1.02 (1.00–1.04, $p = 0.121$)	-
Age ≤ 60 years				
≤60 years	81 (56.6)	62 (43.4)		
>60 years	58 (43.9)	74 (56.1)	1.67 (1.04–2.69, $p = 0.036$) *	1.81 (1.11–2.97, $p = 0.018$) *
Genitals [1]				
No	129 (51.0)	124 (49.0)		
Yes	10 (45.5)	12 (54.5)	1.25 (0.52–3.06, $p = 0.619$)	-
Cutaneous [1]				
No	129 (52.4)	117 (47.6)		
Yes	10 (34.5)	19 (65.5)	2.09 (0.95–4.86, $p = 0.072$)	2.43 (1.04–5.78, $p = 0.036$) *
Scalp [1]				
No	136 (50.7)	132 (64.2)		
Yes	3 (42.9)	4 (57.1)	1.37 (0.30–7.08, $p = 0.681$)	-
Glycemia (mean)	97.4 (22.2)	99.8 (20.4)	1.00 (0.99–1.02, $p = 0.543$)	-

Table 3. *Cont.*

Variable	2 Sites	≥3 Sites	OR Univariate (CI, p Value)	OR Multivariate (CI, p Value)
ADA prediabetes				
No	112 (51.9)	104 (48.1)		
Yes	27 (45.8)	32 (54.2)	1.28 (0.72–2.29, p = 0.408)	-
WHO prediabetes				
No	126 (48.6)	133 (51.4)		
Yes	13 (81.2)	3 (18.8)	0.22 (0.05–0.70, p = 0.020) *	0.20 (0.04–0.65, p = 0.015) *
Diabetes (≥126 mg/dL)				
No	126 (50.2)	125 (49.8)		
Yes	13 (54.2)	11 (45.8)	0.85 (0.36–1.98, p = 0.711)	-
Type of treatment				
No	124 (51.0)	119(49.0)		
Oral antidiabetics	12 (42.9)	16 (57.19)	1.41 (0.64–3.17, p = 0.392)	
Insulin	3 (75.0)	1 (25.0)	0.35 (0.02–2.76, p = 0.363)	

The results of the logistic regression analyses with regard to glycemia, prediabetes and diabetes are shown in Table 4 (Table 4). In the univariate study, patients with ADA prediabetes and those taking oral antidiabetics were more likely to be suffering from OLP, and in the multivariate analysis adjusted for sex and age (≤60), this was corroborated ($OR_{ADAprediabetes}$ = 1.61 (1.04–2.51) and $OR_{oral\ antidiabetic}$ = 2.29 (1.18–4.68)).

Table 4. Association between oral lichen planus and glycemic level adjusted for age (≤60) and sex. Statistically significant associations (p < 0.05 *). OR: Odd ratio. CI: confidence interval. ADA: American Diabetes Association. WHO: World Health Organization.

Variable		OLP	Control	OR Univariate	OR Multivariate
ADA prediabetes (100–125 mg/dL)	No Yes	216 (47.9) 59 (59.6)	235 (52.1) 40 (40.4)	1.60 (1.03–2.51) p = 0.036 *	1.61 (1.04–2.51) p = 0.036 *
WHO prediabetes (110–125 mg/dL)	No Yes	259 (50.6) 16 (42.1)	253 (49.4) 22 (57.9)	0.71 (0.36–1.38) p = 0.315	0.71 (0.36–1.38) p = 0.315
Diabetes (≥126 mg/dL)	No Yes	251 (48.9) 24 (64.9)	262 (51.1) 13 (35.1)	1.93 (0.97–3.98) p = 0.065	1.96 (0.98–4.08) p = 0.062
Oral antidiabetics	No Yes	247 (48.5) 28 (68.3)	262 (51.1) 13 (31.7)	2.28 (1.18–4.65) p = 0.017 *	2.29 (1.18–4.68) p = 0.017 *

4. Discussion

The present study examined a sample of people with OLP whose fasting plasma glucose values and presence of DM were analyzed and compared to a control group. The initial interpretation of our results indicated that prediabetes with glycemia from 100 mg/dL to 125 mg/dL and intake of oral antidiabetics were associated with OLP, which was confirmed when adjusted for sex and age.

There were no statistically significant differences in mean glycemic levels between OLP cases (98.15 mg/dL) and the control group (95.45 mg/dL) as previously reported by Lopez et al. [32]. Although those authors recorded higher levels in the control group (n = 200; 102.5 ± 76.8) than in OLP patients (n = 200; 101.2 ± 22), they found no statistically significant differences [32].

Various case–control studies have not found an association between OLP and DM, but they all failed to note the diagnostic criteria for diabetes [5,18–23]. Other studies have reported statistically significant differences, with DM being more prevalent in OLP patients than in control groups according to medical records [13,14], or electronic medical records [15].

Previous studies have referred to a diagnosis of DM for blood glucose values ≥100 mg/dL, that is, prediabetes and diabetes [24,25]. In our results, the proportion of patients with values greater than ≥100 mg/dL (30.18%) is similar to the figures those studies reported for OLP groups of 30.11% [24] and 35% [25] with samples of 177 and 40 patients, respectively.

The prevalence of blood glucose levels > 125 mg/dL in previous studies with 40 OLP patients has been reported as 28% [16], whereas in our study, 8.73% of the OLP patients had levels ≥ 126 mg/dL. More recently, Chalkoo considered DM when fasting blood glucose >120 mg/dL, reporting a prevalence of DM of 35% in 20 OLP patients, with statistically significant differences compared to the control group [17].

Chung et al. [33], studied treatments for DM using questionnaires and found statistically significantly higher levels of treatment for DM in 32 patients with OLP compared to the control group. From a sample of 156 OLP and 156 controls, Dave et al. [14]. reported statistically significant differences in treatment with biguanides and insulin but not with sulfanyl ureas. In their multivariate model, only DM appeared as a risk factor (OR, 2.8), not treatments for DM. We confirmed that once adjusted for sex and age, the association between oral antidiabetics and OLP was twice as strong as in the control group.

A relationship between the reticular clinical form and treatment of DM in patients with OLP has been described [33], as has a relationship between atrophic-erosive lesions and DM treatment [34]. Our findings confirmed an association between the atrophic-erosive clinical form and oral antidiabetics, and age > 60 years. Although other authors have not found these relationships between the clinical form and DM, Zhao et al. observed that after treatment of erosive OLP with hyperglycaemia, glycemic levels fell [35].

Logistic regression showed that none of the parameters of the diabetes variables were associated with the number of sites of OLP. Instead, being over 60 and having skin involvement were associated with greater increases in the risk of presenting OLP (OR, 1.81, and OR, 2.43, respectively)–risks that we have not seen reported in the literature.

It is interesting to note from our results that between 1.45% (control group) and 3.27% (OLP patients) of the subjects were unaware of having DM. Between 5–10% of patients per year with impaired fasting glucose or impaired glucose tolerance can develop type 2 diabetes, hence the recommendation to screen for diabetes in people who are 35 and over and to test for it annually. The term prediabetes is used to label an elevated glucose level that does not meet the criteria for diabetes, and the ADA recommends identification of individuals with prediabetes, providing an opportunity for intervention [11]. Despite the fact that the progression of OLP does not influence average glucose levels, OLP is a chronic disease, and patients may develop prediabetes or progress from prediabetes to diabetes. Our findings indicate the need to establish preventive measures such as lifestyle modification in patients with OLP to avoid progressing to diabetes. This need is even greater if corticosteroids are used as the first choice of treatment for OLP [36], as they are absorbed following topical application [37], and their adverse effects—hyperglycemia and diabetes—are greater where there is comorbidity [38].

The possible link between diabetes and OLP comes from the presence of common histocompatibility antigens (HLA), especially HLA28, in both diseases [39]. More recently, higher levels of serum interleukins 8 have been observed in OLP and diabetes independently and where they are associated with each other, which would support the relationship between them [40]. Furthermore, the importance of the topic is reinforced from the bidirectional point of view of oral cancer, assuming the association of OLP and oral cancer in patients with diabetes, and oral cancer in patients with OLP [11].

The present study has some limitations. This was a single-center study with a cross-sectional design, so it does not allow causality or pathogenic association to be inferred from our results. Another limitation to consider is that the study did not include variables that could be related to lifestyle or other metabolic diseases associated with DM; however, it does provide interesting future lines for research to confirm the relationship between prediabetes and OLP.

5. Conclusions

In conclusion, we found significantly higher frequencies of glycemia in the order of 100–125 mg/dL in OLP patients. It therefore seems reasonable to test serum fasting plasma glucose values in order to prevent the development of diabetes as well as deal with its possible complications. Our results should be corroborated by new studies to avoid complications of prediabetes in OLP patients.

Supplementary Materials: The following supporting information can be downloaded at: https://www.mdpi.com/article/10.3390/diagnostics13091586/s1, Table S1: STROBE Statement—Checklist of items that should be included in reports of case-control studies.

Author Contributions: Conceptualization, M.G.-P. and L.R.-F.; methodology, M.G.-P., S.L.-P. and L.R.-F.; software, M.G.-P., S.L.-P. and L.R.-F.; formal analysis, M.G.-P., S.L.-P. and L.R.-F..; investigation, M.G.-P., S.L.-P. and L.R.-F.; writing—original draft preparation, M.G.-P., S.L.-P. and L.R.-F.; writing—review and editing, M.G.-P., S.L.-P. and L.R.-F.; visualization, M.G.-P., S.L.-P. and L.R.-F.; supervision, M.G.-P., S.L.-P. and L.R.-F. All authors have read and agreed to the published version of the manuscript.

Funding: This research received no external funding.

Institutional Review Board Statement: The study was conducted in accordance with the Declaration of Helsinki, and approved by the Ethics Committee of the Principality of Asturias (protocol code 127/13).

Informed Consent Statement: Informed consent was obtained from all subjects involved in the study.

Data Availability Statement: The data presented in this study are available on request from the corresponding author.

Acknowledgments: The authors would like to thank Tania Iglesias Cabo, a technician in the statistical consulting unit of the University of Oviedo, who performed the statistical analysis.

Conflicts of Interest: The authors declare no conflict of interest.

References

1. Le Cleach, L.; Chosidow, O. Lichen planus. *N. Engl. J. Med.* **2012**, *366*, 723–732. [CrossRef] [PubMed]
2. González-Moles, M.Á.; Warnakulasuriya, S.; González-Ruiz, I.; González-Ruiz, L.; Ayén, Á.; Lenouvel, D.; Ruiz-Ávila, I.; Ramos-García, P. Worldwide prevalence of oral lichen planus: A systematic review and meta-analysis. *Oral Dis.* **2021**, *27*, 813–828. [CrossRef] [PubMed]
3. van der Meij, E.H.; van der Waal, I. Lack of clinicopathologic correlation in the diagnosis of oral lichen planus based on the presently available diagnostic criteria and suggestions for modifications. *J. Oral Pathol. Med.* **2003**, *32*, 507–512. [CrossRef] [PubMed]
4. Carbone, M.; Arduino, P.G.; Carrozzo, M.; Gandolfo, S.; Argiolas, M.R.; Bertolusso, G.; Conrotto, D.; Pentenero, M.; Broccoletti, R. Course of oral lichen planus: A retrospective study of 808 northern Italian patients. *Oral Dis.* **2009**, *15*, 235–243. [CrossRef]
5. Adamo, D.; Calabria, E.; Coppola, N.; Lo Muzio, L.; Giuliani, M.; Bizzoca, M.E.; Azzi, L.; Croveri, F.; Colella, G.; Boschetti, C.E.; et al. SIPMO (Italian Society of Oral Pathology, Medicine). Psychological profile and unexpected pain in oral lichen planus: A case-control multicenter SIPMO study. *Oral Dis.* **2021**, *29*, 298–414. [CrossRef]
6. De Angelis, L.M.; Cirillo, N.; McCullough, M.J. The immunopathogenesis of oral lichen planus-Is there a role for mucosal associated invariant T cells? *J. Oral Pathol. Med.* **2019**, *48*, 552–559. [CrossRef]
7. El-Howati, A.; Thornhill, M.H.; Colley, H.E.; Murdoch, C. Immune mechanisms in oral lichen planus. *Oral Dis.* **2022**, *28*, 1400–1415. [CrossRef]
8. Liu, W.; Deng, Y.; Shi, H.; Shen, X. Clinical investigation on oral lichen planus and associated comorbidities needs a holistic concept. *Oral Dis.* **2022**, *7*, 327–329. [CrossRef]
9. American Diabetes Association Professional Practice Committee. Classification and Diagnosis of Diabetes: Standards of Medical Care in Diabetes. *Diabetes Care* **2022**, *45* (Suppl. S1), S17–S38. [CrossRef]
10. Ahmad, E.; Lim, S.; Lamptey, R.; Webb, D.R.; Davies, M.J. Type 2 diabetes. *Lancet* **2022**, *400*, 1803–1820. [CrossRef]

11. Ramos-Garcia, P.; Roca-Rodriguez, M.D.M.; Aguilar-Diosdado, M.; Gonzalez-Moles, M.A. Diabetes mellitus and oral cancer/oral potentially malignant disorders: A systematic review and meta-analysis. *Oral Dis.* **2021**, *27*, 404–421. [CrossRef] [PubMed]
12. De Porras-Carrique, T.; Ramos-García, P.; Aguilar-Diosdado, M.; Warnakulasuriya, S.; González-Moles, M.Á. Autoimmune disorders in oral lichen planus: A systematic review and meta-analysis. *Oral Dis.* **2022**, *9*, 1382–1394. [CrossRef]
13. Canjuga, I.; Mravak-Stipetić, M.; Lončar, B.; Kern, J. The Prevalence of Systemic Diseases and Medications in Patients with Oral Lichen Planus. *Acta Stomatol. Croat.* **2010**, *44*, 96–100.
14. Dave, A.; Shariff, J.; Philipone, E. Association between oral lichen planus and systemic conditions and medications: Case-control study. *Oral Dis.* **2021**, *27*, 515–524. [CrossRef] [PubMed]
15. Arduino, P.G.; Karimi, D.; Tirone, F.; Sciannameo, V.; Ricceri, F.; Cabras, M.; Gambino, A.; Conrotto, D.; Salzano, S.; Carbone, M.; et al. Evidence of earlier thyroid dysfunction in newly diagnosed oral lichen planus patients: A hint for endocrinologists. *Endocr. Connect.* **2017**, *6*, 726–730. [CrossRef]
16. Lundström, I.M. Incidence of diabetes mellitus in patients with oral lichen planus. *Int. J. Oral Maxillofac. Surg.* **1983**, *12*, 147–152. [CrossRef]
17. Chalkoo, A.H. Oral Lichen Planus: Relation with Transaminase Levels and Diabetes. Oral Lichen Planus: Relation with Transaminase Levels and Diabetes. *J. Indian Acad. Oral Med. Radiol.* **2010**, *22*, 1–3. [CrossRef]
18. Garg, V.K.; Karki, B.M.; Agrawal, S.; Agarwalla, A.; Gupta, R. A study from Nepal showing no correlation between lichen planus and hepatitis B and C viruses. *J. Dermatol.* **2002**, *29*, 411–413. [CrossRef]
19. Gerayli, S.; Meshkat, Z.; Pasdar, A.; Mosannen Mozafari, P.; Banihashemi, E.; Khajavi, M.A.; Rasekhi, J. The association between oral lichen planus and hepatitis C virus infection; a report from northeast of Iran. *Jundishapur. J. Microbiol.* **2015**, *8*, e16741. [CrossRef]
20. Kats, L.; Goldman, Y.; Kahn, A.; Goldman, V.; Gorsky, M. Oral lichen planus and thyroid gland diseases: Possible associations. *BMC Oral Health* **2019**, *19*, 169. [CrossRef]
21. Nagao, Y.; Sata, M. A retrospective case-control study of hepatitis C virus infection and oral lichen planus in Japan: Association study with mutations in the core and NS5A region of hepatitis C virus. *BMC Gastroenterol.* **2012**, *12*, 31. [CrossRef] [PubMed]
22. Xue, J.L.; Fan, M.W.; Wang, S.Z.; Chen, X.M.; Li, Y.; Wang, L. A clinical study of 674 patients with oral lichen planus in China. *J. Oral Pathol. Med.* **2005**, *34*, 467–472. [CrossRef] [PubMed]
23. Zhou, T.; Li, D.; Chen, Q.; Hua, H.; Li, C. Correlation Between Oral Lichen Planus and Thyroid Disease in China: A Case-Control Study. *Front. Endocrinol.* **2018**, *9*, 330. [CrossRef] [PubMed]
24. Hornstein, O.P.; Stühler, C.; Schirner, E.; Simon, M., Jr. Lichen ruber and diabetes mellitus—Pathogenetic relations? *Hautarzt* **1984**, *35*, 287–291. [PubMed]
25. Ali, A.A.; Suresh, C.S. Oral lichen planus in relation to transaminase levels and hepatitis C virus. *J. Oral Pathol. Med.* **2007**, *36*, 604–608. [CrossRef]
26. Vandenbroucke, J.P.; von Elm, E.; Altman, D.G.; Gøtzsche, P.C.; Mulrow, C.D.; Pocock, S.J.; Poole, C.; Schlesselman, J.J.; Egger, M. Strengthening the Reporting of Observational Studies in Epidemiology (STROBE): Explanation and elaboration. *Int. J. Surg.* **2014**, *12*, 1500–1524. [CrossRef]
27. Kramer, I.R.; Lucas, R.B.; Pindborg, J.J.; Sobin, L.H. Definition of leukoplakia and related lesions: An aid to studies on oral precancer. *Oral Surg. Oral Med. Oral Pathol.* **1978**, *46*, 518–539.
28. Aguirre-Urizar, J.M.; Alberdi-Navarro, J.; Lafuente-Ibáñez de Mendoza, I.; Marichalar-Mendia, X.; Martínez-Revilla, B.; Parra-Pérez, C.; De Juan-Galindez, A.; Echebarria-Goicouria, M.Á. Clinicopathological and prognostic characterization of oral lichenoid disease and its main subtypes: A series of 384 cases. *Med. Oral Patol. Oral Cir. Bucal.* **2020**, *25*, e554–e562. [CrossRef]
29. van der Waal, I. Oral lichen planus and oral lichenoid lesions; a critical appraisal with emphasis on the diagnostic aspects. *Med. Oral Patol. Oral Cir. Bucal.* **2009**, *14*, E310–E314.
30. American Diabetes Association. *Standards of Medical Care in Diabetes-2018 Abridged for Primary Care Providers. Clin. Diabetes* **2018**, *36*, 14–37. [CrossRef]
31. World Health Organization. *Global Report on Diabetes*; World Health Organization: Geneva, Switzerland, 2016; Available online: https://www.who.int/publications/i/item/9789241565257 (accessed on 7 January 2018).
32. López-Jornet, P.; Camacho-Alonso, F.; Rodríguez-Martínes, M.A. Alterations in serum lipid profile patterns in oral lichen planus: A cross-sectional study. *Am. J. Clin. Dermatol.* **2012**, *13*, 399–404. [CrossRef] [PubMed]
33. Chung, C.H.; Yang, Y.H.; Chang, T.T.; Shieh, D.B.; Liu, S.Y.; Shieh, T.Y. Relationship of oral lichen planus to hepatitis C virus in southern Taiwan. *Kaohsiung J. Med. Sci.* **2004**, *20*, 151–159. [CrossRef] [PubMed]
34. Bagan, J.V.; Donat, J.S.; Penarrocha, M.; Milian, M.A.; Sanchis, J.M. Oral lichen planus and diabetes mellitus. A clinico-pathological study. *Bull. Group Int. Rec. Sci. Stomatol. Odontol.* **1993**, *36*, 3–6.
35. Zhao, W.; Yang, Y.; Wu, F.; Zhou, H. The reciprocal association between diabetes mellitus and erosive oral lichen planus. *Oral Dis.* **2019**, *25*, 1235–1236. [CrossRef]
36. Thongprasom, K. Oral lichen planus: Challenge and management. *Oral Dis.* **2018**, *24*, 172–173. [CrossRef]
37. Varoni, E.M.; Molteni, A.; Sardella, A.; Carrassi, A.; Di Candia, D.; Gigli, F.; Lodi, F.; Lodi, G. Pharmacokinetics study about topical clobetasol on oral mucosa. *J. Oral Pathol. Med.* **2012**, *41*, 255–260. [CrossRef]

38. Ceccarelli, E.; Mattaliano, C.; Brazzi, A.; Marinetti, A.C.; Nigi, L.; Chirico, C.; Corallo, C.; Fioravanti, A.; Giordano, N. Hyperglycemia and diabetes induced by glucocorticoids in nondiabetic and diabetic patients: Revision of literature and personal considerations. *Curr. Pharm. Biotechnol.* **2018**, *19*, 1210–1220. [CrossRef]
39. Halevy, S.; Zamir, R.; Gazit, E.; Feuerman, E.J. HLA system in relation to carbohydrate metabolism in lichen planus. *Br. J. Dermatol.* **1979**, *100*, 683–686. [CrossRef]
40. Tavangar, A.; Khozeimeh, F.; Ghoreishian, F.; Boroujeni, M.A. Serum level of Interleukin-8 in subjects with diabetes, diabetes plus oral lichen planus, and oral lichen planus: A biochemical study. *Dent. Res. J.* **2016**, *13*, 413–418. [CrossRef]

diagnostics

MDPI

Article

High-Frequency Ultrasound in Diagnosis and Treatment of Non-Melanoma Skin Cancer in the Head and Neck Region

Tiberiu Tamas [1], Cristian Dinu [1], Lavinia Manuela Lenghel [2,*], Emil Boțan [3], Adela Tamas [4], Sebastian Stoia [1], Daniel Corneliu Leucuta [5], Simion Bran [1], Florin Onisor [1], Grigore Băciuț [1], Gabriel Armencea [1] and Mihaela Băciuț [1]

1 Department of Maxillofacial Surgery and Implantology, Faculty of Dentistry, "Iuliu Hațieganu" University of Medicine and Pharmacy, 400012 Cluj-Napoca, Romania
2 Department of Radiology, "Iuliu Hatieganu" University of Medicine and Pharmacy, 400012 Cluj-Napoca, Romania
3 Department of Pathology, Emergency County Hospital, 400437 Cluj-Napoca, Romania
4 Doctoral School, "Iuliu Hatieganu" University of Medicine and Pharmacy, 400012 Cluj-Napoca, Romania
5 Medical Informatics and Biostatistics Department, "Iuliu Hatieganu" University of Medicine and Pharmacy, 400012 Cluj-Napoca, Romania
* Correspondence: lenghel.manuela@gmail.com; Tel.: +40-740423025

Abstract: Non-melanoma skin cancer is one of the most frequently diagnosed cancers in the human body and unfortunately the incidence continues to increase. NMSC is represented by the basal cell carcinomas (BCCs) and squamous cell carcinomas (SCCs), which are the most prevalent forms, and basosquamous cell carcinomas (BSC) together with Merkel cell carcinoma (MCC), which are rare types but with a very aggressive pattern and poor prognosis. The pathological diagnosis is hard to assess without a biopsy, even by the dermoscopy. Moreover, the staging can be problematic because there is no access clinically to the thickness of the tumor and the depth of the invasion. The aim of this study was to evaluate the role of ultrasonography (US), which is a very efficient imaging method, non-irradiating and cheap, in diagnosis and treatment of non-melanoma skin cancer in the head and neck region. Thirty-one patients with highly suspicious malignant lesions of the head and neck skin were evaluated in the Oral and Maxillo-facial Surgery Department and Imaging Department in Cluj Napoca, Romania. All tumors were measured with three transducers: 13 MHz, 20 MHz and 40 MHz. Doppler examination and elastography were also used. The length, width, diameter, thickness, the presence of necrosis, status of regional lymph nodes, the presence of hyperechoic spots, strain ratio and vascularization were all recorded. After that, all patients were treated by surgical resection of the tumor and reconstruction of the defect. Immediately after surgical resection, all tumors were measured again after the same protocol. The resection margins were evaluated by all three types of transducers in order to detect malignant involvement and the results were compared with the histopathological report. We found that the 13 MHz transducers offered a big picture of the tumor but the level of details, in the form of the presence of the hyperechoic spots, is reduced. We recommend this transducer for evaluation of surgical margins or for the large skin tumors. The 20 and 40 MHz transducers are better for viewing the particularities of malignant lesions and for an accurate measurement; however, in the case of large size lesions, assessing all three dimensions of the tumor can be difficult. The intralesional hyperechoic spots are present in case of BCC and they can be used for differential diagnosis of BCC.

Keywords: non-melanoma skin cancer; basal cell carcinoma; squamous cell carcinoma; ultrasound; echography; high frequency transducer; skin; imaging

1. Introduction

One of the most common diagnosed tumors in the human body is non-melanoma skin cancer (NMSC). Unfortunately, according to GLOBOCAN, the incidence is steadily rising.

Basal cell carcinomas (BCCs) and squamous cell carcinomas (SCCs) are the most prevalent forms, with a good prognosis in general, especially when they are found in the early stages. Other NMSC types are baso-squamous cell carcinomas (BSCs) and Merkel cell carcinomas (MCSs), with a more aggressive behavior and a poorer prognosis overall [1–3]. Because of the constant exposure to ultraviolet (UV) sun radiation, the head and neck region is the most affected anatomical area. Several alternative methods of treatment were documented for NMSC. Topical chemotherapy, cryotherapy, laser or thermocauterization, and superficial radiotherapy have demonstrated modest results. Surgical excision is still the treatment of choice, and achieving favorable cosmetic outcomes will always be difficult [4,5]. Clinical examination, palpation and dermoscopy are used to diagnose skin cancer, which is then followed by an incisional or excisional biopsy. Dermoscopy can be used to evaluate the lesion's longitudinal and horizontal extension, but it cannot examine the lesion's thickness or potential for invasion of adjacent tissues. The depth of invasion (DOI) from the American Joint Committee on Cancer (AJCC) staging system is used for melanoma and oral cavity cancer. In the last edition of the staging system, it was introduced for non-melanoma skin cancer and it modifies the stage when the invasion is deeper than 6 mm directly to T3. Additionally, DOI is correlated with regional metastasis in the case of SCC and other aggressive skin malignancies [6–8].

The imaging's role is not very well defined. Magnetic resonance imaging (MRI) is a non-irradiating imaging method which is considered the gold standard for soft tissue tumors. Unfortunately, due to the small size of the tumors, MRI is not very efficient in this type of pathology. However, it can be a good option for very extensive tumors with local and regional adjacent structures involvement. Computed tomography scan (CT) involves the use of X-rays, and it cannot differentiate between soft tissues as well as the MRI. The advantage of this method is the ability to evaluate the bone involvement. It is reserved for very extensive tumors with high risk of bone tissue invasion [6–9].

Ultrasonography (US) is a very efficient imaging method, non-irradiating and cheap and, for these reasons, has a very high rate of acceptance from the patients. The thickness of the tumor, the depth of invasion, the invaded adjacent structures, the vascularity and the stiffness of the tumor may all be determined with an ultrasound examination. The disadvantage of US is the dependence on the examinator's knowledge. The tumors should be evaluated by a trained practitioner in skin malignancy evaluation [6,10,11].

In recent years, new developments, such as high frequency transducers, have been introduced, as the structures of interest are only a few millimeters below the skin's surface. These technologies can offer a broad range of data, including the precise anatomic position, shape, extent in all directions, blood flow and deeper layers involvement. Higher frequencies come with better resolution but less tissue penetration. In order to measure the thickness of the skin tumors, US transducers with a frequency range of 10 to 40 MHz are frequently utilized in clinical practice. Generally, NMSC are hypoechoic lesions with an inhomogeneous aspect on grayscale US examination [12,13]. In the prospective research by Song et al., high frequency US has shown positive outcomes in determining the DOI prior to surgery. To this end, the distance between the epidermal granulous layer and the deepest point of invasion is measured. An invasion deeper than 6 mm will place the tumor in stage III, even if the diameter of tumor is smaller than 2 cm. The upstaging of tumors means a poorer prognosis and a higher rate of positive margins after surgical treatment [14].

The technique of ultrasonography requires three methods for evaluation. The B-mode ultrasound may be followed by a color or power Doppler US scan of the lesion in order to obtain further details. All diameter measurements, including longitudinal, transverse and deep axe, should be recorded. The peak systolic velocity of the arterial flow, as well as information regarding the size and nature (arteries or veins) of the tumor vessels, are provided. Involvement of deeper layers, including cartilage, muscle and bone should be mentioned, if present. The form, depth, echogenicity (hypoechoic, isoechoic and hyperechoic), homogeneity (homogenous and heterogeneous), size, level of invasion and vascularity must all be included in the ultrasound report. An interesting aspect is the presence of the calcification in some malignant tumors which can be visualized as hyperechoic spots without acoustic shadow. Ulcerated lesions with hemorrhage and crusts can give artifacts and should be carefully evaluated [15,16]. The examinator should be familiarized with the manipulation of the transducers because the US beam is necessary to be perpendicular to the lesion and should find the largest diameter of the tumor. If too much pressure is applied on the tumor, the size can be easily modified and the staging and treatment plan compromised. A high amount of gel and low pressure should be used during the examination [17–19].

Strain elastography is a US examination which measures the stiffness of the tissues. The soft and hard tissues are color-coded differently on the strain elastogram. Usually, malignant tumors are stiffer than normal adjacent tissue, which makes them easily distinguishable. The images can be superposed and compared with those from grayscale examination [6,20].

It has also been suggested that optical coherence tomography (OCT), a high-resolution method based on the characteristics of light, can be used to analyze skin cancers. For thickness assessment in BCC lesions less than 2 mm in size, OCT appears to be more accurate and less biased than high-frequency US, according to the literature [21–24]. Other high-resolution imaging techniques include confocal microscopy, also known as reflectance confocal microscopy, and multiphoton microscopy, which may produce in vivo pictures that are almost as good as those from histologic examination of BCC lesions. In a histologic examination, a grayscale horizontal view of the tumor is displayed, which is in contrast to earlier cross-sectional imaging methods [6,25,26]. The differences between these techniques are summarized in Table 1.

Table 1. Characteristics of different non-invasive tehniques used for evaluation of NMSC.

	Ultrasonography	Dermoscopy	OCT	Confocal Microscopy
Assessing the size of the tumor	It can assess the size in all three dimensions including the thickness [14]	It can assess the size except the thickness [1,6]	It can assess the size in all three dimensions for tumors with a thickness less than 2 mm [21–24]	It can assess the size in all three dimensions for tumors with a thickness less than 3 mm [25,26]
Type of subjacent tissue involvement	It can precisely identify the type of subjacent tissue involved (muscle, fascia, perichondrium)	It cannot precisely identify the type of subjacent tissue involved	Because the depth of penetration is less than 2 mm, we do not recommend using it for evaluation of subjacent tissue involvement	Because the depth of penetration is less than 3 mm, we do not recommend to using it for evaluation of subjacent tissue involvement
Surgical margins involvement detection	Yes (postoperative evaluation of the excised specimen)	Only preoperatively	Only preoperatively. Very limited data in the literature	Yes (Ex vivo confocal microscopy)
Vascularity	Yes	Yes	Yes (Dynamic Optical Coherence Tomography)	Yes
Specificity is diagnosis	85–90% for both BCC and SCC	95–99% for BCC Insufficient data for SCC	80–100% for SCC 70–80% for BCC	92–98% for SCC 93% for BCC
Sensitivity is diagnosis	93–95% for both BCC and SCC	90–93% for BCC 75–77% for SCC	92–93% for SCC 92–95% for BCC	74–77% for SCC 92% for BCC

The gold standard for NMSC treatment is surgical excision and defect reconstruction. Other treatments can be attempted, such as local chemotherapy for very small lesions, or systemic chemotherapy, immunotherapy, Hedgehog Pathway inhibitors, and radiotherapy for highly advanced tumors. In order to attain good quality surgery with negative resection margins, it is necessary to know the diameter and the thickness of the tumor preoperatively. Because the thickness is not accessible without imaging, Mohs surgery should be attempted if it is available. If not, delaying the defect reconstruction until the certification of complete resection is also an option.

The aim of this study is to evaluate the role of ultrasonography in diagnosis, staging and treatment of the NMSC of the head and neck region.

2. Materials and Methods

Thirty-one patients with highly suspicious malignant lesions of the head and neck skin were evaluated in the Oral and Maxillo-facial Surgery Department and Imaging Department in Cluj Napoca, Romania. All patients were informed of the study and agreed to participate. The Ethical Committee of the Emergency County Hospital Cluj-Napoca, Romania (39109/A/0001/UK/R Reg. n. 362/08.01.2020) and the Iuliu Hatieganu University of Medicine and Pharmacy Cluj-Napoca, Romania (433/11.10.2019) approved the study. After clinical examination, the patients were investigated by ultrasonography with high frequency transducers. All tumors were measured with three transducers: 13 MHz, 20 MHz and 40 MHz (Figure 1). Doppler examination and elastography were also used. The length, width, diameter, thickness, the presence of necrosis, status of regional lymph nodes, the presence of hyperechoic spots, strain ratio and vascularization were all recorded. All patients were treated by surgical resection of the tumor and reconstruction of the defect (Figure 2). Immediately after surgical resection, all tumors were placed with the epidermis facing up on sterile gauze and fixed with small-gauge needles. A layer of US gel was applied. The tumors were measured again after the same protocol. Doppler and elastography techniques were not performed. The resection margins were evaluated by all three types of transducers in order to detect malignant involvement of the resection margins or to measure the closest margin to the tumor. The results were compared with the histopathological report.

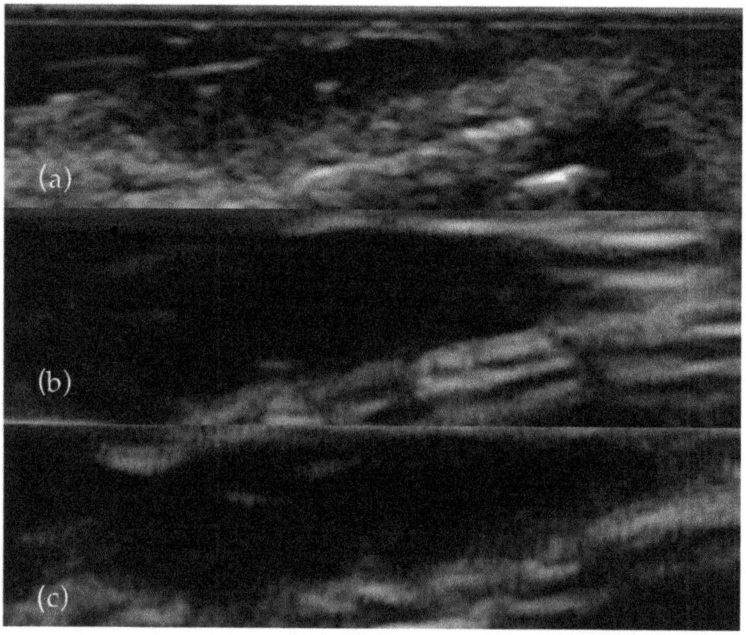

Figure 1. The same BCC examined with 13 MHz (**a**), 20 MHz (**b**) and 40 MHz (**c**). As the frequency increases, the presence of the two hyperechoyc spots is clearer.

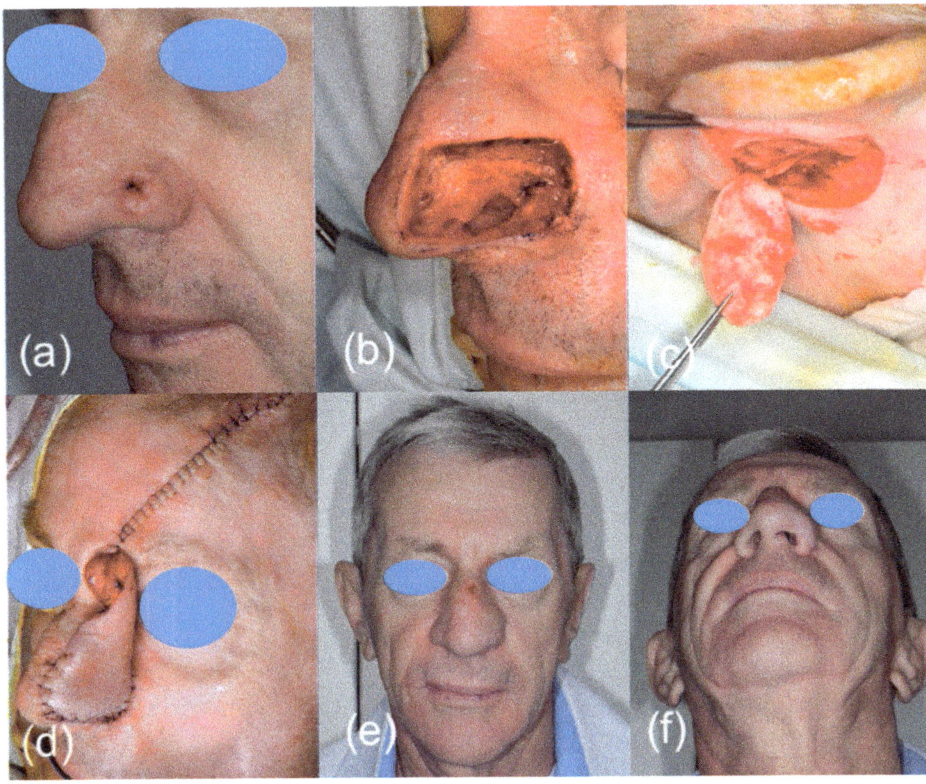

Figure 2. Surgical excision of a nasal BCC: (**a**) clinical presentation of the malignant lesion; (**b**) surgical defect after tumor resection together with alar cartilage and a portion of nasal mucosa; (**c**) auricular cartilage graft for prefabricated frontal flap; (**d**) reconstruction of the nose with prefabricated frontal flap and auricular cartilage; (**e**,**f**) clinical aspect 1 month after surgery.

Statistical Analyses

To choose the interclass correlation coefficient (ICC), we followed the guide of Koo et al. For the type of reliability study, we considered an inter-rater reliability study, with having the same set of raters for all subjects that were not randomized; thus, a two-way mixed effects analysis was chosen. The intended measurement protocol was based on a single rater, and we considered consistency to be the most important quality. Despite that, we did also compute the absolute agreement. We used the interpretation of Portney et al. for ICC: 0.5 suggesting poor reliability, values between 0.5 and 0.75 suggesting moderate reliability, values between 0.75 and 0.9 suggesting good reliability and values greater than 0.90 suggesting excellent reliability. For ICC computations, we used the irr R package [27–29].

Concerning the concordance between the echographic diagnosis, or characteristics (as hyperechoic spots) with the histopathological diagnosis, the inter-rater agreement was sought, as evidenced by Cohen Kappa (following Altman's interpretation: 0.20—poor; 0.21–0.40—fair; 0.41–0.60—moderate; 0.61–0.80—good; 0.81–1.00—very good). Kappa can vary from -1 (lower agreement than expected by chance) to 1 (higher agreement than expected) and is the norm of the difference between the actual rate of agreement and the chance rate of agreement anticipated (almost perfect agreement). However, the kappa coefficient does not account for inter-observer variability (degree of disagreement) or data prevalence (how evenly categories are distributed); thus, we computed the prevalence-adjusted, bias-adjusted Cohen Kappa [30–32].

3. Results

Concerning concordance between the histopathological and echographic measurements of tumor thickness, we found excellent agreement for all transducers, with the best ones being for the preoperative observations and the least favorable ones being the postoperative observations (Figure 3) (Table 2). Within any of the preoperative and postoperative observations, the 40 MHz transducer fared the best, followed by the 20 MHz transducer and finally by the 13 MHz transducer.

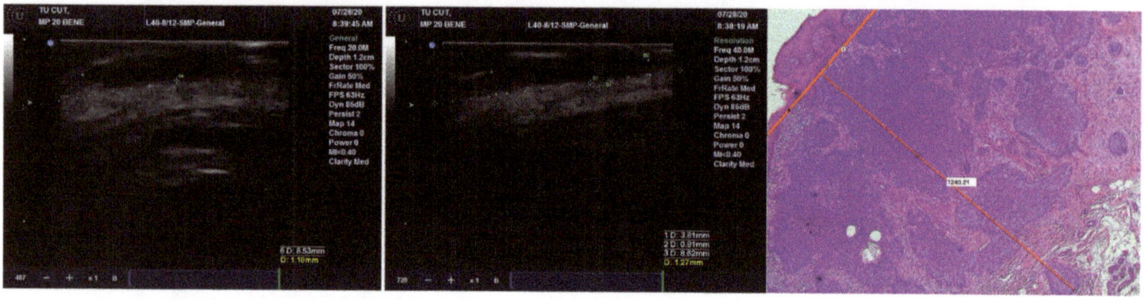

Figure 3. The thickness of the tumor, measured on US with 20 MHz and 40 MHz transducers, compared with the pathological image of the tumor showing very high occurrence of US.

Table 2. Tumor thickness measured pre- and post-operatively, with 13, 20 and 40 MHz transducers, as well as the inter-class correlation coefficient between echography and pathological measurements.

Method	Tumor Thickness (mm), Median (IQR)	Difference (95% CI)	p-Value	ICC Consistency (95% CI)	p-Value	ICC Agreement (95% CI)	p-Value
Pathology	2.5 (1.98–4.5)						
Preoperative							
13 MHz	2.5 (1.75–4.4)	0 (0–0.32)	0.063	0.982 (0.962–0.991)	<0.001	0.98 (0.957–0.99)	<0.001
20 MHz	2.3 (1.75–4.4)	0.2 (−0.09–0.18)	0.574	0.981 (0.96–0.991)	<0.001	0.981 (0.961–0.991)	<0.001
40 MHz	2.2 (1.9–4.5)	0.3 (−0.07–0.15)	0.394	0.992 (0.984–0.996)	<0.001	0.992 (0.985–0.996)	<0.001
Postoperative							
13 MHz	2.3 (1.45–4)	0.2 (0.13–0.65)	0.008	0.927 (0.853–0.964)	<0.001	0.909 (0.776–0.96)	<0.001
20 MHz	2.35 (1.6–4.05)	0.15 (0.1–0.6)	0.011	0.927 (0.855–0.964)	<0.001	0.911 (0.787–0.96)	<0.001
40 MHz	2.4 (1.6–3.95)	0.1 (0.07–0.58)	0.012	0.93 (0.859–0.965)	<0.001	0.916 (0.806–0.962)	<0.001

We assessed the concordance between the final diagnosis and ultrasound diagnosis, and we found a Cohen's Kappa of 0.8897 (95% CI: 0.6783–1.1011), bias-adjusted Kappa of 0.8895 and prevalence- and bias-adjusted Kappa of 0.9355 (Table 3). The concordance between the final diagnosis of BCC and the presence of hyperechoic spots followed an identical pattern. We have associated the presence of the hyperechoic spots with diagnosis of BCC (Figure 4). Only one case of BCC was misdiagnosed by echography due to the missing hyperechoic spots (Figure 5).

Regarding elastography, we can easily see that the tumor tissue is stiffer than the normal adjacent one (Figure 6). We did not find any significant statistical association between the value of strain ratio and the histology of the tumor. The values in the BCC group were also lower than in the SCC group (95% CI −2.98–0.35, $p = 0.126$). Doppler examination (Figure 7) showed increased values for both systolic and diastolic speed in the SCC group, but the results were not statistically significant and no association between the type of histology and vascularization was found (Table 4).

Table 3. Concordance between histopathological diagnosis, echographic diagnosis and hyperechoic spots.

Histopathological Diagnosis	Basal Cell Carcinoma	Squamous Cell Carcinoma	Total
Echographic diagnosis			
Basal cell carcinoma	25	0	25
Squamous carcinoma	1	5	6
Total	26	5	31
Hyperechoic spots			
Present	25	0	25
Absent	1	5	6
Total	26	5	31

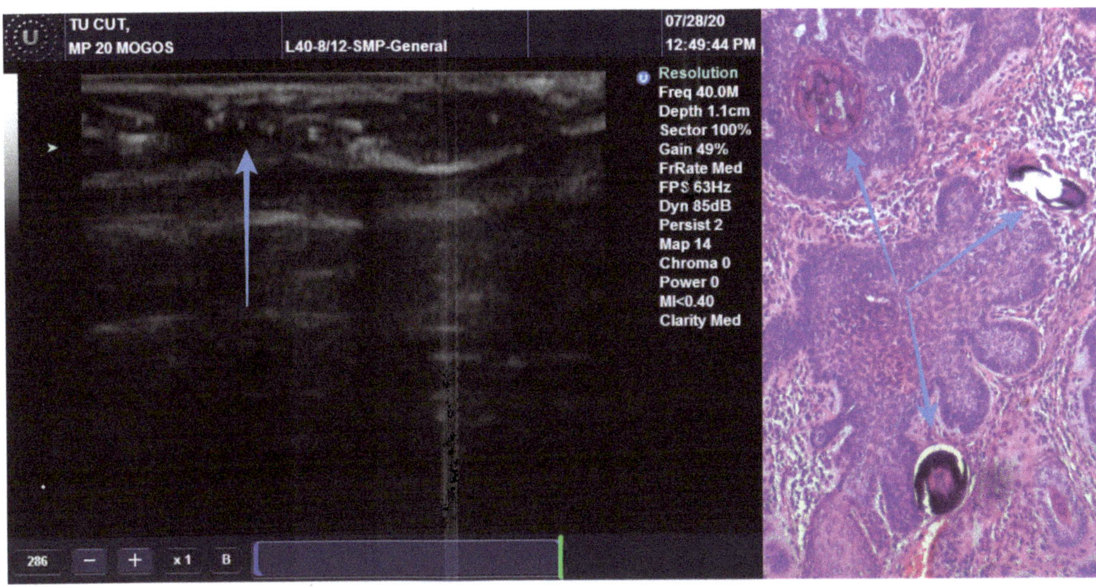

Figure 4. The arrow is showing the presence of the hyperechoic spots in a BCC and the pathological correspondence.

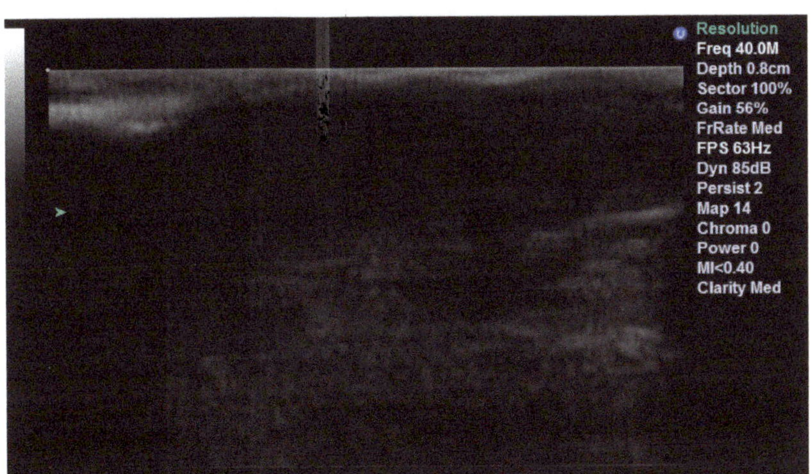

Figure 5. Misdiagnosed BCC due to the absence of hyperechoic spots.

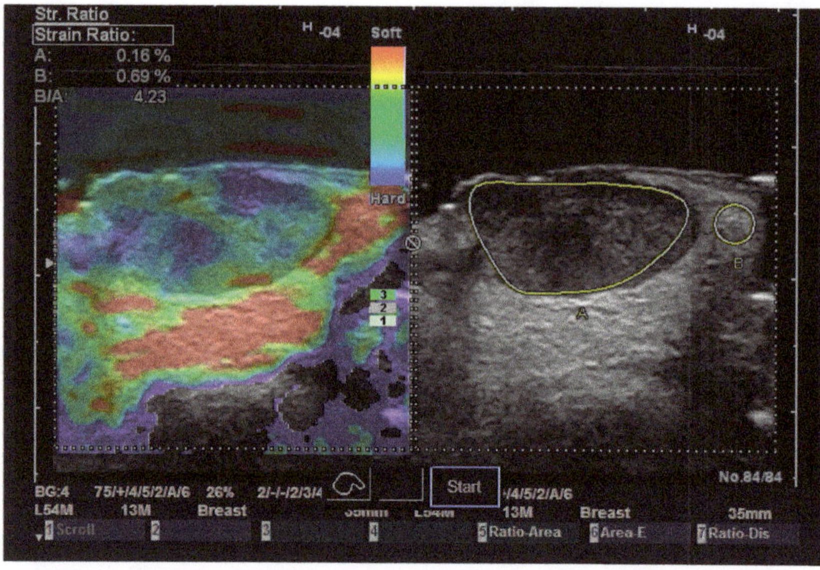

Figure 6. Strain elastography of a SCC.

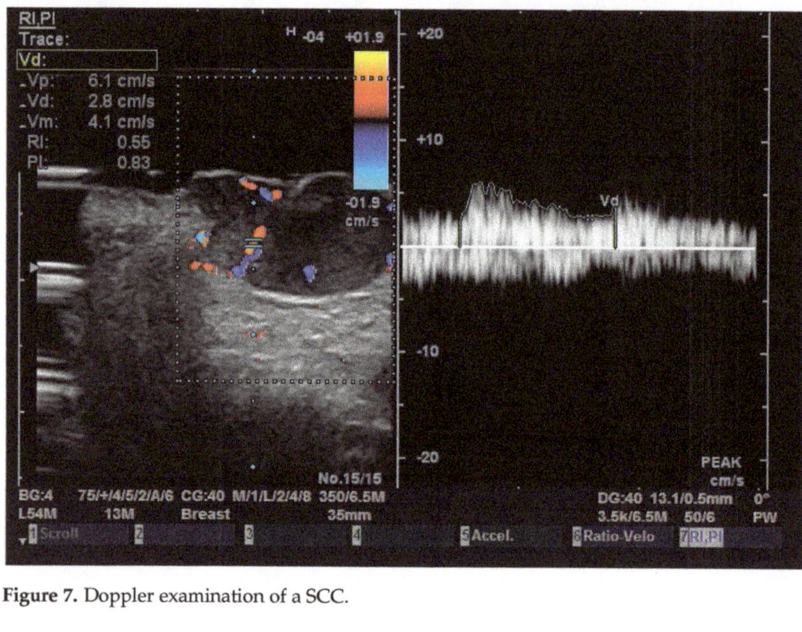

Figure 7. Doppler examination of a SCC.

We found a very high correlation between the ultrasound measurement of the surgical margins and pathological reports, with the best result being achieved with the 13 MHz transducer, followed by the 20 MHz and 40 MHz transducers (Figure 8) (Table 5).

Table 4. Pulsatility index, resistive index, strain ratio, diastolic speed and systolic speed related to histopathological type.

Diagnostic Histopathologic	Cc Basal Cell (n = 25)	Cc Squamous (n = 5)	Difference (95% CI)	p
Pulsatility index, median (IQR)	0.75 (0.6–0.94)	0.82 (0.54–0.95)	0.07 (−0.3–0.31)	0.889• [n1 = 25, n2 = 5]
Resistive index, median (IQR)	0.5 (0.4–0.6)	0.52 (0.42–0.63)	0.02 (−0.13–0.18)	0.933• [n1 = 25, n2 = 5]
Strain ratio, median (IQR)	1.94 (1.38–2.62)	3.2 (2.9–3.6)	1.26 (−2.98–0.35)	0.126• [n1 = 26, n2 = 5]
Diastolic speed, median (IQR)	3.9 (3.9–4.1)	4.3 (2.5–4.9)	0.4 (−1–1.6)	0.822• [n1 = 25, n2 = 5]
Systolic speed, median (IQR)	6.3 (6.1–9)	6.8 (6.4–8.2)	0.5 (−1.9–2.2)	0.758• [n1 = 25, n2 = 5]

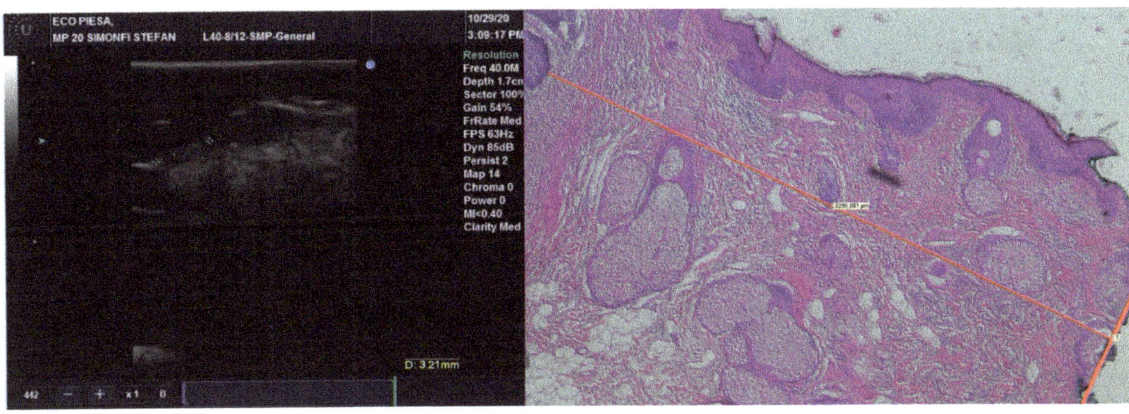

Figure 8. Surgical margins measurement by US and histopathology.

Table 5. Correlation between the ultrasound measurement of the surgical margins and pathological reports.

Method	Tumor Margin (mm), Median (IQR)	Difference (95% CI)	p-Value	ICC Consistency (95% CI)	p-Value	ICC Agreement (95% CI)	p-Value
Pathology	2 (1–2.75)						
13 MHz	2 (1–2.65)	0 (−0.05–0.15)	0.367	0.993 (0.985–0.997)	<0.001	0.993 (0.985–0.997)	<0.001
20 MHz	1.9 (1.15–3)	−0.1 (−0.05–0.25)	0.203	0.961 (0.921–0.981)	<0.001	0.96 (0.92–0.981)	<0.001
40 MHz	2 (1.1–2.8)	0 (−0.05–0.27)	0.195	0.972 (0.942–0.986)	<0.001	0.971 (0.941–0.986)	<0.001

4. Discussion

The skin is defined as the largest organ in the human body and it is also the first line of defense from different external aggressors. This protective line can be harmed by environmental factors, including ultraviolet radiation or chemicals, leading to the development of skin cancer. The incidence of both melanoma and non-melanoma malignancies is rapidly increasing on a worldwide scale. According to GLOBOCAN statistics data, there were more than 1 million new cases globally in 2020, placing NMSC fourth among all malignancies. The death rate is reduced when compared to melanoma or other cancers. There are various subtypes, including Merkel cell carcinoma, cutaneous squamous cell carcinoma, basal cell carcinoma and basosquamous cell carcinoma. There are also different types of presentation among the same histological type as nodular BCC, infiltrative BCC, Morpheus type BCC, etc. There are few studies in the literature that discuss the function of the US in diagnosing, staging and treating of NMSC, particularly in the head and neck area [33,34].

Staging cancer means to start with measurement and assessment of all three dimensions of the tumor (length, width and depth). The updated AJCC staging system uses the invasion measurement in staging of squamous cell carcinoma. In addition, the risk of regional metastases in proportion with the depth of invasion for the aggressive skin cancers as SCC, BSC or MCC [33]. Even if some clinicians choose the biopsy for diagnosis and staging of the highly suspicious lesions, this method can lead to staging errors because the thickness of the tumor is unknown before surgical intervention. This means that the chance to have positive surgical margins and an incorrect staging are higher. Dermoscopy can help the diagnosis process but not the staging process because the thickness of the tumor is not measurable by this method.

The US transducers frequency used in other studies varies between 5–18 MHz. We used three different transducers in order to obtain the closest measurement and to see the advantages and disadvantages of each one. The 13 MHz transducers offered a big picture of the tumor but the level of details, in the form of the presence of the hyperechoic spots, is reduced. We recommend this transducer for evaluation of surgical margins or for large skin tumors. The 20 and 40 MHz transducers are better for viewing the particularities of malignant lesions and for an accurate measurement but in the case of large size lesions, assessing of all three dimensions of the tumor can be difficult. Increasing the resolution of transducers obtains better level of details but, unfortunately, the level of penetrance decreases [6].

Dermoscopy can help the preoperative diagnosis process. Pampena R. et al. have examined retrospectively more than 400 dermoscopy images in order to find if it is possible to differentiate between the types of BCC. They found a prevalence of arborizing telangiectasia in invasive BCC (iBCC), as compared to superficial BCC (sBCC), but there were no significant differences in arborizing telangiectasia between iBCC and nodular BCC (nBCC). They found ulcerations more frequently in iBCC than both in sBCC and nBCC [34]. Because high frequency ultrasound and Doppler examination can evaluate both the vascularization and the presence of ulcerations, its use should be combined with dermoscopy in order to strengthen the results in BCC type differentiation. Knowing that the treatment should be more aggressive for SCC, with wider margins of resection and in some cases completed with neck dissection and even superficial parotidectomy, it is very important to be sure about the preoperative diagnosis.

The incisional biopsy is done for the big lesions but is very difficult or impossible to achieve in the case of small lesions. For these lesions, an excisional biopsy with wide margins is used for diagnosis and also for treatment. Mohs surgery can help for diagnosis and proper treatment of skin cancer. On ultrasound, the BCC and SCC look quite similar, except the BCC presents some hyperechoic spots. They look like a cotton flower but the particularity is the absence of the acoustic shadowing [35]. The spots are histologically represented by microcalcifications, horn cysts and apoptotic cells. Alfageme Roldan F has studied these spots and he found two interesting aspects: the increased number of spots is associated with an aggressive behavior and the micronodular type presents the highest number of spots. These results could be very useful in the diagnosis process and treatment plan [36]. In accordance with the other studies, we found an excellent correlation between the presence of hyperechoic spots and the diagnosis of BCC. We recommend this method for differentiating the BCC from other skin malignancies and for avoiding extending the surgical excision unnecessarily.

The strain value in strain elastography serves as a proximate indicator of tissue stiffness or elasticity. Variable skin layers have different elastic qualities, with subcutaneous tissue being more elastic and the epidermis and dermis being less elastic. Increased marginal stiffness is very suggesting for cancer. The malignant tissue develops a fibrotic environment that makes it stiffer than the surrounding tissues [36,37]. Some researchers came to the conclusion that elastography is a variable evaluation that, when combined with B-mode results, may provide the physician or the surgeon with important knowledge regarding different subtypes of BCC. More specifically, infiltrative and non-infiltrative cutaneous basal cell carcinomas are distinguished by enhanced marginal stiffness [38]. Tanaka et al. showed that the tumor thickness of nodular BCC determined by elastography corresponded better with the histological thickness than the tumor thickness measured by traditional B-mode sonography. These results suggest that elastography can be a dependable approach for planning future surgical safety margins since the infiltrative variety of BCC carries a greater risk of recurrence [39]. We did not find any correlation between the nodular or infiltrative type of NMSC and the stiffness of the tumor. We also did not try to measure the thickness of the tumor on elastography, so we cannot confirm the previous result. A certainty that we observed is that the increased stiffness of the tumor compared with the normal adjacent one, which can help us establish the diagnosis of malignant tumor and guide us in surgical treatment.

The treatment of NMSC is based on surgical excision, regional lymph nodes excision if necessary, and other forms of therapy such as radiation therapy, chemotherapy or immunotherapy in advanced cases. NCCN recommendations are based on a large study by Wolf and Zitelli. This study involved 117 cases of BCCs. The tumors were divided regarding the diameter: greater than 2 cm and smaller than 2 cm. They performed the surgical excision at 4 mm around the tumors. For tumors smaller than 2 cm, the complete removal was obtained in 95% of all cases. For the other tumors, the rate of complete removal was lower and they recommended larger margins than 4 mm [40].

The European Dermatology Forum (EDF) guidelines regarding the surgical margins for BCC are based on the previous EDF guidelines, the British Association of Dermatology guidelines (BAD), and the French guidelines. The recommendations are close to NCCN guidelines. If the BCC is low-risk (less than 2 cm in diameter) the surgical margins can be placed at 3–4 mm around the tumor. They also do not recommend doing the same for larger tumors because the risk of positive margins is higher, so they suggest 5 to 10 mm surgical margins. They confirm the results of previous studies with a complete excision rate of 95% when the mentioned protocol was used [41,42]. Because these tissues are considered natural barriers, the surgical excision should be conducted to the level of fascia perichondrium or periosteum when possible. If the tumor is located in a region with thicker skin, the excision can be done more superficially. Moreover, the type of BCC should guide the surgical excision because the baso-squamous cell carcinoma and morpheus type BCC have a poorer outcome [41,42].

Because a high percentage of the defects are closed immediately with local and regional flaps, finding positive margins based on a pathological report can be problematic. Mohs surgery, named after Frederick Mohs, the surgeon who invented it, is a technique where the surgical margins are microscopically controlled during the surgical intervention. This surgical method provides high cure rates in the treatment of a range of skin malignancies. Unfortunately, not all medical institutions provide it. Evaluation of margin status prior to a histology report would be nearly impossible without this approach [43].

Regarding the role of US in the treatment of NMSC, few studies concluded that US can easily detect malignant involvement of the margins. This may play a role in the treatment of skin cancer and prevention of local recurrence, providing improved overall survival rates. The adjacent normal tissue is measured from the limit of the tumor to the surgical resection margin. It can be easily measured on grayscale examination or elastography [44,45]. Our findings regarding the evaluation of surgical margins postoperatively illustrate the utility of this technique for immediate measurement of surgical margins directly in the operating room. All three transducers (13, 20 and 40 MHz) have shown very good results, but the most accurate was the 13 MHz transducer. This can be explained by how the lower the frequency, the deeper or greater penetrance a sound wave has. If positive or close margins are visualized, the surgery can be completed immediately. This way, the surgeon can avoid insufficient resection of the tumor or unnecessary sacrifice of healthy tissue. All these should reduce the recurrence rate and improve overall survival, but further studies are necessary to ascertain this fact.

Last but not least, US can be very useful for diagnosis and staging of melanoma skin cancer, which is still one of the deadliest types of cancer. A very precise DOI should be measured because the staging and the treatment change from a difference of just 1 mm. The treatment of both melanoma and non-melanoma skin cancer is difficult in the head and neck region due to the multiple anatomical vital elements, and the large excisions recommended for other anatomical regions are frequently not applicable. Comparative studies between these two pathologies could be further directions for skin cancer research.

5. Conclusions

NMSC is one of the most frequent malignancies in the human body and the incidence is quickly rising. The treatment of these tumors is mainly surgical but can differ for BCC, SCC and other types. For this reason, an appropriate preoperative diagnosis is necessary. Clinical examination and dermoscopy can help the diagnosis process, but only the pathological report can offer the final diagnosis. Unfortunately, for small tumors, only excisional biopsy can be done, which is primarily performed at the same time as surgical treatment of the lesion. Ultrasound examination with high frequency transducers can differentiate the BCC from other types of skin malignancy and is very easily accepted by the patient because it is a non-irradiating and non-invasive technique. Unnecessary healthy tissue sacrifice can be avoided with this technique. The staging of the tumor has the same importance as the diagnosis. Dermoscopy and clinical examination cannot evaluate the thickness of the tumor, which is very important for treatment and also for prognosis. Echography with high frequency transducers and elastography can measure precisely the thickness of the tumor and can guide the tumor excision, avoiding the tumor involvement of the surgical margins and reducing the recurrence rate. Involvement of positive margins and detection of close margins can be easily detected with 13, 20 and 40 MHz transducers.

Author Contributions: Conceptualization, T.T. and M.B.; methodology, C.D.; software, S.S. and G.A.; validation, G.B. and S.B.; formal analysis, D.C.L. and L.M.L.; investigation, T.T. and M.B.; resources, L.M.L. and A.T.; writing—original draft preparation, T.T.; writing—M.B. and L.M.L.; visualization, C.D. and F.O.; supervision, E.B.; project administration, T.T. and M.B. All authors have read and agreed to the published version of the manuscript.

Funding: This research received no external funding.

Institutional Review Board Statement: The study was conducted in accordance with the Declaration of Helsinki and was approved by the Ethical Committee of the Emergency County Hospital Cluj-Napoca, Romania (39109/A/0001/UK/R Reg. n. 362/08.01.2020) and the Iuliu Hatieganu University of Medicine and Pharmacy Cluj-Napoca, Romania (433/11.10.2019).

Informed Consent Statement: Informed consent was obtained from all subjects involved in the study. Written informed consent was obtained from the patients to publish this paper.

Data Availability Statement: The data presented in this study are available in the article.

Conflicts of Interest: The authors declare no conflict of interest.

References

1. Altamura, D.; Menzies, S.W.; Argenziano, G.; Zalaudek, I.; Soyer, H.P.; Sera, F.; Avramidis, M.; DeAmbrosis, K.; Fargnoli, M.C.; Peris, K. Dermatoscopy of basal cell carcinoma: Morphologic variability of global and local features and accuracy of diagnosis. *J. Am. Acad. Dermatol.* **2010**, *62*, 67–75. [CrossRef] [PubMed]
2. Didona, D.; Paolino, G.; Bottoni, U.; Cantisani, C. Non Melanoma Skin Cancer Pathogenesis Overview. *Biomedicines* **2018**, *6*, 6. [CrossRef] [PubMed]
3. Nassiri-Kashani, M.; Sadr, B.; Fanian, F.; Kamyab, K.; Noormohammadpour, P.; Shahshahani, M.M.; Zartab, H.; Naghizadeh, M.M.; Sarraf-Yazdy, M.; Firooz, A. Pre-operative assessment of basal cell carcinoma dimensions using high frequency ultrasonography and its correlation with histopathology. *Skin Res. Technol.* **2013**, *19*, e132–e138. [CrossRef] [PubMed]
4. Leiter, U.; Eigentler, T.; Garbe, C. Epidemiology of skin cancer. *Adv. Exp. Med. Biol.* **2014**, *810*, 120–140. [CrossRef] [PubMed]
5. Katalinic, A.; Kunze, U.; Schäfer, T. Epidemiology of cutaneous melanoma and non-melanoma skin cancer in Schleswig-Holstein, Germany: Incidence, clinical subtypes, tumour stages and localization (epidemiology of skin cancer). *Br. J. Dermatol.* **2003**, *149*, 1200–1206. [CrossRef]
6. Tamas, T.; Dinu, C.; Lenghel, M.; Băciuț, G.; Bran, S.; Stoia, S.; Băciuț, M. The role of ultrasonography in head and neck Non-Melanoma Skin Cancer approach: An update with a review of the literature. *Med. Ultrason.* **2021**, *23*, 83–88. [CrossRef]
7. Barcaui, E.d.O.; Carvalho, A.C.; Valiante, P.M.; Barcaui, C.B. High-frequency ultrasound associated with dermoscopy in pre-operative evaluation of basal cell carcinoma. *An. Bras. Dermatol.* **2014**, *89*, 828–831. [CrossRef]
8. Calzavara-Pinton, P.; Ortel, B.; Venturini, M. Non-melanoma skin cancer, sun exposure and sun protection. *G. Ital. Di Dermatol. E Venereol. Organo Uff. Soc. Ital. Di Dermatol. E Sifilogr.* **2015**, *150*, 369–378.
9. Crişan, D.; Badea, A.F.; Crişan, M.; Rastian, I.; Gheuca Solovastru, L.; Badea, R. Integrative analysis of cutaneous skin tumours using ultrasonogaphic criteria. Preliminary results. *Med. Ultrason.* **2014**, *16*, 285–290. [CrossRef]

10. Crisan, M.; Crisan, D.; Sannino, G.; Lupsor, M.; Badea, R.; Amzica, F. Ultrasonographic staging of cutaneous malignant tumors: An ultrasonographic depth index. *Arch. Dermatol. Res.* **2013**, *305*, 305–313. [CrossRef]
11. Badea, R.; Crişan, M.; Lupşor, M.; Fodor, L. Diagnosis and characterization of cutaneous tumors using combined ultrasonographic procedures (conventional and high resolution ultrasonography). *Med. Ultrason.* **2010**, *12*, 317–322.
12. Alfageme Roldán, F. Elastography in Dermatology. Elastografía en dermatología. *Actas Dermo-Sifiliogr.* **2016**, *107*, 652–660. [CrossRef]
13. Wortsman, X. Sonography of facial cutaneous basal cell carcinoma: A first-line imaging technique. *J. Ultrasound Med.* **2013**, *32*, 567–572. [CrossRef]
14. Song, W.J.; Choi, H.J.; Lee, Y.M.; Tark, M.S.; Nam, D.H.; Han, J.K.; Cho, H.D. Clinical analysis of an ultrasound system in the evaluation of skin cancers: Correlation with histology. *Ann. Plast. Surg.* **2014**, *73*, 427–433. [CrossRef]
15. Wortsman, X.; Vergara, P.; Castro, A.; Saavedra, D.; Bobadilla, F.; Sazunic, I.; Zemelman, V.; Wortsman, J. Ultrasound as predictor of histologic subtypes linked to recurrence in basal cell carcinoma of the skin. *J. Eur. Acad. Dermatol. Venereol. JEADV* **2015**, *29*, 702–707. [CrossRef]
16. Schmid-Wendtner, M.H.; Burgdorf, W. Ultrasound scanning in dermatology. *Arch. Dermatol.* **2005**, *141*, 217–224. [CrossRef]
17. Bhatt, K.D.; Tambe, S.A.; Jerajani, H.R.; Dhurat, R.S. Utility of high-frequency ultrasonography in the diagnosis of benign and malignant skin tumors. *Indian J. Dermatol. Venereol. Leprol.* **2017**, *83*, 162–182. [CrossRef]
18. Jasaitiene, D.; Valiukeviciene, S.; Linkeviciute, G.; Raisutis, R.; Jasiuniene, E.; Kazys, R. Principles of high-frequency ultrasonography for investigation of skin pathology. *J. Eur. Acad. Dermatol. Venereol. JEADV* **2011**, *25*, 375–382. [CrossRef]
19. Bezugly, A. High frequency ultrasound study of skin tumors in dermatological and aesthetic practice. *Med. Ultrason.* **2015**, *17*, 541–544. [CrossRef]
20. Dasgeb, B.; Morris, M.A.; Mehregan, D.; Siegel, E.L. Quantified ultrasound elastography in the assessment of cutaneous carcinoma. *Br. J. Radiol.* **2015**, *88*, 20150344. [CrossRef]
21. Markowitz, O.; Schwartz, M.; Feldman, E.; Bienenfeld, A.; Bieber, A.K.; Ellis, J.; Alapati, U.; Lebwohl, M.; Siegel, D.M. Evaluation of Optical Coherence Tomography as a Means of Identifying Earlier Stage Basal Cell Carcinomas while Reducing the Use of Diagnostic Biopsy. *J. Clin. Aesthetic Dermatol.* **2015**, *8*, 14–20.
22. Olmedo, J.M.; Warschaw, K.E.; Schmitt, J.M.; Swanson, D.L. Optical coherence tomography for the characterization of basal cell carcinoma in vivo: A pilot study. *J. Am. Acad. Dermatol.* **2006**, *55*, 408–412. [CrossRef] [PubMed]
23. Gambichler, T.; Orlikov, A.; Vasa, R.; Moussa, G.; Hoffmann, K.; Stücker, M.; Altmeyer, P.; Bechara, F.G. In vivo optical coherence tomography of basal cell carcinoma. *J. Dermatol. Sci.* **2007**, *45*, 167–173. [CrossRef] [PubMed]
24. Sauermann, K.; Gambichler, T.; Wilmert, M.; Rotterdam, S.; Stücker, M.; Altmeyer, P.; Hoffmann, K. Investigation of basal cell carcinoma [correction of carcionoma] by confocal laser scanning microscopy in vivo. *Skin Res. Technol.* **2002**, *8*, 141–147. [CrossRef]
25. Nori, S.; Rius-Díaz, F.; Cuevas, J.; Goldgeier, M.; Jaen, P.; Torres, A.; González, S. Sensitivity and specificity of reflectance-mode confocal microscopy for in vivo diagnosis of basal cell carcinoma: A multicenter study. *J. Am. Acad. Dermatol.* **2004**, *51*, 923–930. [CrossRef] [PubMed]
26. Patalay, R.; Talbot, C.; Alexandrov, Y.; Lenz, M.O.; Kumar, S.; Warren, S.; Munro, I.; Neil, M.A.; König, K.; French, P.M.; et al. Multiphoton multispectral fluorescence lifetime tomography for the evaluation of basal cell carcinomas. *PLoS ONE* **2012**, *7*, e43460. [CrossRef] [PubMed]
27. Koo, T.K.; Li, M.Y. A Guideline of Selecting and Reporting Intraclass Correlation Coefficients for Reliability Research. *J. Chiropr. Med.* **2016**, *15*, 155–163. [CrossRef]
28. Portney, L.G.; Watkins, M.P. *Foundations of Clinical Research: Applications to Practice*, 3rd ed.; Prentice Hall: Upper Saddle River, NJ, USA, 2009; p. 892.
29. Gamer, M.; Lemon, J.; Singh, I.F.P. Various Coefficients of Interrater Reliability and Agreement. 2019. Available online: https://CRAN.R-project.org/package=irr (accessed on 12 October 2022).
30. McHugh, M.L. Interrater reliability: The kappa statistic. *Biochem. Med.* **2012**, *22*, 276–282. [CrossRef]
31. Altman, D.G. *Practical Statistics for Medical Research*, 1st ed.; Chapman and Hall/CRC: Boca Raton, FL, USA, 1990; 624p.
32. Sim, J.; Wright, C.C. The kappa statistic in reliability studies: Use, interpretation, and sample size requirements. *Phys. Ther.* **2005**, *85*, 257–268. [CrossRef]
33. Tamas, T.; Baciut, M.; Nutu, A.; Bran, S.; Armencea, G.; Stoia, S.; Manea, A.; Crisan, L.; Opris, H.; Onisor, F.; et al. Is miRNA Regulation the Key to Controlling Non-Melanoma Skin Cancer Evolution? *Genes* **2021**, *12*, 1929. [CrossRef]
34. Pampena, R.; Parisi, G.; Benati, M.; Borsari, S.; Lai, M.; Paolino, G.; Cesinaro, A.M.; Ciardo, S.; Farnetani, F.; Bassoli, S.; et al. Clinical and Dermoscopic Factors for the Identification of Aggressive Histologic Subtypes of Basal Cell Carcinoma. *Front. Oncol.* **2021**, *10*, 630458. [CrossRef]
35. Uhara, H.; Hayashi, K.; Koga, H.; Saida, T. Multiple hypersonographic spots in basal cell carcinoma. *Dermatol. Surg. Off. Publ. Am. Soc. Dermatol. Surg.* **2007**, *33*, 1215–1219. [CrossRef]
36. Alfageme Roldán, F. Ultrasound skin imaging. *Actas Dermo-Sifiliogr.* **2014**, *105*, 891–899. [CrossRef]
37. Morris, M.A.; Ring, C.M.; Managuli, R.; Saboury, B.; Mehregan, D.; Siegel, E.; Dasgeb, B. Feature analysis of ultrasound elastography image for quantitative assessment of cutaneous carcinoma. *Skin Res. Technol.* **2018**, *24*, 242–247. [CrossRef]

38. Alfageme, F.; Salgüero, I.; Nájera, L.; Suarez, M.L.; Roustan, G. Increased Marginal Stiffness Differentiates Infiltrative from Noninfiltrative Cutaneous Basal Cell Carcinomas in the Facial Area: A Prospective Study. *J. Ultrasound Med.* **2019**, *38*, 1841–1845. [CrossRef]
39. Tanaka, T.; Tada, Y.; Ohnishi, T.; Watanabe, S. Usefulness of real-time tissue elastography for detecting the border of basal cell carcinomas. *J. Dermatol.* **2017**, *44*, 438–443. [CrossRef]
40. National Comprehensive Cancer Network. NCCN Clinical Practice Guidelines in Oncology; Merkel Cell Carcinoma. Available online: http://www.nccn.org (accessed on 15 January 2017).
41. Tai, P. A practical update of surgical management of Merkel cell carcinoma of the skin. *ISRN Surg.* **2013**, *2013*, 850797. [CrossRef]
42. European Dermatology Forum. Guideline on the Diagnosis and Treatment of Merkel Cell Carcinoma. Available online: http://www.euroderm.org (accessed on 3 January 2016).
43. Prickett, K.A.; Ramsey, M.L. Mohs Micrographic Surgery. In *StatPearls*; StatPearls Publishing: Treasure Island, FL, USA, 2022. Available online: https://www.ncbi.nlm.nih.gov/books/NBK441833/ (accessed on 25 July 2022).
44. Nolan, G.S.; Wormald, J.C.R.; Kiely, A.L.; Totty, J.P.; Jain, A. Global incidence of incomplete surgical excision in adult patients with non-melanoma skin cancer: Study protocol for a systematic review and meta-analysis of observational studies. *Syst. Rev.* **2020**, *9*, 83. [CrossRef]
45. Vilas-Sueiro, A.; Alfageme, F.; Salgüero, I.; De Las Heras, C.; Roustan, G. Ex Vivo High-Frequency Ultrasound for Assessment of Basal Cell Carcinoma. *J. Ultrasound Med.* **2019**, *38*, 529–531. [CrossRef]

diagnostics

MDPI

Article

Autofluorescence Imaging of the Skin Is an Objective Non-Invasive Technique for Diagnosing Pseudoxanthoma Elasticum

Klára Farkas [1], Szabolcs Bozsányi [1], Dóra Plázár [1], András Bánvölgyi [1], Luca Fésűs [1], Pálma Anker [1], Sára Zakariás [1], Ilze Lihacova [2], Alexey Lihachev [2], Marta Lange [2], Tamás Arányi [3,4], Norbert M. Wikonkál [1], Márta Medvecz [1,†] and Norbert Kiss [1,*,†]

1 Department of Dermatology, Venereology and Dermatooncology, Semmelweis University, 1085 Budapest, Hungary; farkas.klara@phd.semmelweis.hu (K.F.); bozsanyi.szabolcs@med.semmelweis-univ.hu (S.B.); plazar.dora@phd.semmelweis.hu (D.P.); banvolgyi.andras@med.semmelweis-univ.hu (A.B.); fesus.luca@med.semmelweis-univ.hu (L.F.); anker.palma@phd.semmelweis.hu (P.A.); zakarias.sara@phd.semmelweis.hu (S.Z.); wikonkal.norbert@med.semmelweis-univ.hu (N.M.W.); medvecz.marta@med.semmelweis-univ.hu (M.M.)

2 Biophotonics Laboratory, Institute of Atomic Physics and Spectroscopy, University of Latvia, LV-1004 Riga, Latvia; ilze.lihacova@gmail.com (I.L.); aleksejs.lihacovs@gmail.com (A.L.); marta.lange.rtu@gmail.com (M.L.)

3 Department of Molecular Biology, Semmelweis University, 1085 Budapest, Hungary; aranyi.tamas@ttk.mta.hu

4 Institute of Enzymology, Research Center for Natural Sciences, 1117 Budapest, Hungary

* Correspondence: kiss.norbert@med.semmelweis-univ.hu

† Authors share senior authorship.

Abstract: Pseudoxanthoma elasticum (PXE) is a rare multisystemic autosomal recessive connective tissue disease. In most cases, skin manifestations of PXE are the first to develop, followed later by severe ocular and cardiovascular complications. In our present study, in addition to dermoscopy, we introduced novel techniques, autofluorescence (AF) and diffuse reflectance (DR) imaging for the assessment of affected skin sites of five PXE patients. PXE-affected skin areas in most skin sites showed a previously observed pattern upon dermoscopic examination. With the novel imaging, PXE-affected skin lesions displayed high AF intensity. During our measurements, significantly higher mean, minimum and maximum AF intensity values were found in areas of PXE-affected skin when compared to uninvolved skin. Conversely, images acquired with the use of 660 and 940 nm illumination showed no mentionable difference. Our results demonstrate that AF imaging may be used in the in vivo diagnostics and quantification of the severity of the skin lesions of PXE patients. In addition, it is a safe, fast and cost-effective diagnostic method. AF imaging may be also used to objectively monitor the efficacy of the possible novel therapeutic approaches of PXE in the future.

Keywords: pseudoxanthoma elasticum; autofluorescence imaging; LED; dermoscopy; quantitative analysis; calcification; diffuse reflectance imaging; diagnosis

1. Introduction

Rare diseases may occur infrequently, yet, taken together, they affect a significant proportion of the population and place an increasing burden on society and healthcare [1]. Pseudoxanthoma elasticum (PXE, OMIM#264800) is a rare multisystemic autosomal recessive connective tissue disease. Its prevalence is estimated to be between 1 in 25,000 and 1 in 50,000, with a 2:1 female predominance [2]. PXE can be caused by over 300 mutations, located at the *ABCC6* (ATP-binding cassette subfamily C member 6) gene [3]. These mutations lead to decreased serum levels of inorganic pyrophosphate (PPi), which is an anti-mineralization factor. As a consequence, fragmentation of elastic fibers and deposition of calcium salts develop in the mid-dermis of the skin, in the Bruch's membrane of the retina and in blood vessels [4]. Histopathological features of PXE are also characteristic for PXE-like conditions that include PXE-like syndrome with multiple coagulation factor deficiency [5] and generalized arterial calcification of infancy (GACI, OMIM#614473), which should be considered in the differential diagnosis [6]. Other dermal elastic tissue disorders, such as perforating calcific elastosis, late-onset focal dermal elastosis or papillary dermal elastolysis may clinically and histologically resemble PXE, while they mainly occur in the elderly and lack systemic manifestations [7,8]. Clinical signs of PXE usually occur during the first two decades of life. In most cases, skin lesions appear first as small, 1–5 mm yellowish papules on the neck, and similar lesions progressively develop at the periumbilical region and the flexural areas of the extremities, in particular, the axillar, popliteal and inguinal regions. Later in life, the papules coalesce into plaques and the skin becomes redundant and loose [9–11].

Ocular manifestations of PXE present most frequently with angioid streaks, that lead to choroidal neovascularization (CNV) which may associate with oedema, bleeding, chorioretinal atrophy and scars, and consequently loss of central vision [12,13]. The fragmentation of the elastic fibers and the calcification of the lamina elastica of the medium-sized arteries are behind the cardiovascular manifestations of PXE. These include premature arteriosclerosis, intermittent claudication, increased risk of early acute myocardial infarctions and stroke [14].

Dermoscopy (DS) is a widely available clinical tool, yet only a few recent studies investigated the dermoscopic features of PXE. These include yellowish-white clods, a light purplish-red background, and reticulated vessels. Yellow or white areas show differently arranged globules. The pattern of globules can present as dots, irregular broad or narrow mesh networks or parallel lines that coalesces to plaques. The vascular pattern appears as a fine reticular network or a pink background [15]. A previous study assumed that the dermal elastolysis results in the vascular realignment with a linear superficial vessel appearance [16]. Kawashima et al. described the appearance of DS images and observed the differences of dermoscopic features between PXE and PXE-like conditions. They found that the colors of the clods and the background are different, so the differential diagnosis may be possible based on DS [17]. In pilot studies, high-frequency ultrasonography (HFUS) and reflectance confocal microscopy (RCM) could detect certain characteristic features of PXE [18,19]. Ex Vivo nonlinear optical microscopy was also previously introduced to visualize calcification and fragmentation of elastin fibers in PXE [20,21], among investigations on other rare disorders [22,23]. However, in vivo use of this technique was not yet reported in these indications. In a recent study, inflammation and calcification were detected in PXE-affected skin with 18-FluroDeoxyGlucose and 18F-Sodium Fluoride positron emission tomography-computed tomographic (PET-CT) imaging [24]. Limitations of this technique are the radiation exposure, low resolution and high cost.

Autofluorescence (AF) and diffuse reflectance (DR) imaging using light emitting diode (LED) illumination has been in use since the 2010s [25] and has previously been utilized to visualize different skin conditions [26–28]. A multispectral LED-based device prototype was successfully tested in the assessment of skin cancer and the detection of recurrence in post-operative scars [27,29]. Furthermore, it was capable to distinguish melanoma from benign lesions, including seborrheic keratosis and pigmented nevi [26]. Compared to other imaging techniques of the skin, it is a fast, cost-effective method which can be applied conveniently by any physician. The morphological and optical properties of skin lesions render PXE a strong candidate for successful use of AF imaging. In particular, the calcium salt deposits in the mid-dermis are expected to give significant AF signal based on previous spectrometric and ophthalmologic studies [30–32]. In our present study, in addition to DS, we set out to assess the clinical relevance of AF and DR imaging to visualize and objectify the skin lesions in PXE patients.

2. Materials and Methods

2.1. Patient Data

Three female and two male patients were included in this study, with a mean age of 55 ± 8.1 years. All patients were diagnosed and managed with PXE at Semmelweis University, Budapest, Hungary. Definitive diagnosis of the patients was established based on the revised diagnostic criteria of PXE [33], including skin biopsy and histological analysis. The diagnosis of PXE was verified by molecular genetic analysis of the *ABCC6* gene in all patients by the Center for Medical Genetics, Ghent University Hospital, Ghent, Belgium. We evaluated the severity of the disease with the Phenodex score based on clinical findings of six organ systems [34]. The patients' data are summarized in Table 1. This study was approved by the Ethics Committee of Semmelweis University, Budapest, Hungary (SE RKEB No. 228/2018, approval date: 12 December 2018). All involved patients were informed about the study and signed the required consent.

Table 1. Demographic data and Phenodex score of the PXE patients.

Pt. No.	Sex	Age (y)	Phenodex Score
1	F	53	S3 E3 G0 V1 C0 R0
2	F	53	S2 E3 G0 V1 C1 R0
3	M	69	S3 E3 G0 V0 C0 R0
4	M	52	S2 E3 G0 V0 C1 R0
5	F	48	S2 E2 G0 V0 C0 R0

Pt, patient; F, female; M, male; y, years; S, skin; E, eye; G, gastrointestinal; V, vascular; C, cardiac; R, renal; PXE, pseudoxanthoma elasticum.

2.2. Dermoscopic and Clinical Image Acquisition

Dermoscopic (Heine Delta 20, Heine Optotechnik GmbH, Herrsching, Germany) and clinical photos were collected from all PXE-affected skin sites (the neck, the periumbilical region and the flexural areas, in particular, the axillar, popliteal and inguinal regions) of the patients.

2.3. Autofluorescence and Diffuse Reflectance Imaging with Narrow-Band LED Excitation

A multispectral LED device prototype was used, as described earlier [35]. AF signal was excited at 405 nm, while DR images were acquired with 660 and 940 nm LEDs (SML-LXL8047UVC, Lumex Inc., Ronkonkoma, NY, USA) arranged circularly for uniform illumination with a power density of 20 mW/cm^2. A >515 nm long pass filter was applied in order to prevent detection of 405 nm LED emission. Measurements were controlled by the direct skin contact of the device in order to block outside illumination to achieve consistent data collection. Images were collected with a color CMOS 5 megapixel IDS camera (MT9P006STC, IDS uEye UI3581LE-C-HQ, Obersulm, Germany) fixed at 60 mm distance from the illuminated skin [36]. The acquired images were automatically transferred to a cloud server for further data processing and analysis [37].

All affected skin sites of the patients were captured by this multispectral LED device. Quantitative analyses were carried out using the 405 nm AF channel and 660 and 940 nm DR images by ImageJ v1.52a software (NIH, Bethesda, MD, USA). Seven different skin sites of each patient were analyzed using the images acquired with the three different illumination wavelength settings. Altogether, 105 images were analyzed. Using the AF images, the PXE-affected areas were selected manually as regions of interest (ROI) based on their DS morphology according to patterns described earlier [15,17]. Adjacent to the affected areas, the uninvolved background skin served as the control group. Then, we transferred the selected ROI of AF images to DR images. Images were analyzed using intensity descriptors in comparison to uninvolved skin. We analyzed the mean, minimum and maximum AF intensity values within the selected ROIs. Intensity values are expressed as arbitrary units (A.U.) as measured by ImageJ software.

2.4. Statistical Analysis

Statistical analyses were performed with GraphPad Prism v9.0.0 software (GraphPad Software Inc., La Jolla, CA, USA) using unpaired two-tailed Student's t-test. We considered p-values less than 0.05 statistically significant. All results are expressed as mean $\pm$ standard deviation.

3. Results

3.1. Clinical Presentation and Histopathological Features

The clinical photos of the patients showed the characteristic skin changes of PXE (Figure 1). In the case of Patient 2 or Patient 5, we found moderate skin involvement, while in the rest of the cases, more extensive, prominent skin manifestations were present.

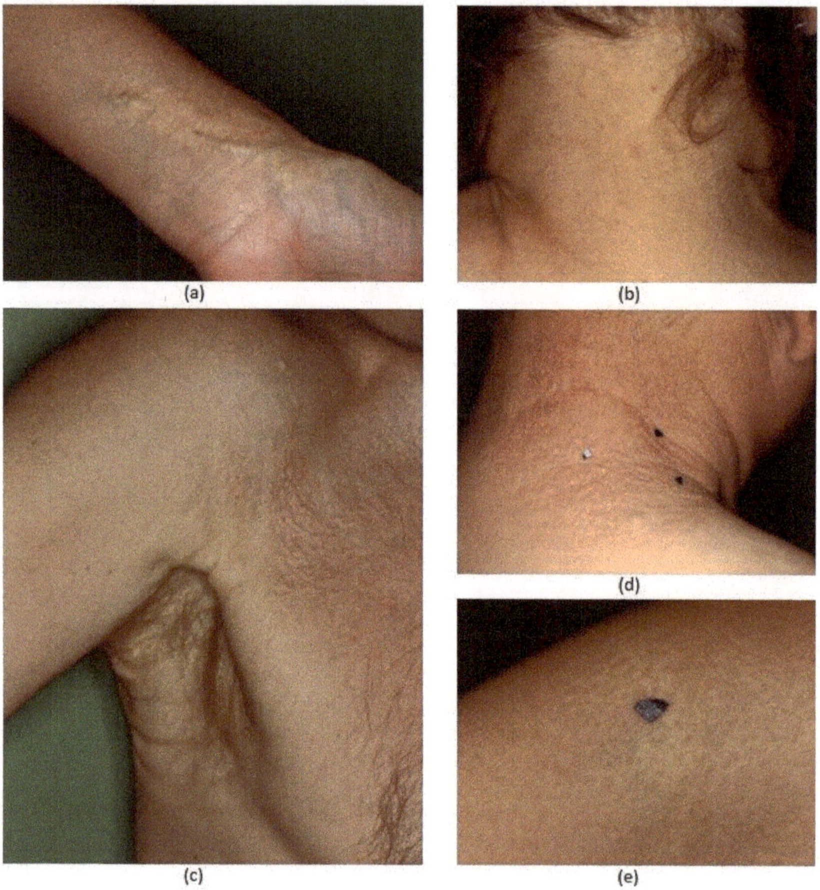

Figure 1. Clinical photographs that display the typical cutaneous manifestations of the patients with pseudoxanthoma elasticum. (**a**) Patient 1, large coalescent plaques on the wrist (**b**) Patient 2, 1–2 mm yellowish papules on the right side of the neck (**c**) Patient 3, redundant, loose skin in the axilla and 2–5 mm papules on the axilla (**d**) Patient 4, 2–6 mm papules on the right and dorsal side of the neck (**e**) Patient 5, right antecubital fossa, yellow-white flat plaques.

The location of the skin lesions varied between patients. Particularly, while the neck was affected in all cases, yellowish papules on the periumbilical region were only visible in two cases and the chest was affected only in Patient 1. The localization of the patients' skin manifestations is summarized in Table 2.

Table 2. Affected skin sites of the pseudoxanthoma elasticum patients.

	Pt 1	Pt 2	Pt 3	Pt 4	Pt 5
neck	+	+	+	+	+
axilla	+	(+)	+	+	(+)
antecubital fossa	+	(+)	+	(+)	+
popliteal fossa	+	(+)	+	(+)	(+)
inguinal	+	+	+		(+)
periumbilical	+				+
wrist	+			(+)	
chest	(+)				

Pt, patient; +, clearly visible papules or plaques; (+) mild, poorly visible skin lesions.

Skin biopsy samples were collected from clinically affected skin lesions. The histopathologic findings confirmed the diagnosis of PXE. Typical histological features were observed with specific stains by light microscopy. Hematoxylin and eosin staining showed disorganized connective tissue structure. Extensive calcium deposits were identified in the mid-dermis with Von Kossa staining. Weigert's elastic staining revealed the fragmentation of mineralized elastin, and van Gieson staining showed abnormal, disrupted collagen fibers surrounding calcified areas (Figure 2).

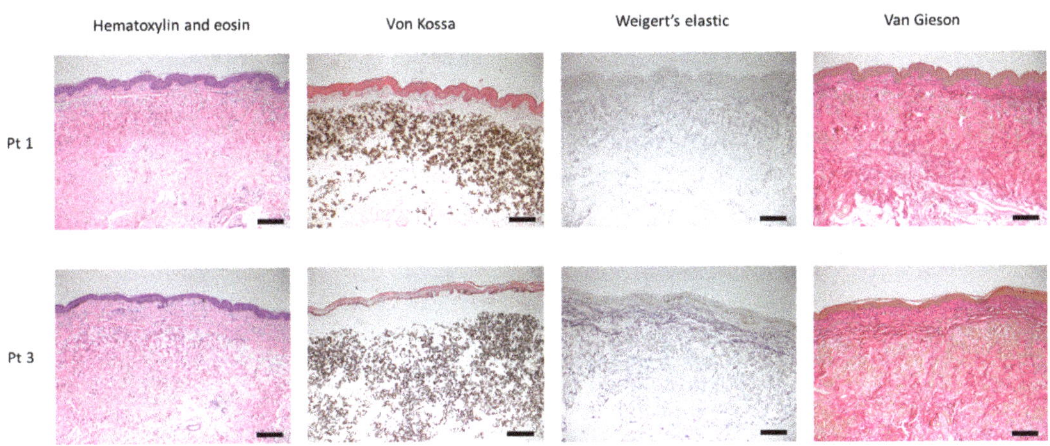

Figure 2. Histopathological characteristics of skin biopsies of the pseudoxanthoma elasticum patients (Pt) with hematoxylin and eosin (H&E), von Kossa (VK) staining for calcium deposits, Weigert's elastic (WE) staining for elastin and van Gieson (VG) staining for collagen. Scale bars display 100 μm.

3.2. Dermoscopy Images

In the acquired DS images, in most cases, we identified yellowish clods on a light purplish-red background and reticulated vessels. The pattern of globules appeared as dots, irregular mesh networks or parallel lines (Figure 3).

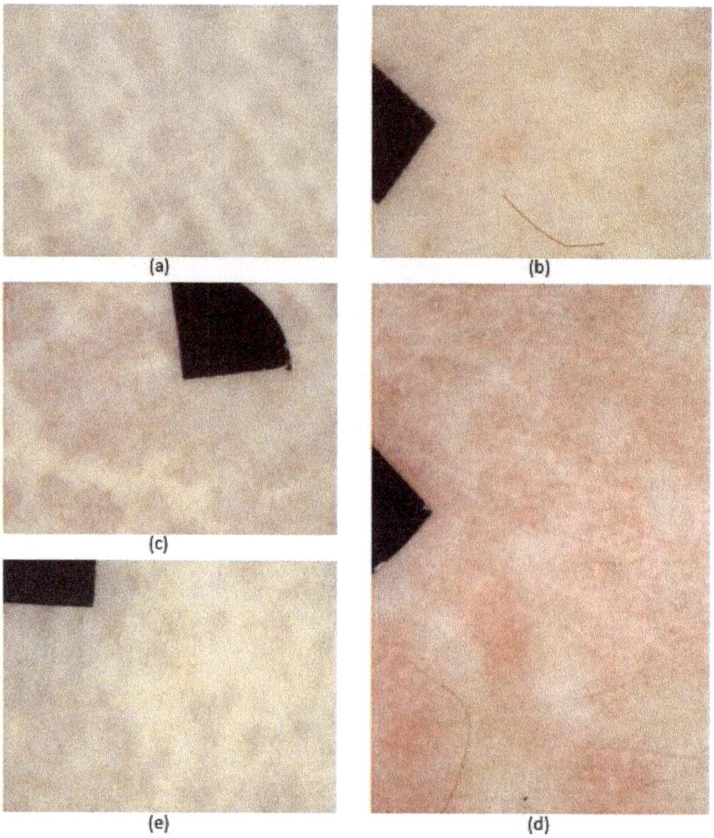

Figure 3. Representative dermoscopic images of the skin lesions of the patients with pseudoxanthoma elasticum. (**a**) Patient 1, linear pattern on the cubital fossa. (**b**) Patient 2, dots on the neck. (**c**) Patient 3, narrow mesh network on the axilla. (**d**) Patient 4, broad mesh network on the neck. (**e**) Patient 5, plaques of confluent dots and linear patterns in the antecubital fossa. Black markers do not point to areas of interest, they are used for image alignment (area: 0.125 cm^2).

However, there were some clinically affected skin sites where the specific DS features were absent (Table 3). Namely, in inguinal regions, the typical pattern appeared only in Patient 1. At the periumbilical region of Patient 1, slightly visible reticulated vessels and not obviously distinguishable globules could be detected. There were no visible DS patterns on the chest of Patient 1, the wrist of Patient 4 and the periumbilical regions of Patient 5.

Table 3. Characteristic dermoscopic features of the affected skin sites of the pseudoxanthoma elasticum patients according to the nomenclature of Berthin et al. [15] and Kawashima et al. [17].

	Structure	Pt 1	Pt 2	Pt 3	Pt 4	Pt 5
neck	globules background vessels	plaques red reticulated, linear	dots, linear brownish reticulated	mesh network purple-red reticulated	linear red reticulated	dots, plaques brownish no
axilla	globules background vessels	plaques purple linear	dots brownish no	mesh network purple-red reticulated	dots pink no	dots brownish no

Table 3. *Cont.*

	Structure	Pt 1	Pt 2	Pt 3	Pt 4	Pt 5
antecubital fossa	globules background vessels	linear purple-red reticulated	(non-typical)	linear purple reticulated	linear brownish no	plaque purple no
popliteal fossa	globules background vessels	linear purple-red several reticulated	linear brownish no	linear purple reticulated	dots pink reticulated	linear purple reticulated
inguinal	globules background vessels	dots purple reticulated	(non-typical)	(non-typical)		(non-typical)
periumbilical	globules background vessels	dots purple-red reticulated				(absent)
wrist	globules background vessels	plaques purple reticulated			(absent)	
chest	globules background vessels	(absent)				

Pt, patient. Blank cells indicate that DS images were not captured in that region, due to the lack of cutaneous manifestations.

3.3. *Autofluorescence and Diffuse Reflectance Imaging*

The PXE-affected skin sites of each patient selected for AF and DR imaging are summarized in Table 4. AF, DR images with narrow-band LED excitation and DS findings are displayed in Figure 4. PXE-affected skin areas appeared as fields with high AF intensity with 405 nm LED excitation. With 660 nm illumination, high background skin DR signal was visible. Images acquired with 940 nm illumination showed unremarkable, low-contrast DR patterns, or in certain cases, no additional signal was detected.

AF images showed high-contrast signal of the skin lesions even in those sites of the skin where non-typical or normal DS pattern was visible. AF images acquired from the antecubital fossa and inguinal region (Figure 4d) of Patient 2 showed a well-visible affected area compared to DS images where a non-typical pattern was apparent. The AF image from the inguinal region of Patient 3 showed clearly visible lesions, but the DS image was non-typical. AF image from the wrist of Patient 4 displayed higher intensity area, while DS pattern was completely absent (Figure 4h). AF images from the inguinal (Figure 4i) and periumbilical regions of Patient 5 revealed remarkable high intensity areas, but the DS images presented non-typical or normal pattern.

Table 4. The investigated skin sites of the pseudoxanthoma elasticum patients.

	Pt 1	Pt 2	Pt 3	Pt 4	Pt 5
Site 1	popliteal fossa	popliteal fossa	popliteal fossa	popliteal fossa	periumbilical
Site 2	Inguinal	inguinal	inguinal	axilla	inguinal
Site 3	periumbilical	axilla	inguinal	axilla	axilla
Site 4	axilla	axilla	axilla	axilla	antecubital fossa
Site 5	wrist	antecubital fossa	antecubital fossa	wrist	antecubital fossa
Site 6	antecubital fossa	neck	neck	antecubital fossa	neck
Site 7	neck	neck	neck	neck	neck

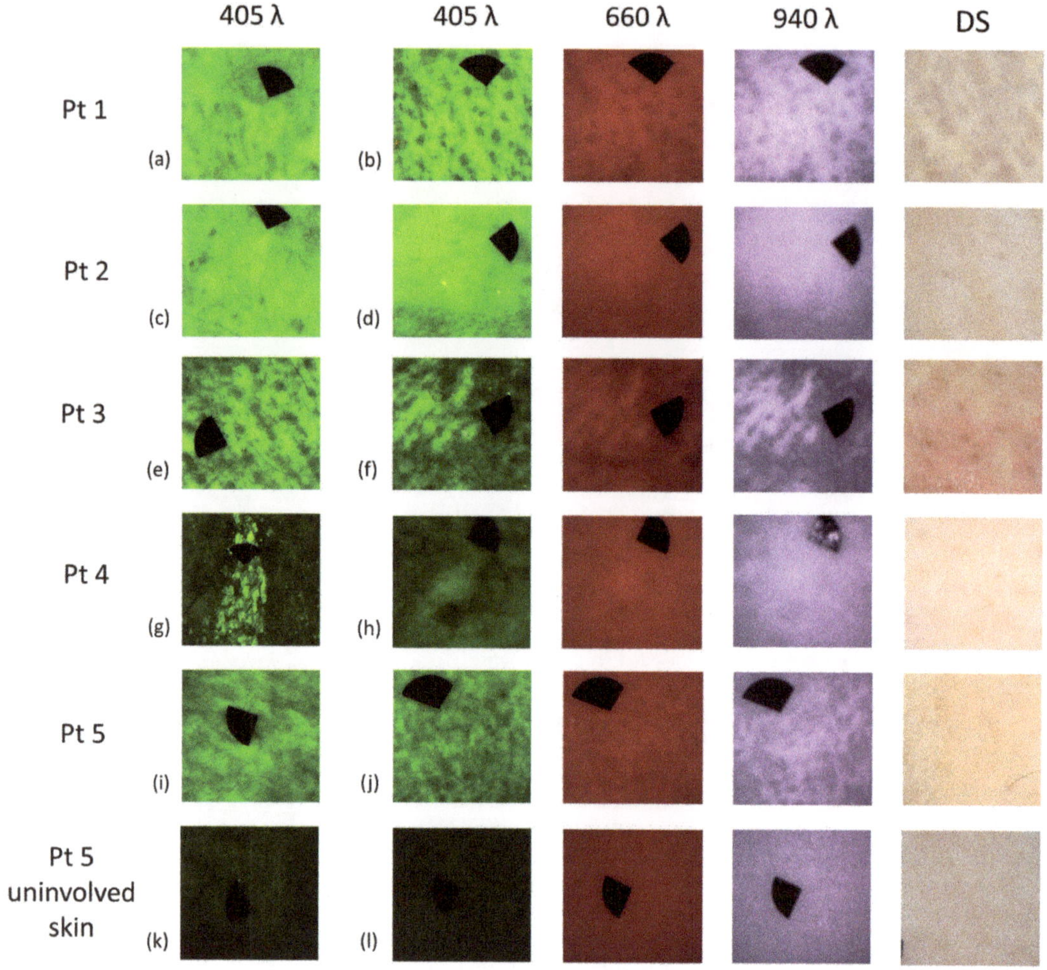

Figure 4. Representative autofluorescence (AF), diffuse reflectance (DR) and dermoscopy (DS) images of the affected skin sites of the pseudoxanthoma elasticum patients. (**a**) AF image of the axilla of Patient (Pt) 1. (**b**) AF, DR and DS images of the antecubital fossa of Pt 1, showing PXE-specific morphologic structures. (**c**) AF image of the axilla of Pt 2. (**d**) AF, DR and DS images of the inguinal region of Pt 2. AF image reveals a well-visible extensive area with high AF signal. DR images show barely noticeable pattern. DS image shows no typical pattern. (**e**) AF image of the antecubital fossa of Pt 3. (**f**) AF, DR and DS images of the neck of Pt 3, showing PXE-specific morphologic structures. (**g**) AF image from popliteal fossa of Pt 4. (**h**) AF, DR and DS images from wrist of Pt 4. AF image gives high-contrast signal. DR images are less informative. (**i**) AF image of the inguinal region of Pt 5. (**j**) AF, DR and DS images of the axilla of Pt 5, displaying PXE-specific morphologic structures. (**k**) AF images of Pt 5, uninvolved normal skin. (**l**) AF, DR and DS images of uninvolved normal skin, upper arm of Pt 5. The size of the images is 2×2 cm^2. Black markers do not point to areas of interest, they are used for image alignment (area: 0.125 cm^2).

Quantitative analyses revealed significantly higher mean AF intensity values in the PXE-affected skin areas compared to uninvolved skin (60.74 ± 23.86 vs. 43.40 ± 16.76 A.U.; $p < 0.0008$, Figure 5a,b). Upon measuring the minimum (36.14 ± 15.29 vs. 25.17 ± 11.66 A.U.; $p < 0.0012$) and the maximum values (109.8 ± 30.87 vs. 79.69 ± 23.76 A.U.; $p < 0.0001$) of AF intensity, we found significantly higher rates in the affected areas (Figure 5b). Analyzing DR images with 660 and 940 nm illumination of the same field of views, no significant differences were found.

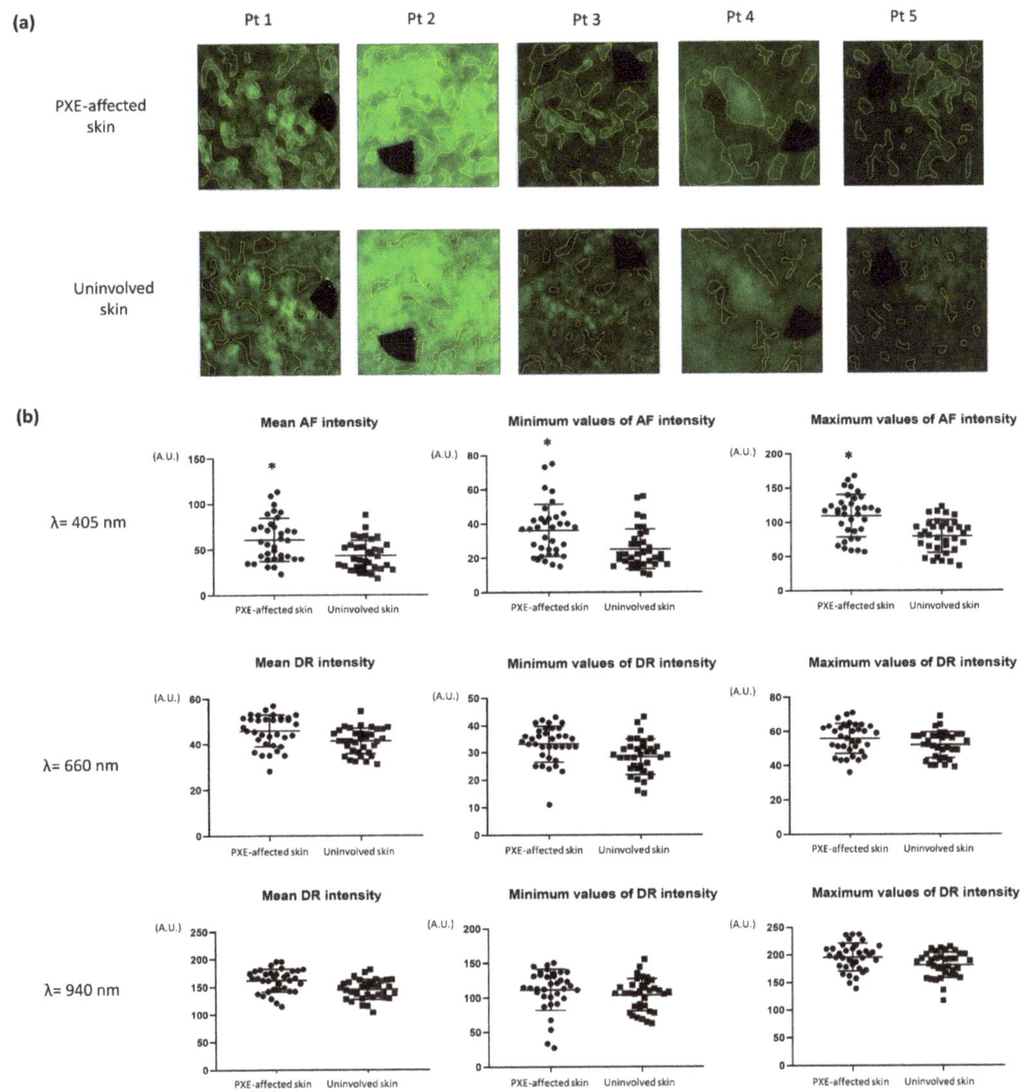

Figure 5. Regions of interest (ROI) selection and results of quantitative analyses of autofluorescence (AF) and diffuse reflectance (DR) images of the skin of the pseudoxanthoma elasticum (PXE) patients. (**a**) Marked PXE-affected and uninvolved skin areas as ROI. Patient (Pt) 1, neck; Pt 2, antecubital fossa; Pt 3, axilla; Pt 4, wrist; Pt 5, antecubital fossa. (**b**) In the AF images, the investigated parameters were significantly increased in PXE-affected skin compared to uninvolved skin. Analyzing DR images, no significant differences were found between PXE-affected and healthy skin. * $p < 0.05$. A.U., arbitrary unit.

4. Discussion

Early diagnosis of PXE is crucial to start the immediate management of the disorder which can delay or stop the progression of the systemic complications. However, to date, the diagnosis of PXE is usually delayed and often missed. The fact that most commonly the first manifestations are subtle, barely noticeable skin lesions poses a further challenge to diagnose PXE. Therefore, at the time of the diagnosis, ocular or cardiovascular complications are often already present [38].

For this reason, using a tool that enables an objective detection of the skin manifestations may open new perspectives in the diagnostic algorithm of PXE. DS is a widely available non-invasive diagnostic method, but it is difficult to notice the DS pattern of PXE, even for trained dermatologists, and requires practice. Although we also identified the previously published characteristic DS features in the PXE patients, not all affected regions show the typical patterns. Furthermore, in some PXE-affected skin sites, DS signs were completely absent. Also, DS does not allow quantification of the severity of skin involvement. The other non-invasive diagnostic modalities that may be used for the assessment of PXE also have major limitations. HFUS and RCM are both expensive techniques and require special expertise and training, and they are only scarcely available, mostly at large dermatology centers.

In our images with narrow-band LED excitation, the skin lesions of PXE patients appeared with different degrees of intensity with the different wavelengths. AF images were significantly more detailed, while DR images were generally less informative, and they did not seem to provide a significant benefit compared to dermoscopy. It was not always possible to distinguish the PXE-affected skin in DR images. In accordance, our quantitative analyses did not find significant differences between the DR images of PXE-affected skin compared to uninvolved skin.

AF imaging provides an easy and effective non-invasive diagnostic method, and probably enables the visualization of skin lesions in PXE. In skin sites which were only slightly affected by PXE, DS showed atypical pattern or DS signs were not visible. However, with the LED device, increased AF signal could still be detected. Utilizing an AF imaging device, we could markedly differentiate the PXE-affected areas from the uninvolved skin.

We suppose that the higher AF intensity of PXE-affected lesional skin could be correlated with the light absorption of calcium deposits. The peak of the absorption spectrum of ionized calcium was shown to be at 422.7 nm findings [30]. However, calcium deposits in PXE mainly consist of calcium hydrogen phosphate, calcium hydroxyapatite and, to a minor extent, iron precipitates [39,40]. Among these components, in one report, calcium hydroxyapatite was shown to harbour fluorescence properties, with a maximum excitation intensity at 405 nm [41]. However, in the paper, calcium hydroxyapatite was investigated in the form of nanoparticles. Also, it was not verified that the exact source of AF signal was calcium hydroxyapatite. As we used identical illumination wavelength of 405 nm, we hypothesise that it may be possible that the AF intensity was also emitted by calcium hydroxyapatite when present in the skin as a component of calcium deposits. In addition, calcium phosphate was reported to emit AF at 345 and 470 nm excitation wavelength using fluorescence microscopy. This could suggest that calcium phosphate should also give AF signal with 405 nm illumination [42].

Even though we could not measure the exact concentration of calcium deposits in the skin, we could recognize characteristic pattern of PXE based on the AF signal with 405 nm illumination. Moreover, DR images of the same field of views could be used to estimate relative absorption and scattering properties of the examined skin lesions. As seen, the DR images showed high reflection in the exact same areas that showed high AF signal. According to the optical properties of the skin (including fluorophores and chromophores), absorbers and scatters have minimal absorption and scattering in the red and near-infrared (NIR) region [43,44]. In the DR channels of PXE-affected skin, we could detect increased reflection which presumably is not related to the skin major chromophores and fluorophores, as they have a minimum absorption and scattering in the red and NIR region. By comparing the images (AF and DR channels), we could clearly see that PXE lesions produce high AF intensity signal with 405 nm excitation, as well as demonstrate high optical reflection visible in DR images with 660 and 940 nm illumination. Analyzing the histopathological images, the structures responsible for high AF signal should be calcium deposits. However, the origin of high AF and reflection signal could be attributed to main tissue compounds present among calcium deposits visualized by histology. The mineralization of elastin fibers and alterations of the collagen structure are also histopathological features of PXE [21]. Therefore, in addition to calcium compounds, elastin and collagen as endogenous fluorophores could also be responsible for the increased AF signal [45]. Additional studies would be necessary to clarify the exact origin of high AF signal in PXE lesions under 405 nm excitation. Also, identification of the optimal LED illumination wavelength for AF could further increase the selectivity of the imaging for skin lesions of PXE.

In this work, in addition to DS, we performed quantitative AF imaging and successfully distinguished PXE-affected skin from uninvolved skin areas using intensity descriptors with the aim to objectify the degree of skin involvement. A recent retrospective clinical study of 125 PXE patients found a significant correlation between the number of the PXE-affected skin sites of and the occurrence of severe cardiovascular events and/or ophthalmological complications [46]. In addition, in a smaller cohort of 14 patients, similar results were reported regarding the cardiovascular involvement [47]. These studies indicate that the accurate assessment of the skin manifestations could also indicate the degree of systemic involvement in PXE and could harbor great prognostic value in the future.

Here, we introduced a LED-based fluorescence imaging device as a potential non-invasive diagnostic tool for the assessment of the skin of PXE patients. Our results demonstrate that AF imaging is capable of the in vivo detection of PXE-affected skin and the quantification of the severity of the skin lesions of PXE patients. Moreover, it is a safe, fast and cost-effective diagnostic approach. AF imaging may be also used to objectively monitor the efficacy of the possible novel therapeutic approaches of PXE in the future.

Author Contributions: Conceptualization, K.F., S.B., D.P., A.B., N.M.W., M.M. and N.K.; methodology, K.F., S.B., D.P., A.B., I.L., A.L., M.L., M.M. and N.K.; software, K.F., S.B., I.L., A.L., M.L. and N.K.; validation, K.F., S.B., D.P., A.B., L.F., P.A., S.Z., M.M. and N.K.; formal analysis, K.F. and S.B.; investigation, K.F., S.B., D.P., A.B., L.F., P.A., S.Z., M.M. and N.K.; resources, I.L., A.L., M.L., N.M.W., M.M. and N.K.; data curation, K.F., S.B., A.L., M.M. and N.K.; writing—original draft preparation, K.F., S.B., M.M. and N.K.; writing—review and editing, K.F., S.B., D.P., L.F., P.A., S.Z., I.L., A.L., M.L., T.A., N.M.W., M.M. and N.K.; visualization, K.F., S.B., D.P., M.M. and N.K.; supervision, I.L., A.L., T.A., N.M.W., M.M. and N.K.; project administration, K.F., S.B., M.M. and N.K.; funding acquisition, S.B., P.A., S.Z., A.L., I.L., T.A., N.M.W., M.M. and N.K. All authors have read and agreed to the published version of the manuscript.

Funding: This work was supported by grants from the National Research, Development and Innovation Office of Hungary—NKFIH [FK_131916, 2019 (Semmelweis University, M.M.); K_132695, 2019 (Semmelweis University, T.A.)], EFOP-3.6.3-VEKOP-16-2017-00009 (P.A., S.B.) the ÚNKP-20-4-II-SE-7 (N.K.) and ÚNKP-20-3-I-SE-24 (S.Z.) New National Excellence Program of the Ministry For Innovation and Technology from the source of the National Research, Development and Innovation Fund of Hungary and the European Regional Development Fund projects "Time-resolved autofluorescence methodology for non-invasive skin cancer diagnostics" [No. 1.1.1.2/16/I/001, agreement No. 1.1.1.2/VIAA/1/16/014 (A.L.)] and "Development and clinical validation of a novel cost effective multi-modal methodology for early diagnostics of skin cancers" [No. 1.1.1.2/16/I/001 agreement No. 1.1.1.2/VIAA/1/16/052 (I.L.)] and COST action CA16115 EuroSoftCalcNet.

Institutional Review Board Statement: The study was conducted according to the guidelines of the Declaration of Helsinki, and approved by the Ethics Committee of Semmelweis University, Buda-pest, Hungary (SE RKEB No. 228/2018, approval date: 12 December 2018).

Informed Consent Statement: Informed consent was obtained from all subjects involved in the study. Written informed consent has been obtained from the patients to publish this paper.

Data Availability Statement: The data presented in this study are available on request from the corresponding author. The data are not publicly available due to ethical considerations.

Acknowledgments: The Department of Dermatology, Venereology, and Dermatooncology, Semmelweis University, is a Reference Centre of the ERN-Skin: European Reference Network on Rare and Undiagnosed Skin Disorders. Authors thank Olivier Vanakker and Matthias Van Gils from the Center for Medical Genetics, Ghent University Hospital, Ghent, Belgium, for the molecular genetic analyses of the patients. We thank Viktória Szabó and the Department of Ophthalmology, for the ophthalmological examinations and Anikó Ilona Nagy and the Heart and Vascular Center, Semmelweis University, for the cardiovascular assessments. Histopathologic evaluation was performed by Enikő Kuroli at the Department of Dermatology, Venereology, and Dermatooncology, Semmelweis University.

Conflicts of Interest: The authors declare no conflict of interest. The funders had no role in the design of the study; in the collection, analyses, or interpretation of data; in the writing of the manuscript, or in the decision to publish the results.

References

1. Ferreira, C.R. The burden of rare diseases. *Am. J. Med. Genet. A* **2019**, *179*, 885–892. [CrossRef] [PubMed]
2. Köblös, G.; Andrikovics, H.; Prohászka, Z.; Tordai, A.; Váradi, A.; Arányi, T. The R1141X loss-of-function mutation of the ABCC6 gene is a strong genetic risk factor for coronary artery disease. *Genet. Test. Mol. Biomark.* **2010**, *14*, 75–78. [CrossRef]
3. Verschuere, S.; Navassiolava, N.; Martin, L.; Nevalainen, P.I.; Coucke, P.J.; Vanakker, O.M. Reassessment of causality of ABCC6 missense variants associated with pseudoxanthoma elasticum based on Sherloc. *Genet. Med.* **2021**, *23*, 131–139. [CrossRef] [PubMed]
4. Jansen, R.S.; Duijst, S.; Mahakena, S.; Sommer, D.; Szeri, F.; Váradi, A.; Plomp, A.; Bergen, A.A.; Oude Elferink, R.P.; Borst, P.; et al. ABCC6-mediated ATP secretion by the liver is the main source of the mineralization inhibitor inorganic pyrophosphate in the systemic circulation-brief report. *Arterioscler. Thromb. Vasc. Biol* **2014**, *34*, 1985–1989. [CrossRef] [PubMed]
5. Uitto, J.; Jiang, Q.; Váradi, A.; Bercovitch, L.G.; Terry, S.F. Pseudoxanthoma elasticum: Diagnostic features, classification and treatment options. *Expert Opin. Orphan Drugs* **2014**, *2*, 567–577. [CrossRef]
6. de Vilder, E.Y.; Vanakker, O.M. From variome to phenome: Pathogenesis, diagnosis and management of ectopic mineralization disorders. *World J. Clin. Cases* **2015**, *3*, 556–574. [CrossRef]
7. Hosen, M.J.; Lamoen, A.; de Paepe, A.; Vanakker, O.M. Histopathology of Pseudoxanthoma Elasticum and Related Disorders: Histological Hallmarks and Diagnostic Clues. *Scientifica* **2012**, *2012*, 598262. [CrossRef]
8. Panagou, E.; Ratynska, M.; Heelan, K. Pseudoxanthoma elasticum-like papillary dermal elastolysis: A case report and review of literature. *Int. J. Dermatol.* **2019**, *58*, 93–97. [CrossRef]
9. Aydin, E.; Demir, H.D.; Batioglu, F.; Sezer, E. Atypical Fundus Lesions in Juvenile Pseudoxanthoma Elasticum. *Ophthalmic Res.* **2007**, *39*, 344–347. [CrossRef] [PubMed]
10. Bertamino, M.; Severino, M.; Grossi, A.; Rusmini, M.; Tortora, D.; Gandolfo, C.; Pederzoli, S.; Malattia, C.; Picco, P.; Striano, P.; et al. ABCC6 mutations and early onset stroke: Two cases of a typical Pseudoxanthoma Elasticum. *Eur. J. Paediatr. Neurol.* **2018**, *22*, 725–728. [CrossRef] [PubMed]
11. Dibi, A.; El Fahime, E.L.; Mouane, N.; Dafiri, R.; Bentahila, A. Pseudoxanthoma elasticum: A rare cause of gastrointestinal bleeding in children. *Arch. Pediatr.* **2016**, *23*, 591–594. [CrossRef]
12. Gliem, M.; Birtel, J.; Müller, P.L.; Hendig, D.; Faust, I.; Herrmann, P.; Holz, F.G.; Adamus, G.; Issa, P.C. Acute Retinopathy in Pseudoxanthoma Elasticum. *JAMA Ophthalmol.* **2019**, *137*, 1165–1173. [CrossRef] [PubMed]
13. Risseeuw, S.; van Norel, J.O.; Klaver, C.C.W.; Colijn, J.M.; Imhof, S.M.; van Leeuwen, R. Visual acuity in pseudoxanthoma elasticum. *Retina* **2019**, *39*, 1580–1587. [CrossRef]
14. Lefthériotis, G.; Omarjee, L.; Saux, O.L.; Henrion, D.; Abraham, P.; Prunier, F.; Willoteaux, S.; Martin, L. The vascular phenotype in Pseudoxanthoma elasticum and related disorders: Contribution of a genetic disease to the understanding of vascular calcification. *Front. Genet.* **2013**, *4*, 4. [CrossRef] [PubMed]
15. Berthin, C.; Phan, A.; Navasiolava, N.; Michalak, S.; Humeau, H.; Grimaux, X.; Martin, L. Dermoscopic phenotype of pseudoxanthoma elasticum skin lesions: A study of 16 patients. *J. Eur. Acad. Dermatol. Venereol.* **2019**, *33*, e262–e265. [CrossRef] [PubMed]
16. Lacarrubba, F.; Verzì, A.E.; Caltabiano, R.; Micali, G. Dermoscopy of pseudoxanthoma elasticum. *J. Am. Acad. Dermatol.* **2017**, *76*, S69–S70. [CrossRef]
17. Kawashima, S.; Togawa, Y.; Miyachi, H.; Matsue, H. Dermoscopic features of pseudoxanthoma elasticum. *Clin. Exp. Dermatol.* **2018**, *43*, 175–179. [CrossRef]
18. Mandel, V.; Boraldi, F.; Pellacani, G.; Ciardo, S.; Mazzaglia, G.; Farnetani, F. Pseudoxanthoma elasticum and reflectance confocal microscopy: Report of two affected young sisters. *J. Pediatr. Neonatal Individ. Med.* **2015**, *4*, e040125. [CrossRef]
19. Guérin-Moreau, M.; Leftheriotis, G.; Corre, Y.L.; Etienne, M.; Amode, R.; Hamel, J.F.; Croué, A.; Saux, O.L.; Machet, L.; Martin, L. High-frequency (20–50 MHz) ultrasonography of pseudoxanthoma elasticum skin lesions. *Br. J. Dermatol.* **2013**, *169*, 1233–1239. [CrossRef]

20. Murata, T.; Honda, T.; Miyachi, Y.; Kabashima, K. Morphological character of pseudoxanthoma elasticum observed by multiphoton microscopy. *J. Dermatol. Sci.* **2013**, *72*, 199–201. [CrossRef]
21. Kiss, N.; Fésűs, L.; Bozsányi, S.; Szeri, F.; van Gils, M.; Szabó, V.; Nagy, A.I.; Hidvégi, B.; Szipőcs, R.; Martin, L.; et al. Nonlinear optical microscopy is a novel tool for the analysis of cutaneous alterations in pseudoxanthoma elasticum. *Lasers Med. Sci.* **2020**, *35*, 1821–1830. [CrossRef]
22. Kiss, N.; Haluszka, D.; Lorincz, K.; Kuroli, E.; Harsing, J.; Mayer, B.; Karpati, S.; Fekete, G.; Szipocs, R.; Wikonkal, N.; et al. Ex Vivo nonlinear microscopy imaging of Ehlers-Danlos syndrome-affected skin. *Arch. Dermatol. Res.* **2018**, *310*, 463–473. [CrossRef]
23. Cui, J.Z.; Tehrani, A.Y.; Jett, K.A.; Bernatchez, P.; van Breemen, C.; Esfandiarei, M. Quantification of aortic and cutaneous elastin and collagen morphology in Marfan syndrome by multiphoton microscopy. *J. Struct. Biol.* **2014**, *187*, 242–253. [CrossRef] [PubMed]
24. Omarjee, L.; Mention, P.J.; Janin, A.; Kauffenstein, G.; Pabic, E.L.; Meilhac, O.; Blanchard, S.; Navasiolava, N.; Leftheriotis, G.; Couturier, O.; et al. Assessment of Inflammation and Calcification in Pseudoxanthoma Elasticum Arteries and Skin with 18F-FluroDeoxyGlucose and 18F-Sodium Fluoride Positron Emission Tomography/Computed Tomography Imaging: The GOCAPXE Trial. *J. Clin. Med.* **2020**, *9*, 3448. [CrossRef]
25. Lesins, J.; Lihachev, A.; Rudys, R.; Bagdonas, S.; Spigulis, J. Skin Autofluorescence Photo-Bleaching and Photo-Memory. In Proceedings of the Medical Laser Applications and Laser-Tissue Interactions V, Munich, Germany, 22–26 May 2011; p. 80920N.
26. Lihachev, A.; Lihacova, I.; Plorina, E.V.; Lange, M.; Derjabo, A.; Spigulis, J. Differentiation of seborrheic keratosis from basal cell carcinoma, nevi and melanoma by RGB autofluorescence imaging. *Biomed. Opt. Express* **2018**, *9*, 1852–1858. [CrossRef] [PubMed]
27. Lange, M.; Kiss, N.; Fesus, L.; Plorina, E.; Derjabo, A.; Spigulis, J. Non-Invasive LED-Based Screening Solution for Skin Cancer. In *Clinical and Preclinical Optical Diagnostics*; OSA Publishing: Washington, DC, USA, 2019; p. 17. [CrossRef]
28. Lihacova, I.; Cibulska, E.; Lihachev, A.; Lange, M.; Plorina, E.V.; Bliznuks, D.; Derjabo, A.; Kiss, N. Challenges of Automatic Processing of Large amount of Skin Lesion Multispectral Data. *Proc. SPIE* **2020**, *11585*, 115850C. [CrossRef]
29. Marta, L.; Szabolcs, B.; Emilija, V.P.; Lihachev, A.; Derjabo, A. Spectral Imaging as A Tool for the Evaluation of Skin Cancer Post-Operative Scars. *Proc. SPIE* **2020**, *11585*, 1158506. [CrossRef]
30. Rust, J.A.; Nóbrega, J.A.; Calloway, C.P.; Jones, B.T. Fraunhofer Effect Atomic Absorption Spectrometry. *Anal. Chem.* **2005**, *77*, 1060–1067. [CrossRef]
31. Siddiqui, A.A.; Paulus, Y.M.; Scott, A.W. Use of Fundus Autofluorescence to Evaluate Retinal Artery Occlusions. *Retina* **2014**, *34*, 2490–2491. [CrossRef]
32. Rajesh, B.; Hussain, R.; Giridhar, A. Autofluorescence and Infrared Fundus Imaging for Detection of Retinal Emboli and Unmasking Undiagnosed Systemic Abnormalities. *J. Ophthalmic Vis. Res.* **2016**, *11*, 449–451. [CrossRef]
33. Plomp, A.S.; Toonstra, J.; Bergen, A.A.; van Dijk, M.R.; de Jong, P.T. Proposal for updating the pseudoxanthoma elasticum classification system and a review of the clinical findings. *Am. J. Med. Genet. A* **2010**, *152a*, 1049–1058. [CrossRef]
34. Legrand, A.; Cornez, L.; Samkari, W.; Mazzella, J.M.; Venisse, A.; Boccio, V.; Auribault, K.; Keren, B.; Benistan, K.; Germain, D.P.; et al. Mutation spectrum in the ABCC6 gene and genotype-phenotype correlations in a French cohort with pseudoxanthoma elasticum. *Genet. Med.* **2017**, *19*, 909–917. [CrossRef] [PubMed]
35. Tamošiūnas, M.; Plorina, E.V.; Lange, M.; Derjabo, A.; Kuzmina, I.; Bļizņuks, D.; Spigulis, J. Autofluorescence imaging for recurrence detection in skin cancer postoperative scars. *J. Biophotonics* **2020**, *13*, e201900162. [CrossRef]
36. Lihachev, A.; Derjabo, A.; Ferulova, I.; Lange, M.; Lihacova, I.; Spigulis, J. Autofluorescence imaging of basal cell carcinoma by smartphone RGB camera. *J. Biomed. Opt.* **2015**, *20*, 120502. [CrossRef]
37. Bliznuks, D.; Jakovels, D.; Saknite, I.; Spigulis, J. Mobile Platform for Online Processing of Multimodal Skin Optical Images: Using Online Matlab Server for Processing Remission, Fluorescence and Laser Speckle Images, Obtained by Using Novel Handheld Device. In Proceedings of the 2015 International Conference on BioPhotonics (BioPhotonics), Florence, Italy, 20–22 May 2015; pp. 1–4.
38. Michael, N.; Boisseau, C.; Bonicel, P.; Daudon, P.; Bonneau, D.; Chassaing, N.; Martin, L. Manifestations of pseudoxanthoma elasticum in childhood. *Br. J. Dermatol.* **2009**, *161*, 635–639. [CrossRef]
39. Jensen, O.A. Bruch's membrane in pseudoxanthoma elasticum. Histochemical, ultrastructural, and x-ray microanalytical study of the membrane and angioid streak areas. *Graefe Arch. Clin. Exp. Ophthalmol.* **1977**, *203*, 311–320. [CrossRef] [PubMed]
40. Walker, E.R.; Frederickson, R.G.; Mayes, M.D. The mineralization of elastic fibers and alterations of extracellular matrix in pseudoxanthoma elasticum. Ultrastructure, immunocytochemistry, and X-ray analysis. *Arch. Dermatol.* **1989**, *125*, 70–76. [CrossRef]
41. Machado, T.R.; Leite, I.S.; Inada, N.M.; Li, M.S.; da Silva, J.S.; Andrés, J.; Beltrán-Mir, H.; Cordoncillo, E.; Longo, E. Designing biocompatible and multicolor fluorescent hydroxyapatite nanoparticles for cell-imaging applications. *Mater. Today Chem.* **2019**, *14*, 100211. [CrossRef]
42. Uchida, Y.; Uchida, Y.; Kawai, S.; Kanamaru, R.; Sugiyama, Y.; Tomaru, T.; Maezawa, Y.; Kameda, N. Detection of vulnerable coronary plaques by color fluorescent angioscopy. *JACC Cardiovasc. Imaging* **2010**, *3*, 398–408. [CrossRef] [PubMed]
43. Cao, Q.; Zhegalova, N.; Wang, S.; Akers, W.; Berezin, M. Multispectral imaging in the extended near-infrared window based on endogenous chromophores. *J. Biomed. Opt.* **2013**, *18*, 101318. [CrossRef] [PubMed]
44. Spigulis, J.; Rubins, U.; Kviesis-Kipge, E.; Rubenis, O. SkImager: A concept device for in-vivo skin assessment by multimodal imaging. *Proc. Est. Acad. Sci.* **2014**, *63*, 213. [CrossRef]

45. Giovannacci, I.; Magnoni, C.; Vescovi, P.; Painelli, A.; Tarentini, E.; Meleti, M. Which are the main fluorophores in skin and oral mucosa? A review with emphasis on clinical applications of tissue autofluorescence. *Arch. Oral Biol* **2019**, *105*, 89–98. [CrossRef] [PubMed]
46. Navasiolava, N.; Gnanou, M.; Douillard, M.; Saulnier, P.; Aranyi, T.; Ebran, J.M.; Henni, S.; Humeau, H.; Leftheriotis, G.; Martin, L. The extent of pseudoxanthoma elasticum skin changes is related to cardiovascular complications and visual loss: A cross-sectional study. *Br. J. Dermatol.* **2019**, *180*, 207–208. [CrossRef] [PubMed]
47. Utani, A.; Tanioka, M.; Yamamoto, Y.; Taki, R.; Araki, E.; Tamura, H.; Miyachi, Y. Relationship between the distribution of pseudoxanthoma elasticum skin and mucous membrane lesions and cardiovascular involvement. *J. Dermatol.* **2010**, *37*, 130–136. [CrossRef] [PubMed]

diagnostics

MDPI

Article

AK-DL: A Shallow Neural Network Model for Diagnosing Actinic Keratosis with Better Performance than Deep Neural Networks

Liyang Wang [1], Angxuan Chen [2], Yan Zhang [2], Xiaoya Wang [1], Yu Zhang [2], Qun Shen [1] and Yong Xue [1,*]

[1] Beijing Advanced Innovation Center for Food Nutrition and Human Health, Key Laboratory of Plant Protein and Grain Processing, National Engineering and Technology Research Center for Fruits and Vegetables, College of Food Science and Nutritional Engineering, China Agricultural University, Beijing 100083, China; 18259800533@163.com (L.W.); 15384665858@163.com (X.W.); shenqun@cau.edu.cn (Q.S.)

[2] College of Information and Electrical Engineering, China Agricultural University, Beijing 100083, China; angxuan.chen@foxmail.com (A.C.); edison_yan@icloud.com (Y.Z.); 18810760793@163.com (Y.Z.)

* Correspondence: xueyong@cau.edu.cn

Received: 16 March 2020; Accepted: 9 April 2020; Published: 13 April 2020

Abstract: Actinic keratosis (AK) is one of the most common precancerous skin lesions, which is easily confused with benign keratosis (BK). At present, the diagnosis of AK mainly depends on histopathological examination, and ignorance can easily occur in the early stage, thus missing the opportunity for treatment. In this study, we designed a shallow convolutional neural network (CNN) named actinic keratosis deep learning (AK-DL) and further developed an intelligent diagnostic system for AK based on the iOS platform. After data preprocessing, the AK-DL model was trained and tested with AK and BK images from dataset HAM10000. We further compared it with mainstream deep CNN models, such as AlexNet, GoogLeNet, and ResNet, as well as traditional medical image processing algorithms. Our results showed that the performance of AK-DL was better than the mainstream deep CNN models and traditional medical image processing algorithms based on the AK dataset. The recognition accuracy of AK-DL was 0.925, the area under the receiver operating characteristic curve (AUC) was 0.887, and the training time was only 123.0 s. An iOS app of intelligent diagnostic system was developed based on the AK-DL model for accurate and automatic diagnosis of AK. Our results indicate that it is better to employ a shallow CNN in the recognition of AK.

Keywords: actinic keratosis; AK-DL; intelligent diagnostic app; mainstream deep model

1. Introduction

Actinic keratosis (AK), also called senile keratosis, is usually caused by long-term exposure to sunlight or ionizing radiation [1,2]. It is one of the most common precancerous skin lesions and presents as reddish brown or yellow macules or small plaques with clear boundary. The theory of field cancerization postulates that the skin encircling AK is also at increased risk of possible malignant transformation because it has suffered from the same chronic ultraviolet light [3]. Early medical diagnosis and treatment are believed to delay the development of AK and the cancerization of the skin surrounding it [3–5]. However, AK can be easily confused with benign keratosis (BK), such as seborrheic keratosis (SK). At present, the clinical diagnosis is mainly based on histopathological examination with a lengthy and complicated process [6–9].

With the development of computer vision, image recognition based on deep learning is increasingly being used in medical diagnostics [10–15], including cancer identification, histocyte classification, tumor image segmentation, etc. Kawahara et al. [16] used convolution neural network (CNN) early on to classify dermatologic images and achieved good results. Kaur et al. [17] demonstrated that the classification accuracy of hybrid deep learning method could reach up to 82.0% among six kinds of skin images, which was better than the traditional CNN models. Romero-Lopez et al. [18] exploited VGGNet to classify benign and malignant types of skin cancer and achieved a classification accuracy of 78.66% based on the International Skin Imaging Collaboration (ISIC) dataset. Harangi et al. [19] incorporated softmax layer outputs of four different neural architectures for intelligent diagnosis of melanoma and eventually found that this method was superior to single neural network. Generally, research on the use of machine learning to identify skin diseases is dominated by mainstream deep CNNs, which are characterized by long duration of training and high requirement of arithmetic facility. Recent studies have found that the performance of shallow CNNs are not worse, but even better, than deep CNNs in some cases [20,21]. However, such results have not been found in dermatology, especially in recognition of AK. The excellent image identification of actinic keratosis based on a shallow CNN should have high accuracy, high efficiency, and low requirement for operating equipment so that it can improve the efficiency of clinical diagnosis and facilitate self-monitoring and prevention of skin cancer.

In this study, we designed a new CNN model named actinic keratosis deep learning (AK-DL), which includes two convolutional layers and a max pooling layer, to recognize actinic keratosis. We then compared it with the current mainstream deep CNNs of skin mirror diagnosis, such as AlexNet, GoogLeNet V1, and ResNet-50, as well as traditional medical image processing methods. Finally, an iOS app of intelligent diagnostic system was developed based on the AK-DL model.

2. Materials and Methods

2.1. Data Acquisition

A public dataset HAM10000 (collected by the following two departments over a period of 20 years: the Department of Dermatology at the Medical University of Vienna, Austria, and the skin cancer practice of Cliff Rosendahl in Queensland, Australia) from the ISIC dataset was employed [22], consisting of 100,015 skin images belonging to 7 common skin diseases. It is worth noting that each mirror in this dataset has been strictly confirmed by dermatologists. A part of the data was used in our study, including 327 skin images diagnosed as AK and 1099 skin images diagnosed as BK (Table S1). To our knowledge, there has been no study using keratosis images alone from the ISIC dataset.

2.2. Data Preprocessing

Preprocessing, an indispensable step in computer vision, transforms the initial image into a form that can be computerized. The operations of preprocessing in this study included data augmentation, image enhancement, and color space conversion.

Due to the imbalance of categories in this dataset, the data was augmented to improve the robustness of the model and enhance its generalization ability. The methods of data augmentation usually contain image rotation, image flip, random cropping, adding noise, etc. Considering the fact that the number of AK images was limited, the images were rotated 90 degrees, horizontally flipped, vertically flipped, and 1308 derived samples were finally produced. Data augmentation can not only avoid the poor training results caused by unbalanced data but also prevent overfitting caused by a small dataset.

To improve the image quality and easily identify key features, images are often enhanced by several methods, such as histogram equalization, contrast enhancement, Gamma correction, etc. In our study, histogram equalization was performed on the amplified images to highlight the texture region of keratosis. Furthermore, the color space of images was converted into Lab space from RGB space for the expansion of color gamut. Lab space has a wider color gamut than RGB and CMYK and is thus a great compensation for the uneven distribution of RGB and CMYK color modes. Additionally, Lab space separates brightness from color; the L channel has no color, while the a and b channels have color, so the adjustment operation is simple and easy for computer processing. The preprocessing of images is shown in Figure 1.

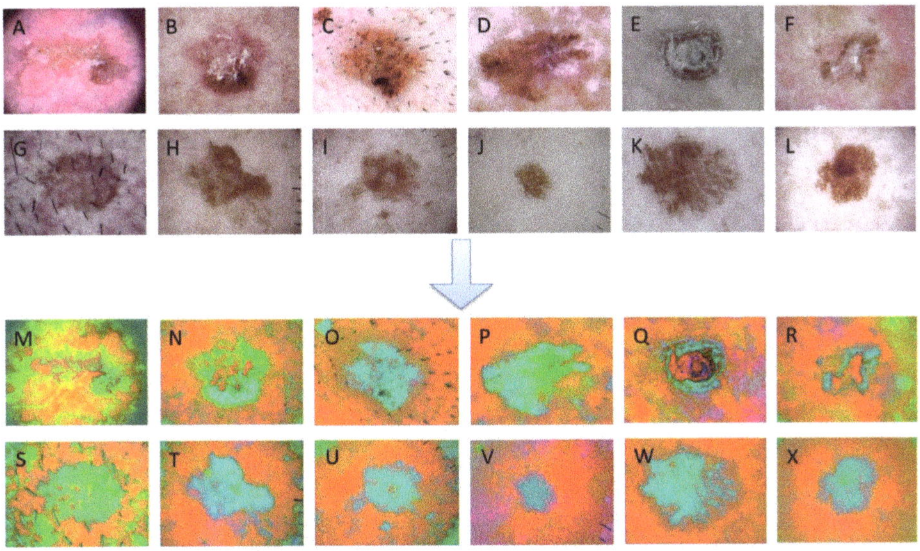

Figure 1. Comparisons of skin images before and after preprocessing. (**A–F**) Actinic keratosis (AK) images before preprocessing, (**G–L**) benign keratosis (BK) images before preprocessing, (**M–R**) actinic keratosis images after preprocessing, (**S–X**) benign keratosis images after preprocessing.

2.3. Models Building

Since Lecun et al. [23] first proposed the classic CNN model LeNet-5 in 1999, more and more mainstream deep learning algorithms have been applied in the field of medical images. AlexNet appeared in 2012, followed by VGG, GoogLeNet, ResNet, and other deep CNNs. In this study, we investigated a new deep learning model named AK-DL and compared it with the current mainstream algorithms of medical diagnosis (AlexNet, GoogLeNet, and ResNet). All building and training processes of the models were carried out in Windows 10 operating system with CPU Intel Core I7-6700HQ (3.5 GHz, 4 GB memory).

2.3.1. AK-DL Model

As a classic CNN architecture, the input layer size of the LeNet-5 model is $28 \times 28 \times 1$. The input image is handwritten numeral, and its texture information is simple. However, the texture information of the keratosis image in the present study was relatively complex, so the input layer size was modified to $64 \times 64 \times 3$. The aim of the convolutional layer is to extract different characteristics of the input, and two convolutional layers were set in the AK-DL model. The core parameters of the convolution operation were as follows: the size of the convolution kernel was set as 5×5, the convolution stride length was set as 1, and the padding was set as 2. The number of types on the two convolutional layers were set as 6 and 16, respectively. The pooling layer, which is mainly used for feature dimensionality reduction, can reduce the number of parameters, prevent overfitting, and improve the fault tolerance of the model. In this research, a max pooling layer was adopted after the first convolutional layer to construct a neural network, and a 2×2 sampling window was used to control the reduction rate of the feature map. Usually, the fully connected layer is used to highly purify the features extracted by the convolutional layer, and it can thus play an important role in mapping the distributed feature with representation of the sample space. In our research, a fully connected layer after the last convolutional layer was added in the AK-DL model.

The traditional shallow CNN model has a certain shortcoming in generalization ability, resulting in overfitting in the case of insufficient data. Considering the relatively small amount of keratosis images, the dropout layer was introduced following the fully connected layer to prevent overfitting. During training, dropout can discard some neurons in the hidden layer with a certain probability, resulting in changes of the network structure. The whole process of dropout is equivalent to taking the average value of multiple different neural networks. In other words, different networks have different overfitting, and some "inverse" overfitting can cancel each other out to reduce overall overfitting; to some extent, it can also reduce the complex coadaptation among neurons to achieve regularization effect. After several tests, it was found that dropout was performed the best when set as 0.5.

The activation function of traditional CNN models is the sigmoid function, which is prone to having the vanishing gradient problem and easily leads to failure because of its soft saturation. Instead, the ReLU function is often employed in CNN models as the activation function, which could alleviate the vanishing gradient problem. Compared with the sigmoid function, it is not necessary to calculate the exponent for the ReLU function when calculating the activation value, so it is faster than the sigmoid function. However, the ReLU function is "fragile" in the training process when applied at a rapid learning rate, resulting in unsuccessful training. To solve this problem, the LeakyReLU function was finally used in the AK-DL model as the activation function. This function introduces a Leaky value into the negative half interval, and the function output has a small slope to the negative value input. After several experiments, it was found that the recognition accuracy of actinic keratosis was the highest when the Leaky value was set as 0.01. The network architecture of the AK-DL model is shown in Figure 2.

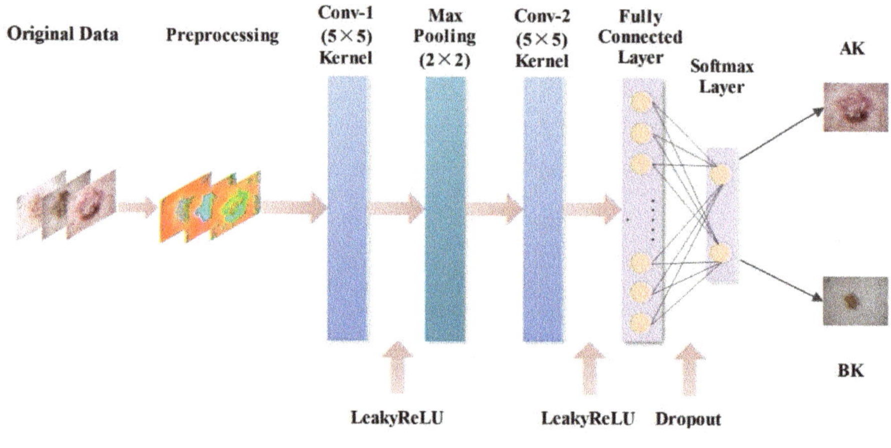

Figure 2. The network architecture of the actinic keratosis deep learning (AK-DL) model.

2.3.2. AlexNet Transfer Learning

The AlexNet model was proposed by Alex Krizhevsky in 2012 and won the 2012 ILSVRC competition. This model has 8 neural network layers, including 5 convolutional layers (3 convolutional layers are followed by the max pooling layer) and 3 fully connected layers, with 630 million links, 60 million parameters, and 650,000 neurons [24]. In this study, AlexNet was used for transfer learning. First, 1000 kinds of natural images from ImageNet were pretrained to get the initialization model. Then, the number of neurons in the output layer was changed to 2, and the last layer was updated. Transfer learning can alleviate overfitting caused by insufficient data and improve training efficiency.

2.3.3. GoogLeNet Transfer Learning

The GoogLeNet model was the champion of 2014 ILSVRC challenge. Unlike AlexNet, GoogLeNet has a total of 22 layers. In order to avoid the problem of vanishing gradient in deep network, two losses at different depths are added into GoogLeNet to prevent gradient vanishing. In addition, GoogLeNet possesses modular unit inception, which not only increases the accuracy of classification but also greatly reduces the number of inputs. The modified inception unit was used in this study, including 9 inception modules (the GoogLeNet V1 architecture). In order to improve training efficiency, transfer learning was also applied to GoogLeNet in this study. After pretraining based on ImageNet, three layers (fully connected layer, softmax layer, and classification output layer) were added to the last three layers of the original network. Meanwhile, the learning factor of the fully connected layer was increased to make the learning rate of the new layer higher than the original layer. Finally, the last transport layer in the network (pool5-drop_7 × 7_s1) was connected to the new layer.

2.3.4. ResNet Transfer Learning

The ResNet model, which was first proposed in 2015 and won first place in the classification task in the ImageNet contest, has been extensively used for medical image processing [25–27]. In this study, classical architecture ResNet-50 was employed for transfer learning. After pretraining 1000 kinds of natural images from ImageNet, the initialization model was obtained, and the transfer process was then completed following the method mentioned above.

2.4. Traditional Machine Learning

The traditional algorithms of medical image processing were also introduced and compared in order to verify the superiority of AK-DL. The Histogram of Oriented Gradient (HOG) of the image was employed as the feature vector of the classifier input. The image resolution was modified to 200 × 200, the cell size was set as 32 × 32 pixels, and each block contained 2 × 2 cells. Finally, a 900-dimension feature vector was extracted from each sample. In this study, the classical machine learning algorithm classifiers employed included support vector machine (SVM), random forest (RF), and *k*-nearest Neighbor (KNN).

2.5. Parameter Optimization

Four parameters, namely, Momentum, InitialLearnRate, MiniBatchSize, and L2Regularization, were optimized with the Momentum optimizer in the CNN training process. A weighted average of the movement index based on the gradient was calculated by the Momentum optimizer, which could accelerate the learning process when facing small and continuous gradients with a lot of noise. The optimized parameters are shown in Table 1. In addition, optimization of the traditional classifiers were also performed in the training process.

Table 1. Convolutional neural network (CNN) parameters.

Parameters	AK-DL	AlexNet Transfer	GoogLeNet Transfer	ResNet Transfer
Momentum	0.9	0.6	0.9	0.7
InitialLearnRate	0.0001	0.0001	0.0001	0.0001
MiniBatchSize	10	20	15	20
L2Regularization	0.0005	0.0001	0.0001	0.00001

2.6. Evaluation Indexes

In this study, accuracy (Acc), sensitivity (Sens), specificity (Spec), precision (Prec), Matthews correlation coefficient (MCC), and area under the receiver operating characteristic curve (AUC) were included as evaluation indexes. The calculations for Acc, Sens, Spec, Prec, and MCC are given below. In addition, considering the impact of training time of the model, the training time was also investigated. A 10-fold cross-validation was applied in the present study, meaning that 10% of the data was chosen randomly as the test data and the rest as the training data for 10 times. The mean values of the evaluation indexes from the 10-fold validation were calculated as the final values.

$$\text{Acc} = \frac{TN + TP}{TN + TP + FN + FP} \quad (1)$$

$$\text{Sens} = \frac{TP}{TP + FN} \quad (2)$$

$$\text{Spec} = \frac{TN}{TN + FP} \quad (3)$$

$$\text{Prec} = \frac{TP}{TP + FP} \quad (4)$$

$$\text{MCC} = \frac{TP \times TN - FP \times FN}{\sqrt{(TP + FP)(TP + FN)(TN + FP)(TN + FN)}} \quad (5)$$

where *TP* (true positive) is the proportion of the positive samples correctly classified, *FP* (false positive) is the proportion of negative samples incorrectly classified, *TN* (true negative) is the proportion of the negative samples correctly classified, and *FN* (false negative) is the proportion of positive samples incorrectly classified.

3. Results

3.1. Comparison of AK-DL with Machine Learning Models

Based on the dataset of keratosis, although AK-DL had a relatively simple architecture, its performance was better than the mainstream deep CNN models, including AlexNet, GoogLeNet, and ResNet. The accuracy of AK-DL was the highest (92.5%) compared with the other models (Table 2), and its AUC was also the highest among all the CNN models (the corresponding receiver operating characteristic curves (ROC) are shown in Figure 3). It is noteworthy that the average training time of AK-DL (123.0 s) was far less than AlexNet (2426.0 s), GoogLeNet (13,761.0 s), and ResNet (15,488.0 s), suggesting that this shallow CNN model had the highest practical value in commercialization, and it was thus possible to develop a convenient and rapid diagnostic system of keratosis.

Table 2. CNN model comparison.

CNN Models	Acc	Sens	Spec	Prec	MCC	Training Time (s)
AK-DL	0.925	0.938	0.909	0.924	0.848	123.0
AlexNet	0.862	0.908	0.815	0.832	0.727	2426.0
GoogLeNet	0.874	0.874	0.875	0.901	0.746	13,761.0
ResNet	0.774	0.829	0.721	0.740	0.553	15,488.0

We also calculated the evaluation indexes of traditional processing algorithms and compared them with the results of the AK-DL model (shown in Table 3 and Figure 4). Although SVM had the best classification performance among the traditional algorithms with an accuracy of 79.1% and AUC of 0.749, the AK-DL still had obvious advantages according to its higher accuracy and AUC. Remarkably, the training time of the traditional processing algorithms was less than the AK-DL model due to the simplicity of the model architecture.

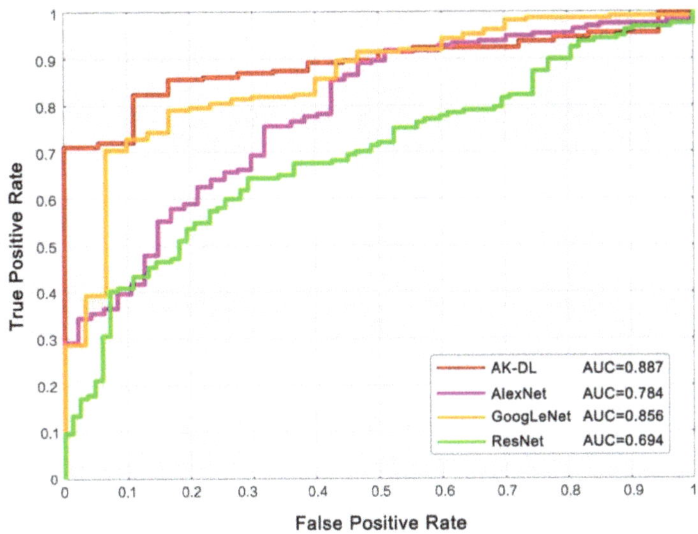

Figure 3. Receiver operating characteristic curves (ROC) of CNN models.

Table 3. Comparison of AK-DL and traditional machine learning algorithms.

Models	Acc	Sens	Spec	Prec	MCC	Training Time (s)
AK-DL	0.925	0.938	0.909	0.924	0.848	123.0
HOG+K-NN	0.713	0.954	0.511	0.619	0.506	12.4
HOG+RF	0.778	0.796	0.763	0.735	0.557	20.1
HOG+SVM	0.791	0.780	0.800	0.766	0.579	20.2

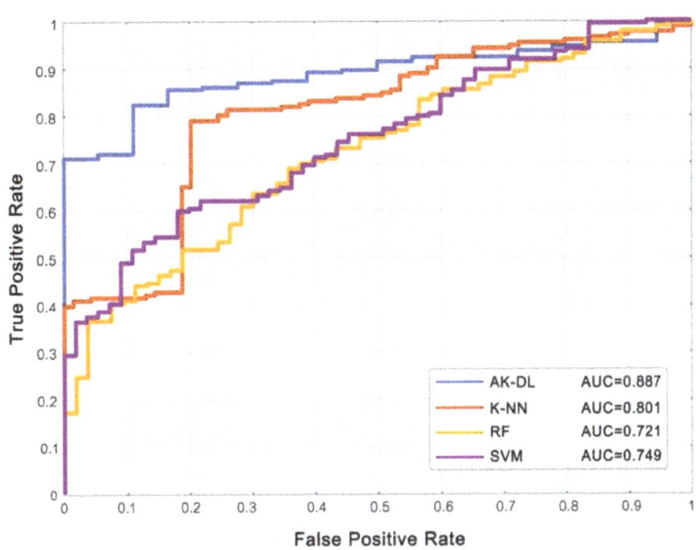

Figure 4. ROC curves of AK-DL and traditional machine learning algorithms.

3.2. Intelligent Diagnostic System

To achieve the end-to-end mode of the keratosis diagnostic system and to make it convenient for people to easily examine their skin condition in real time, we developed an intelligent diagnostic system on iOS using the AK-DL model. The development language was objective-C, and the iOS SDK was used to build the development environment. The workflow of the system is as follows (Figure 5). Firstly, the video input stream from the iPhone camera is obtained, and representative frames are then extracted and sent to the server. Secondly, the server employs PHP to execute Python commands, and the obtained frames are sent to the trained AK-DL model via tube (the AK-DL model requires the input format to be RGB image, and all images are scaled to the unified image size of 64 × 64 × 3). Thirdly, the server transforms the background into binary data and sends it to the trained model. Finally, the output value predicted by the model will feed back to the users. The diagnostic interface is shown in Figure 6.

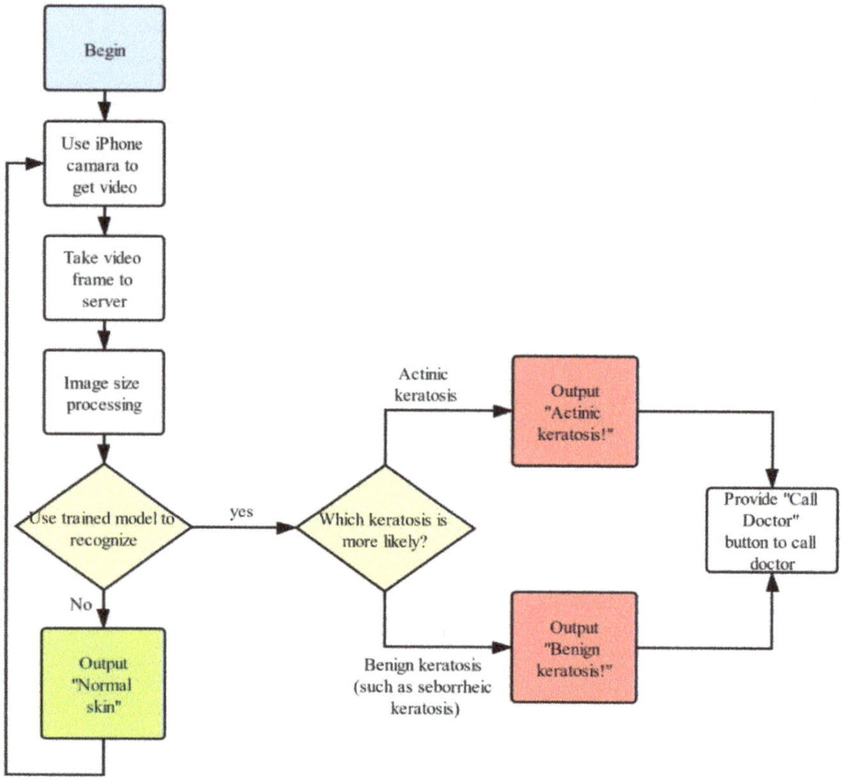

Figure 5. Flow chart of intelligent diagnostic system of keratosis.

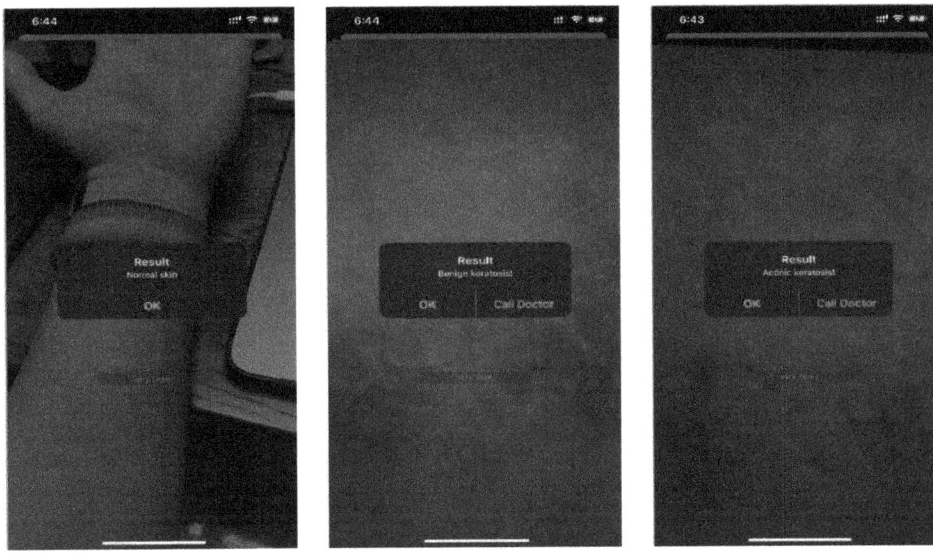

(a) (b) (c)

Figure 6. Diagnosis system interface of keratosis built on iOS. (**a**) Normal skin, (**b**) benign keratosis, (**c**) actinic keratosis.

4. Discussion

In this study, a new shallow CNN named AK-DL was designed to recognize AK instead of adopting mainstream CNN architecture. The network depth was compressed in the AK-DL model to shorten training time, and specific choices in optimizer, activation function, and other aspects could improve the performance of the model. In order to prevent overfitting caused by insufficient and unbalanced image data, data augmentation and dropout layer were used in this study. Our results showed that this network possessed an outstanding performance in recognizing AK.

As there are few related studies on using deep learning algorithms to diagnose AK, we compared the results of our study with some latest studies of other skin diseases. Li et al. [28] used the deep CNN Inception-V3 for training and then made a classification test on 2000 melanoma mirror images, resulting in a final AUC value of 0.8000. Similarly, Codella et al. [29] combined deep CNNs, sparse coding, and SVM to identify melanoma and some atypical lesions and made a cross-validation on 2624 data samples, eventually obtaining an accuracy of 0.7390, a sensitivity of 0.7380, and a specificity of 0.7430. However, not all deep CNNs perform well in the recognition of skin diseases. Mirunalini et al. [30] employed GoogLeNet transfer learning based on a melanoma dataset of 2000 samples from ISIC, and only an overall accuracy of 0.6580 was obtained on the validation set. It is worth noting that the AK-DL model, a shallow CNN, was more effective than the abovementioned studies based on deep CNNs, with the accuracy reaching up to 0.925, while the sensitivity, specificity, and AUC were 0.938, 0.909, and 0.887, respectively. A comparison of AK-DL with the current mainstream deep CNNs, such as AlexNet, GoogLeNet V1, and ResNet-50, based on the same dataset of skin keratosis also reflected the superior performance of AK-DL. Theoretically speaking, deep CNN can extract more feature information for recognition. However, the textural features of skin keratosis are relatively simpler compared with the pathological images of breast cancer, lung cancer, colorectal cancer, etc. It is possible that, based on the feature extraction of special images with sample texture, the capacity of shallow CNN is not lower, and may be even higher, than deep neural networks. In contrast, considering the fact that deep CNN models have more convolution layers, the characteristics extracted are more complex and prone to overfitting when the image dataset is relatively small.

Additionally, we compared the result of AK-DL with other studies using traditional machine learning algorithms, and it also supported the superiority of the AK-DL model. Vasconcelos et al. [31] proposed a new color evaluation method and then applied feature selection and machine learning classification, resulting in an identification accuracy of 0.7775 and 0.8138 on two smaller public datasets, respectively. Ohki et al. [32] used the RF model, a traditional classifier, to automatically classify 1148 skin images (including 980 melanocytic nevi and 168 melanomas) and acquired sensitivity of 0.7980 and specificity of 0.8070 after taking a 10-fold cross-validation. It is worth noting that the performance of AK-DL, a shallow CNN, was better than the abovementioned traditional machine learning methods. This might be due to the inherent shortcomings of traditional machine learning, including the dependence of manual feature extraction and the incomplete reflection of data. On the contrary, the convolution layer was employed in AK-DL to extract features, which could extract image details with end-to-end recognition patterns.

It is worth noting that the efficiency of the AK-DL model (123.0 s for training time) was significantly lower than the deep neural networks, including the AlexNet (2426.0 s), GoogLeNet (13,761.0 s), and ResNet (15,488.0 s) models. Considering the fact that a high recognition accuracy (0.925) was achieved, our results show that shallow CNNs, such as AK-DL, can reduce the training time as well as ensure effectiveness of the model, thus contributing to a reduction in training costs. Therefore, it is possible to develop this real-time image recognition technique with high-accuracy as a supplementary diagnostic method for skin keratosis. It should be pointed out that the AK-DL model was only confirmed in the keratosis dataset, and the application of a shallow CNN thus needs more research and further exploration.

Actinic keratosis is prone to occur in sun-exposed areas in middle-aged and elderly people. It is a type of precancerous skin lesion caused by sun damage. If not actively treated, it may turn into squamous cell carcinoma [33,34]. According to statistics, the prevalence rate of actinic keratosis in Europe, the United States, and Australia has reached 4.5% to 60.0% [35]; therefore, early and active interventions should be advocated. In many cases, this disease is easily confused with BK, which is easily ignored, and diagnosis needs to be observed through pathological examination with a relatively complex process. In view of the situation above, it is of great significance to develop a portable app for accurate automatic diagnosis of AK. A portable app would not only provide real-time monitoring, such as in family doctor services for middle-aged and elderly groups, but also help complete the task of clinical diagnosis with high efficiency and low cost. Based on this, we applied our own AK-DL model to complete the development of the software and tried to put it into service. The AK-DL model performed best in the practical application of this app due to its high diagnostic accuracy and low training cost. In future, we will continue to improve the app service functions and actively promote it gradually.

In order to highlight the characteristics of deep learning for AK diagnosis, we only selected the mature and mainstream technologies for comparison. Actually, several novel methods have recently been developed to diagnose AK. Spectral diagnosis of skin precancerous lesions possesses noninvasive characteristics, which draws conclusions by analyzing and summarizing the spectroscopy information of skin components [36,37]. Compared with neural networks, this method is more accurate, although its analysis is more complicated. Additionally, employing the Markov chain model [38] or traditional statistical methods based on the high-throughput sequencing data of human skin microbial community is also one of the methods for diagnosing AK, but the cost is obviously much higher. Moreover, several other technologies, including Antera 3D technology [39,40], in-vivo or ex-vivo reflectance confocal microscopy (RCM) [41,42], and dermoscopy analysis [43], have been employed for the diagnosis of skin diseases and monitor them after treatment. The detection effect of these abovementioned methods have been verified in actual application, which are considered relatively effective, but their equipment cost and operation complexity are higher than the model proposed in our study. The requirements of equipment for the AK-DL model are relatively lower and the equipment operates simply, which provides the possibility of portable diagnosis.

5. Conclusions

Considering the complexity of diagnosis and the lack of attention during the early stage of AK, a shallow CNN named AK-DL was designed to promote the rapid recognition of AK with high accuracy. Our findings showed that the AK-DL model had the best performance in recognition of AK compared with mainstream deep CNNs (AlexNet, GoogLeNet, and ResNet) and traditional machine learning algorithms (SVM, RF, and KNN). Based on the AK-DL model, an intelligent diagnosis system of AK was developed to satisfy real-time and portable diagnosis. The results of the present study indicate that it is better to employ a shallow CNN in the recognition of AK, and further studies are warranted to determine whether the application of a shallow CNN may promote real-time and automatic diagnosis of other skin diseases. One limitation of this study is that we only validated the detection performance of AK-DL on the images of AK and BK. In real-life diagnosis of skin diseases, only a fraction of precancerous skin lesions are AK. Thus, we would not be able to detect precancerous skin in areas not yet visibly affected by AK due to the lack of relevant training images. In order to carry this work into real clinical setting, a large-scale database that covers different precancerous skin lesions must be built to further train and evaluate the different algorithms.

Supplementary Materials: Supplementary materials can be found at http://www.mdpi.com/2075-4418/10/4/217/s1.

Author Contributions: Conceptualization, L.W. and Y.X.; methodology, L.W.; software, L.W.; validation, A.C. and Y.Z. (Yan Zhang); formal analysis, A.C.; data curation, X.W.; writing—original draft preparation, L.W., Y.Z. (Yu Zhang), and A.C.; writing—review and editing, A.C.; visualization, Y.X. and Q.S.; project administration, Y.X.; funding acquisition, Y.X. All authors have read and agreed to the published version of the manuscript.

Funding: This research was funded by the National Natural Science Foundation of China (Grant No. 81803234).

Conflicts of Interest: The authors declare no conflict of interest.

References

1. Lebwohl, M. Actinic keratosis: Epidemiology and progression to squamous cell carcinoma. *Br. J. Dermatol.* **2003**, *149*, 31–33. [CrossRef] [PubMed]
2. Cantisani, C.; Paolino, G.; Melis, M.; Faina, V.; Romaniello, F.; Didona, D.; Cardone, M.; Calvieri, S. Actinic Keratosis Pathogenesis Update and New Patents. *Recent Pat. Inflamm. Allergy Drug Discov.* **2016**, *10*, 40–48. [CrossRef] [PubMed]
3. Jetter, N.; Chandan, N.; Wang, S.; Tsoukas, M. Field Cancerization Therapies for Management of Actinic Keratosis: A Narrative Review. *Am. J. Clin. Dermatol.* **2018**, *19*, 543–557. [CrossRef] [PubMed]
4. Lanoue, J.; Chen, C.; Goldenberg, G. Actinic keratosis as a marker of field cancerization in excision specimens of cutaneous malignancies. *Cutis* **2016**, *97*, 415–420. [PubMed]
5. Didona, D.; Paolino, G.; Bottoni, U.; Cantisani, C. Non Melanoma Skin Cancer Pathogenesis Overview. *Biomedicines* **2018**, *6*, 6. [CrossRef]
6. Braun, R.P.; Ludwig, S.; Marghoob, A.A. Differential Diagnosis of Seborrheic Keratosis: Clinical and Dermoscopic Features. *J. Drugs Dermatol. JDD* **2017**, *16*, 835–842.
7. Quaedvlieg, P.; Tirsi, E.; Thissen, M.; Krekels, G. Actinic keratosis: How to differentiate the good from the bad ones? *Eur. J. Dermatol.* **2006**, *16*, 335–339.
8. Ortonne, J.P. From actinic keratosis to squamous cell carcinoma. *Br. J. Dermatol.* **2002**, *146*, 20–23. [CrossRef]
9. Lebwohl, M.; Swanson, N.; Anderson, L.L.; Melgaard, A.; Xu, Z.; Berman, B. Ingenol mebutate gel for actinic keratosis. *N. Engl. J. Med.* **2012**, *366*, 1010–1019. [CrossRef]
10. Sirinukunwattana, K.; Raza, S.E.A.; Tsang, Y.-W.; Snead, D.R.; Cree, I.A.; Rajpoot, N.M. Locality sensitive deep learning for detection and classification of nuclei in routine colon cancer histology images. *IEEE Trans. Med. Imaging* **2016**, *35*, 1196–1206. [CrossRef]

11. Albarqouni, S.; Baur, C.; Achilles, F.; Belagiannis, V.; Demirci, S.; Navab, N. Aggnet: Deep learning from crowds for mitosis detection in breast cancer histology images. *IEEE Trans. Med. Imaging* **2016**, *35*, 1313–1321. [CrossRef] [PubMed]
12. Li, W.; Li, J.; Sarma, K.V.; Ho, K.C.; Shen, S.; Knudsen, B.S.; Gertych, A.; Arnold, C.W. Path R-CNN for prostate cancer diagnosis and gleason grading of histological images. *IEEE Trans. Med. Imaging* **2018**, *38*, 945–954. [CrossRef] [PubMed]
13. Jalal Deen, K.; Ganesan, R.; Merline, A. Fuzzy-C-means clustering based segmentation and CNN-classification for accurate segmentation of lung nodules. *Asian Pac. J. Cancer Prev. APJCP* **2017**, *18*, 1869.
14. Gao, F.; Wu, T.; Li, J.; Zheng, B.; Ruan, L.; Shang, D.; Patel, B. SD-CNN: A shallow-deep CNN for improved breast cancer diagnosis. *Comput. Med. Imaging Graph.* **2018**, *70*, 53–62. [CrossRef] [PubMed]
15. Zhang, J.; Hu, H.; Chen, S.; Huang, Y.; Guan, Q. Cancer cells detection in phase-contrast microscopy images based on Faster R-CNN. In Proceedings of the 2016 9th International Symposium on Computational Intelligence and Design (ISCID), Hangzhou, China, 10–11 December 2016; pp. 363–367.
16. Kawahara, J.; Hamarneh, G. Multi-resolution-tract CNN with hybrid pretrained and skin-lesion trained layers. In Proceedings of the International Workshop on Machine Learning in Medical Imaging, Athens, Greece, 17 October 2016; pp. 164–171.
17. Kaur, P.; Dana, K.J.; Cula, G.O.; Mack, M.C. Hybrid deep learning for reflectance confocal microscopy skin images. In Proceedings of the 2016 23rd International Conference on Pattern Recognition (ICPR), Cancún, Mexico, 4–8 December 2016; pp. 1466–1471.
18. Lopez, A.R.; Giro-i-Nieto, X.; Burdick, J.; Marques, O. Skin lesion classification from dermoscopic images using deep learning techniques. In Proceedings of the 2017 13th IASTED International Conference on Biomedical Engineering (BioMed), Innsbruck, Austria, 20–21 February 2017; pp. 49–54.
19. Harangi, B. Skin lesion detection based on an ensemble of deep convolutional neural network. *arXiv* **2017**, arXiv:1705.03360.
20. Tajbakhsh, N.; Suzuki, K. Comparing two classes of end-to-end machine-learning models in lung nodule detection and classification: MTANNs vs. CNNs. *Pattern Recognit.* **2017**, *63*, 476–486. [CrossRef]
21. Zhao, X.; Liu, L.; Qi, S.; Teng, Y.; Li, J.; Qian, W. Agile convolutional neural network for pulmonary nodule classification using CT images. *Int. J. Comput. Assist. Radiol. Surg.* **2018**, *13*, 585–595. [CrossRef]
22. Tschandl, P.; Rosendahl, C.; Kittler, H. The HAM10000 dataset, a large collection of multi-source dermatoscopic images of common pigmented skin lesions. *Sci. Data* **2018**, *5*, 180161. [CrossRef]
23. LeCun, Y.; Bottou, L.; Bengio, Y.; Haffner, P. Gradient-based learning applied to document recognition. *Proc. IEEE* **1998**, *86*, 2278–2324. [CrossRef]
24. Krizhevsky, A.; Sutskever, I.; Hinton, G.E. Imagenet classification with deep convolutional neural networks. In Proceedings of the Advances in Neural Information Processing Systems, Lake Tahoe, NV, USA, 3–6 December 2012; pp. 1097–1105.
25. Han, S.S.; Lim, W.; Kim, M.S.; Park, I.; Park, G.H.; Chang, S.E. Interpretation of the Outputs of a Deep Learning Model Trained with a Skin Cancer Dataset. *J. Investig. Dermatol.* **2018**, *138*, 2275. [CrossRef]
26. Wu, E.; Wu, K.; Cox, D.; Lotter, W. Conditional infilling GANs for data augmentation in mammogram classification. In *Image Analysis for Moving Organ, Breast, and Thoracic Images*; Springer: Berlin/Heidelberg, Germany, 2018; pp. 98–106.
27. Mendes, D.B.; da Silva, N.C. Skin lesions classification using convolutional neural networks in clinical images. *arXiv* **2018**, arXiv:1812.02316.
28. Li, Y.; Shen, L. Skin lesion analysis towards melanoma detection using deep learning network. *Sensors* **2018**, *18*, 556. [CrossRef] [PubMed]
29. Codella, N.; Cai, J.; Abedini, M.; Garnavi, R.; Halpern, A.; Smith, J.R. Deep learning, sparse coding, and SVM for melanoma recognition in dermoscopy images. In Proceedings of the International workshop on machine learning in medical imaging, Nice, France, 1 October 2012; pp. 118–126.
30. Mirunalini, P.; Chandrabose, A.; Gokul, V.; Jaisakthi, S. Deep learning for skin lesion classification. *arXiv* **2017**, arXiv:1703.04364.
31. Vasconcelos, M.J.M.; Rosado, L.; Ferreira, M. A new color assessment methodology using cluster-based features for skin lesion analysis. In Proceedings of the 2015 38th International Convention on Information and Communication Technology, Electronics and Microelectronics (MIPRO), Opatija, Croatia, 25–29 May 2015; pp. 373–378.

32. Ohki, K.; Celebi, M.E.; Schaefer, G.; Iyatomi, H. Building of readable decision trees for automated melanoma discrimination. In Proceedings of the International Symposium on Visual Computing, Las Vegas, NV, USA, 14–16 December 2015; pp. 712–721.
33. Lober, B.A.; Lober, C.W. Actinic keratosis is squamous cell carcinoma. *South. Med. J.* **2000**, *93*, 650–655. [CrossRef] [PubMed]
34. Dinehart, S.M.; Nelson-Adesokan, P.; Cockerell, C.; Russell, S.; Brown, R. Metastatic cutaneous squamous cell carcinoma derived from actinic keratosis. *Cancer* **1997**, *79*, 920–923. [CrossRef]
35. Zhao, Y.; Li, C.; Wen, C.; Wei, Y.; Li, R.; Wang, G.; Tu, P. The prevalence of actinic keratosis in patients visiting dermatologists in two hospitals in China. *Br. J. Dermatol.* **2016**, *174*, 1005–1010. [CrossRef]
36. Gagniuc, P.A.; Ionescu-Tirgoviste, C.; Gagniuc, E.; Militaru, M.; Nwabudike, L.C.; Pavaloiu, B.I.; Vasilăţeanu, A.; Goga, N.; Drăgoi, G.; Popescu, I. Spectral forecast: A general purpose prediction model as an alternative to classical neural networks. *Chaos Interdiscip. J. Nonlinear Sci.* **2020**, *30*, 033119. [CrossRef]
37. Lanzanò, L.; Scordino, A.; Privitera, S.; Tudisco, S.; Musumeci, F. Spectral analysis of Delayed Luminescence from human skin as a possible non-invasive diagnostic tool. *Eur. Biophys. J.* **2007**, *36*, 823–829. [CrossRef]
38. Gagniuc, P.A. *Markov Chains: From Theory to Implementation and Experimentation*; John Wiley & Sons: Hoboken, NJ, USA, 2017.
39. Cantisani, C.; Paolino, G.; Pellacani, G.; Didona, D.; Scarno, M.; Faina, V.; Gobello, T.; Calvieri, S. MAL daylight photodynamic therapy for actinic keratosis: Clinical and imaging evaluation by 3D camera. *Int. J. Mol. Sci.* **2016**, *17*, 1108. [CrossRef]
40. Cantisani, C.; Paolino, G.; Corsetti, P.; Bottoni, U.; Didona, D.; Calvieri, S. Evaluation of Ingenol mebutate efficacy for the treatment of actinic keratosis with Antera 3D camera. *Eur. Rev. Med. Pharm. Sci.* **2015**, *19*, 92–97.
41. Mazzella, C.; Greco, V.; Costa, C.; Scalvenzi, M.; Russo, D.; Savastano, R.; Staibano, S.; Fabbrocini, G. Management of clinical and subclinical actinic keratoses with histological and immunohistochemical assessments by confocal microscopy. *Dermatol. Ther.* **2018**, *31*, e12672. [CrossRef] [PubMed]
42. Mercuri, S.R.; Rizzo, N.; Bellinzona, F.; Pampena, R.; Brianti, P.; Moffa, G.; Colombo, F.L.; Bearzi, P.; Longo, C.; Paolino, G. Digital ex-vivo confocal imaging for fast Mohs surgery in nonmelanoma skin cancers: An emerging technique in dermatologic surgery. *Dermatol. Ther.* **2019**, *32*, e13127. [CrossRef] [PubMed]
43. Carbone, A.; Ferrari, A.; Paolino, G.; Buccini, P.; De Simone, P.; Eibenschutz, L.; Piemonte, P.; Silipo, V.; Sperduti, I.; Catricalà, C. Lentigo maligna of the face: A quantitative simple method to identify individual patient risk probability on dermoscopy. *Australas. J. Dermatol.* **2017**, *58*, 286–291. [CrossRef] [PubMed]

diagnostics

MDPI

Review

MRI Findings in Axial Psoriatic Spondylarthritis

Loredana Sabina Pascu [1,†], Nicolae Sârbu [1,2,*], Andrei Vlad Brădeanu [2,3,†], Daniela Jicman (Stan) [2], Madalina Nicoleta Matei [1,2,†], Mihaela Ionela Sârbu [2,*], Doina Carina Voinescu [2,3], Aurel Nechita [1,2] and Alin Laurențiu Tatu [2,4,5]

1 "Sf. Ioan" Clinical Emergency Children Hospital, 800487 Galati, Romania
2 Faculty of Medicine and Pharmacy, "Dunarea de Jos" University of Galati, 800008 Galati, Romania
3 "Sf. Andrei" Emergency County Clinical Hospital, 177 Brailei st, 800578 Galati, Romania
4 Infectious Diseases Dermatology Department, "Sf. Parascheva" Infectious Diseases Clinical Hospital, 800179 Galati, Romania
5 Multidisciplinary Integrated Center of Dermatological Interface Research MIC-DIR, 800008 Galati, Romania
* Correspondence: dnsarbu@gmail.com (N.S.); sarbumihaelaionela@gmail.com (M.I.S.)
† These authors contributed equally to this work.

Abstract: Psoriatic arthritis is a significant medical condition with a high prevalence, a wide variety of non-specific symptoms, and a high degree of overlap with other spondylarthritis disorders, particularly ankylosing spondylitis. Hence, knowledge of the magnetic resonance imaging (MRI) manifestations and a multidisciplinary strategy are required for the better management of these patients. We searched publications from the last 10 years and focused on the most relevant ones which discussed the classification criteria, the MRI characteristics of axial psoriatic arthritis, the importance of MRI for follow up, and the reliability of skin and synovial biopsy. Axial spondylarthritis can be diagnosed and followed up on using the well-established MRI technique and, additionally, a biopsy. The analysis and concordance between them can provide new directions for future studies.

Keywords: psoriatic arthritis; magnetic resonance imaging; ankylosing spondylitis; skin biopsy

1. Introduction

Spondylarthritis (SpA) represents a group of rheumatic diseases that include ankylosing spondylitis (AS), axial and peripheral spondylarthritis, psoriatic arthritis (PsA), inflammatory bowel disease-associated spondylarthritis (IBDSpA), reactive arthritis and, non-radiographic axial SpA. They all have some standard features, including pain, stiffness, axial involvement, arthritis, dactylitis, enthesitis, and a high prevalence of the HLA-B27 allele. There are multiple terms to indicate patients with axial PsA, such as psoriatic spondylitis (PS), axial psoriatic arthritis (axPsA), or psoriatic spondyloarthritis [1].

Initially, PsA was considered part of rheumatoid arthritis (RA) until psoriasis (PsO) was linked to aggressive forms of inflammatory arthritis that had a different articular pattern from RA, affecting the distal interphalangeal joints and the sacroiliac joints. After discovering the rheumatoid factor, they were classified into two different entities. Even so, the rheumatoid factor is not specific to RA, and a false-positive test might mislead the diagnosis of PsA [2]. It is a potentially debilitating disease, affecting 0.4 to 2% of the general population and up to 19.7% of patients with skin or scalp psoriasis. It has a higher prevalence in Europe—22.7%—with the mean age of individuals with axial PsA being around 51 years old [3]. It does not have a male or female predilection, compared with other arthropathies. On the other hand, axPsA symptoms can develop in older patients, unlike for AS. SpA develops at a younger age and is more prevalent in males who are HLA-B27-positive [4–7]. There are two phenotypes of PsO: type I (85% of patients) is characterized by the beginning of symptoms in young adults before the age of 40, often with a severe evolution; and type 2 (15%) beginning in patients after the age of 40 with a milder evolution [8].

A multicenter European study involving 1560 subjects in 2006 showed that 31% of patients with psoriasis might develop PsA after 30 years, with the prevalence increasing over time [9], unlike the pediatric population where arthritis occurs prior to psoriatic skin lesions up to 10 years, typically in children before the age of 10, while only 2% of children with psoriasis will encounter juvenile idiopathic arthritis (JIA) [10,11]. Psoriatic arthritis is an immune-mediated, inflammatory, multifactorial disorder in genetically susceptible patients. In a genetically predisposed individual, factors such as obesity, infections, articular trauma (sprain or dislocation), or skin trauma (triggering the Koebner phenomenon) can lead to a self-sustained immune response and clinical manifestations [12]. The spondyloarthropathies are characterized by articular and periarticular inflammatory modifications, with a destructive evolution, different from mechanical changes in non-inflammatory conditions [13].

Radiological involvement can be observed late in the evolution of inflammatory changes, so MRI evaluation has gained more and more ground in recent years in the early diagnosis of PsA. However, there needs to be more international consensus regarding the definition of axPsA [14].

In recent years, papers concerning PsA and its extra-articular manifestations have been published, but the features of axPsA are still a matter for debate. Most axPsA patients have associated peripheral articular involvement, which has been widely studied. However, axPsA is still challenging to diagnose due to its unspecific symptoms and biological findings [15].

Collaboration between dermatologists and rheumatologists is essential, as they have an important role in diagnosing PsA, since psoriasis might precede arthritis by 7–12 years [16]. Screening patients with psoriasis without initial articular rheumatic features and identifying the early signs is fundamental for preventing irreversible evolution, joint damage, and function loss.

MRI allows for the assessment of morphological details in axial and peripheral articulations without ionizing radiation. Its main limitations are high costs and the fact that it is a time-consuming procedure [16].

We reviewed the literature from the past 10 years for publications about MRI findings in axPsA. MRI is a well-established technique for diagnosis and follow-up in axial spondylarthritis but has not been studied extensively in axial PsA.

2. Materials and Methods

We performed searches from January 2013 to May 2022 to identify articles referring to MRI findings in axial PsA, as detailed in Figure 1. We used keywords such as "axial psoriatic arthritis," "MR", "MRI," "spondylarthritis", "biopsy", "psoriasis", "psoriasiform", "sacroiliitis", "inflammatory back pain", "juvenile idiopathic arthritis", "juvenile idiopathic psoriatic arthritis", "dactylitis", "treatment", "side effects", and "DMARDs" in the PubMed, ScienceDirect, and SpringerLink electronic databases. We selected 96 studies and reports that we have cited in our review. Among these, a list of the most relevant articles is shown in Table 1.

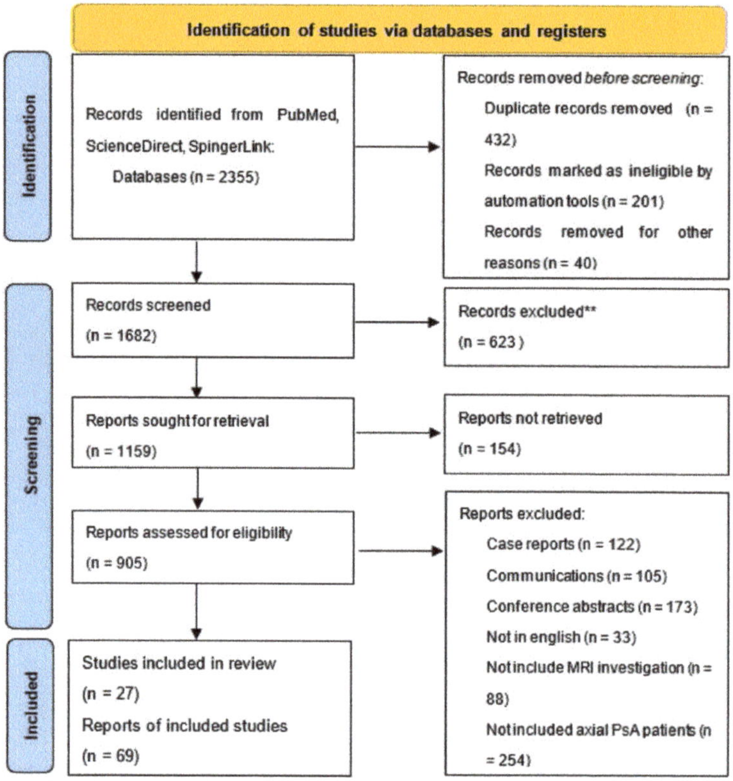

**reports referring to genes deletions, particular drugs treatment responses, changes in paraclinical analyses, book chapters

Figure 1. PRISMA flow chart of our study research.

Table 1. Most relevant selected articles cited in this review.

Author	Year	Study Population	Provided Information
Braga et al. [17]	2020	45 patients	MRI sacroiliitis evaluation
Benavent et al. [18]	2021	3684 patients	
Feld et al. [19]	2021	1354 patients	
Furer et al. [20]	2021	107 patients	
Diaz et al. [21]	2022	93 patients	
Castillo-Gallego et al. [22]	2013	76 patientes	Assement of inflammatory and structural damage
Ye et al. [23]	2019	121 patients	
Poggenborg et al. [24]	2015	48 patients	Comparison with other SpA entities
Jadon et al. [25]	2017	402 patients	
Gensler et al. [26]	2020	203 patients	
Salinas et al. [27]	2021	139 patients	
Abdelaziz et al. [28]	2021	100 patients	
Arnbak et al. [29]	2016	1037 patients	Inflammatory back pain prevalence, MRI findings, and clinical correlations
Yap et al. [30]	2018	171 patients	
Aydin et al. [31]	2017	1195 patients	
Kivity et al. [32]	2018	224 patients	
Zisman et al. [33]	2017	361 patients	Juvenile idiopatic arthritis, risks of psoriasis association, and other comorbidities
Ekelund et al. [34]	2017	440 patients	
Brandon et al. [11]	2018	26710 patients	
Lamot et al. [35]	2021	71 patients	
Jawad et al. [36]	2022	335 patients	Treatment side effects

3. Results

3.1. Clinical Manifestations

PsA might be a challenge in terms of diagnosis, mainly because of the absence of any specific biomarker. Patients often complain of long-standing back pain with an insidious onset before any other symptoms. The rheumatologist needs to differentiate between inflammatory back pain (IBP) and mechanical back pain (MBP), as both of them may occur in axPsA patients and their accurate differentiation allows for appropriate treatment administration [30,37] (Table 2). Patients with IBP usually complain of lower back stiffness in the morning for at least 30 min, with no improvement at rest, sometimes waking patients at night or early in the morning, with an improvement in symptoms after starting physical activity and a good response to non-steroidal anti-inflammatory drugs (NSAIDs) [15]. Typically, IBP last over 3 months, with a gradual onset in middle-aged adults, under 40 years old; however, studies have revealed that 45–55% of axial PsA patients do not report IBP symptoms [30]. In contrast, MBP may occur at any age, typically over 50 years old, with an improvement in symptoms at rest and worsening with movement, and an insidious/acute onset.

Nonetheless, IBP cannot be used as a specific criterion since it is typical of other subsets of SpA and even more frequent in AS patients [4]. AS patients report more frequent and intense spinal pain, while axPsA is associated more frequently with dactylitis, enthesitis and peripheral arthritis, and worse peripheral destruction, and only 4% of the patients have pure axial disease [25,28,38]. Additionally, Abdelaziz et al. revealed that psoriatic nail dystrophy and dactylitis are exclusively seen in axPsA patients, which may help in the differentiation of these two entities [28]. Most of the classification criteria use clinical, genetic, laboratory, and imagistic information, but many signs and symptoms are not specific to a particular subset of SpA. The classifications have a sensitivity and specificity that do not generally exceed 80% [39]. They define a homogenous group of patients that are easier to study [40].

Table 2. Differences between IBP and MBP.

Pain Features	IBP	MBP
Age	<40 yo	>50 yo
Onset	Insidious	Variable
Timing	During the night, can wake the patient.	Daytime
Improvement	With physical activity, no improvement at rest.	With rest, movement may worsen the pain.
Diagnosis	NO history of back injury or heavy work load; Alternating buttock pain; Favorable response to NSAIDs.	History of back injury or heavy work load; No or minimal response to NSAIDs *.
Duration	>3 months	<4 weeks; persisting pain needs additional tests and imaging (HLA-B27, acute phase reactants, MRI, clinical examination).
Location	Anywhere; the proximal/distal third or the posterior region are highly suggestive.	Anterior mid-third of sacroiliac joints, often bilateral.
Treatment	See Figure 2.	Physiotherapy, NSAIDs, corticosteroids, when necessary.

* NSAIDs, nonsteroidal anti-inflammatory drugs.

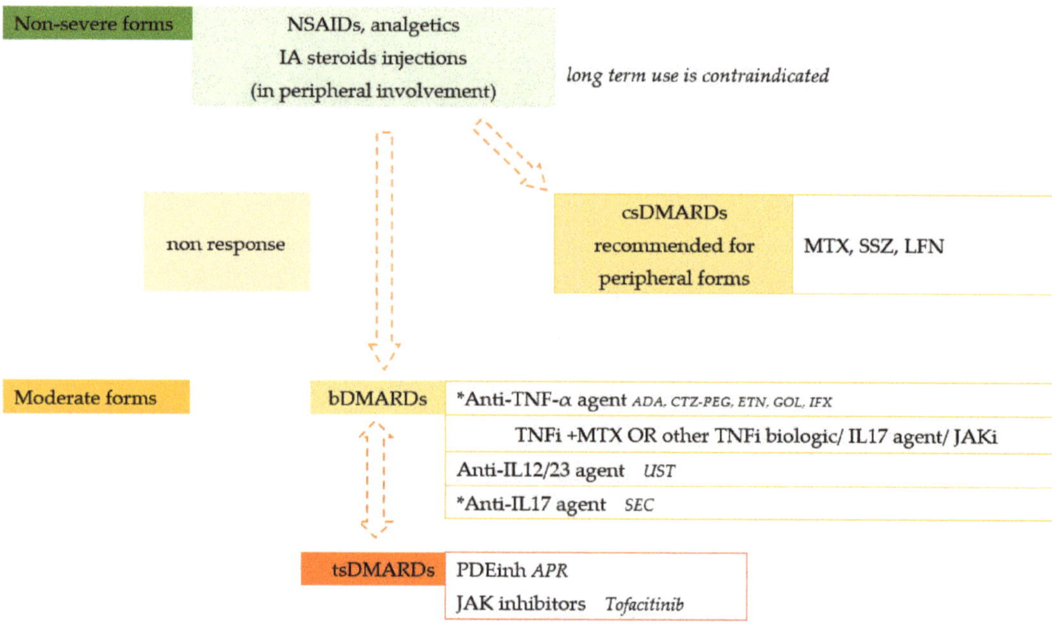

ADA, adalimumab; *APR*, apremilast; *bDMARDs*, biologic disease-modifying antirheumatic drugs; *csDMARDs*, conventional synthetic disease-modifying antirheumatic drugs; *CTZ-PEG*, certolizumab pegol; *ETN*, etanercept; *GOL*, golimumab; *IA*, intraarticular; *IFX*, infliximab; *IL*, interleukin; *LFN*, leflunomide; *MTX*, methotrexate; *PDEinh*, phosphodiesterase inhibitors; *PEbG*, polyethylene glycol; *SEC*, secukinumab; *SSZ*, sulfasalazine; *TNFi*, tumor necrosis factor inhibitor; *tsDMARDs*, targeted synthetic disease-modifying antirheumatic drugs; *UST*, ustekinumab.
*First-line treatment

Figure 2. Treatment guidelines for PsA.

MRI turned out to be the cornerstone in the early detection of minor signal modifications, such as bone marrow edema or fatty lesions in the spine and sacroiliac joints (SIJ), to estimate the probability of a suspected disease in patients with specific symptoms. Given the overlaps between PsA, SpA, and AS, currently, the most used classification for the SpA group is Assessment of SpondyloArthritis international Society (ASAS) classification (98.4% sensitivity, 77% sensitivity), which includes both MRI characteristics for diagnosis, and the modified New York classification [41]. The ASAS criteria are often used as diagnostic criteria in both research and clinical practice [39].

A small percentage of patients who present with isolated spine inflammatory changes, with negative HLA-B27 and no evidence of sacroiliitis on MRI, will not meet the ASAS criteria [42]. Fortunately, the ASAS and the North American Spondyloarthritis Research and Treatment Network (SPARTAN) have recently started a new study that aims to improve ASAS criteria specificity [43].

Today's most widely accepted and applied classification criteria for PsA is the CASPAR criteria set (Classification of Psoriatic Arthritis), which includes dermatological, clinical, and radiological criteria [40,44], developed for a homogeneous group of patients in clinical trials and providing guidance for practitioners, with a high sensitivity of 91.4% [40,44] and 98.7% specificity [2]. Even though CASPAR criteria is a little less sensitive than the Vasey and Espinoza criteria, which include the family history and is applied to both RF-negative and -positive patients, it presents better clinical applicability over other criteria sets, such as Vasey and Espinoza or Moll and Wright criteria [38,39]. Recent studies have revealed that the CASPAR classification is more sensitive in terms of the diagnostic classification assessment of both early and late PsA disease compared to the ASAS criteria [45]. The

Brazilian Society of Rheumatology 2020 guidelines for psoriatic arthritis recommend clinical and imaging criteria for diagnosis and the CASPAR criteria for disease classification [46]. Still, it has limitations in cases of recent onset (Table 3).

Table 3. List of PsA classifications.

Classification	Details	Pros	Cons
CASPAR	Inflammatory arthritis of joint/spine/enthesis + ≥3 points from the following: 1. Psoriasis of the scalp/skin - Current psoriasis, 2p - Personal history, 1p - Family history, 1p 2. Nail dystrophy, 1p 3. Negative rheumatoid factor, 1p 4. Dactylitis, 1p - Present - Personal history 5. Radiological evidence of juxta-articular syndesmophytes, 1p	- Most applied in clinical practice. - Easy to use. - High specificity (98%) and sensitivity (91%). - Also includes patients without psoriasis features or patients with a positive RF blood test. - Includes juxta-articular new bone formation as a feature of PsA, unlike other previous criteria. - Developed via multiple international cohort studies by PsA experts and now it has been widely adopted and implemented.	- The weakest aspect is the initial qualification criterion, since the inflammatory articular disease is not yet well-defined. - Does not include MRI findings, which are often more specific in early diagnosis.
Moll and Wright	Arthritis + Psoriasis + Negative rheumatoid factor. Divides PsA into five categories: - DIP joint only; - Asymmetrical oligoarthritis; - Polyarthritis; - Spondylitis; - Arthritis mutilans.	Historically, it was the simplest and most used criteria.	- Patients without skin psoriasis or those who have a positive RF blood test are not included. - Poor discrimination between PsA and RA. - Is no longer sustained over time and treatment.
ESSG (European SpA Study Group)	Synovitis/inflammatory back pain + Current psoriasis or personal history. Other features: - Arthritis; - Buttock pain; - Enthesitis; - Sacroiliitis; - Inflammatory bowel disease; - Episodes of acute diarrhea; - Urethritis [8].	- Easy applicability. - Allows for a diagnosis without evidence of skin psoriasis.	Lower sensitivity.

Table 3. *Cont.*

Classification	Details	Pros	Cons
Vasey and Espinoza	Psoriasis/psoriatic nail dystrophy + 1. Peripheral disease: - >4 weeks arthritis of DIP joint; - Asymmetrical peripheral arthritis (dactylitis); - Absent RF or rheumatoid nodule; - Radiographic changes (pencil-in-cup deformity, whittling of the terminal phalanges, fluffy periostitis, and bony ankylosis). 2. Axial disease - >4 weeks of spinal pain and stiffness associated with motion restriction; - Symmetric sacroiliitis—grade 2; - Unilateral sacroiliitis—grade 3/4 (according to the New York criteria).	- Easy applicability. - Describes only two patterns of PsA. - Good specificity.	- Does not include patients without skin psoriasis or nail dystrophy. - Does not have enough statistical validation. - Lower sensitivity.
ASAS	Sacroiliitis in imaging studies (active inflammatory lesions in an MRI examination or X-ray changes defined according to the modified New York criteria) + $\geq$1 sign of spondyloarthropathy, or HLA-B27 antigen present + $\geq$2 signs of spondyloarthropathy SpA features: - Inflammatory back pain (IBP) - Arthritis - Enthesitis (heel) - Uveitis - Dactylitis - Psoriasis - Crohn's disease/colitis - Good response to NSAIDs - Family history of SpA - HLA-B27 - Elevated CRP	- Easy to use. - High sensitivity (98.4%). - Includes MRI findings for the better assessment of early disease. - Developed via multiple international cohort studies by PsA experts and now it has been widely adopted and implemented.	Low specificity (77%) for PsA, as it has common features with other axSpA.

Table 3. *Cont.*

Classification	Details	Pros	Cons
ILAR [47]	Definite jPsA. Arthritis and psoriasis or Arthritis with at least two of the following: - Dactylitis; - Nail pitting or onycholysis; - Psoriasis in a first-degree relative. Exclusion criteria: - Arthritis in an HLA-B27-positive male with arthritis onset after 6 years of age; - Ankylosing spondylitis, enthesitis-related arthritis, sacroiliitis with IBD, Reiter's syndrome, or acute anterior uveitis in a first-degree relative; - Presence of the IgM rheumatoid factor on at least two occasions more than 3 months apart; - Presence of systemic JIA. Definite ERA. Arthritis and enthesitis, or arthritis or enthesitis with at least two of the following: - Presence or history of sacroiliac joint tenderness and/or inflammatory lumbosacral pain; - Presence of the HLA-B27 antigen; - Onset of arthritis in a male over 6 years of age; - Acute (symptomatic) anterior uveitis; - History of ankylosing spondylitis, enthesitis-related arthritis, sacroiliitis with inflammatory bowel disease, Reiter's syndrome, or acute anterior uveitis in a first-degree relative. Exclusion criteria: - Psoriasis or a history of psoriasis in the patient or a first-degree relative; - Presence of an IgM rheumatoid factor on at least two occasions at least 3 months apart; - Presence of systemic JIA in the patient.	- Provides a single classification scheme accepted worldwide. - Better than previous criteria.	- Lower sensitivity and specificity compared to the CASPAR criteria. - Restrictive criteria, so patients risk being included in other subgroups. - Difficult to use in assessing patients that have psoriasis, or a psoriasis-like rash, or have relatives with psoriasis. - Classification has had some modifications over time, making the comparision of studies difficult.

On the other hand, for the pediatric population with juvenile arthritis, the most common classification system is that of the International League of Associations of Rheumatology (ILAR). ILAR classifies JIA into seven subgroups:

1. Systemic arthritis;
2. Olygoarthritis (persistent/extended);
3. Polyarthritis RF-negative;
4. Polyarthritis RF-positive;
5. Psoriatic arthritis;
6. Enthesitis-related arthritis (ERA);
7. Undifferentiated arthritis (UA).

Despite the differences between adult and juvenile PsA (jPsA), the CASPAR criteria set may diagnose more JIA patients with juvenile PsA [33], therefore it may be useful for further revisions and a more specific classification.

The COVID-19 pandemic improved access to remote medical consultations through telemedicine. Patients could benefit from consultations, a diagnosis of PsA, and faster treatment. The better detection of PsA was observed through telemedicine, since it helped clinicians use proper diagnostic criteria and screening tools and better understand the characteristics of PsA. Screening tools via telemedicine included validated ones, such as Psoriasis Epidemiology Screening Tool (PEST), and non-validated ones, such as mnemonic PSA (Pain, Stiffness/Sausage finger, and Axial spine involvement) [48].

Post-pandemic studies have revealed that some patients developed chronic arthritis, including psoriatic SpA with bilateral sacroiliitis modifications, validated on MRI, after COVID-19 infection, in the absence of a family history of PsO or HLA-B27 [49,50]. In addition, a paper published in 2022 [51] claimed that axial PsA might undergo exacerbation after the COVID-19 vaccination.

3.2. Imaging Protocols

The primary imaging tool widely used in diagnosing inflammatory joint disease has been conventional radiography, included in the modified New York criteria. Its sensitivity is low for early disease. Nevertheless, it remains a valuable investigation tool in evaluating structural disease progression and is widely available, fast, cost-effective, and requires a low dose of ionizing radiations compared to computer tomography (CT) [52].

However, MRI is the best imaging tool for assessing preclinical PsA, identifying and monitoring inflammatory and structural damage in terms of both axial and peripheral changes, and facilitating better visualization of the peculiar changes involved in PsA. Some studies claim ultrasonography is more sensitive in evaluating peripheral PsA (metacarpophalangeal—MCP, proximal interphalangeal—PIP, and distal interphalangeal—DIP joints) [2,53–55] compared to radiography or MRI. In contrast, other recent studies assert that MRI is more confident in detecting morphological details, but is more expensive, time-consuming, not widely available, and might require contrast agents [52]. In axial disease, the best imaging modality for both inflammatory (bone marrow edema (BME), osteitis) and structural damage, such as erosions or fat deposits, remains MRI investigation [52]. Nevertheless, MRI has limitations in characterizing bone remodeling, the morphology of syndesmophytes, or appreciating sclerosis [15], and it is contraindicated in patients with pacemakers or ferromagnetic metal implants [16].

The MRI examination protocol includes the following steps [41,56,57].

3.2.1. Axial PsA

- Sagittal T1-weighted (T1w) sequences—best for the characterization of fat content or to assess structural bone lesions.
- Sagittal T2-weighted Fat-Sat fast spin echo sequences and short tau inversion recovery (STIR) sequences (short tau inversion recovery) sequences in two planes for water content evaluation/BMO.
- For a better assessment of the costovertebral and costotransverse joints and the facet joints, coronal sequences can be added to the protocol [58].
- Sagittal T1w Fat-Sat sequences with gadolinium enhancement—used for the detection of osteitis; increased perfusion; rarely used, in cases of doubt and high suspicion, to differentiate exudate from synovitis or to assess the activity of bone erosions [59].

3.2.2. Sacroiliac Joints Sequences

- T1-weighted axial oblique and semi-coronal sequences.
- T1-weighted Fat-Sat (TIWFS) spin echo/T1 Dixon/3D gradient echo such as VIBE—for the better evaluation of erosions; VIBE sequences showed a better ability to detect cartilage erosions before extending to the underlying bone [60].

- Axial oblique semi-coronal STIR and T2-weighted fat-suppressed (T2WFS) sequences (coronal plane tilted parallel to the long axis of the sacroiliac joint) with 4-mm slice thickness.
- Apparent diffusion coefficient (ADC) map—STIR or T2WFS sequences may be substituted or supplemented for the better assessment of SIJ inflammation.
- Coronal and axial oblique T1w Fat-Sat with gadolinium—detects osteitis; increased perfusion; recommended in cases of doubt and high suspicion; T1WFS pre- and postcontrast administration can differentiate active inflammation from a simple fluid.

Intravenous contrast can be omitted if it is aimed at detecting bone erosions, edema, and bone productions. However, in active PsA, in addition to bone edema, post-contrast sequences show periosteal enhancement and synovitis, and differentiate active bone erosions from remission [59]. Similar MRI protocols are used in spondylarthritis for diagnosis and follow-up.

3.3. MRI Findings

MRI evaluation of the spine and sacroiliac joints indicates the location of lesions and the distinction between the cartilaginous and ligamentous joint compartments. It can assess bone marrow edema and structural lesions, such as sclerosis, squaring, erosion, spine syndesmophytes/joint space width, or the ankylosis of sacroiliac joints [59].

However, there needs to be a universal consensus regarding MRI scores. There is a broad range of studies, including classifications such as the Berlin modification ankylosing arthritis spine MRI activity score, and the Ankylosing Spondylitis Spine Magnetic Resonance Imaging-activity (ASspiMRI-a), Spondylarthritis Research Consortium of Canada (SPARCC), and Psoriatic Arthritis Magnetic Resonance Image scoring system (PsAMRIS), to assess disease activity in spondylarthritis group diseases [61,62]. Based on a multireader experiment, it is difficult to select one of the three methods for inflammation assessing of the spine and SIJs. The ASspiMRI-a and Berlin methods offer a better overall picture of spine inflammation; however, the SPARCC approach may have advantages in terms of reliability due to its greater consistency in this area [61]. Two large working groups from Canada and Denmark developed and validated an MRI scoring system called CANDEM for better evaluating inflammatory and structural lesions over time, as well as the treatment response. Unfortunately, it has limited clinical applicability since it requires a particular acquisition protocol with 4-mm-thickness T1-weighted turbo spin-echo and STIR sequences on the sagittal plane. Separating the cervical, thoracic, and lumbar spine acquisitions with central and lateral sagittal sequences is helpful for a better assessment of the posterior elements of the spine [15]. This scoring method may be useful in an evaluation of how various medications affect the particular elements of inflammation and damage in the spine, the relationships between lesions, and the progression of inflammation and damage throughout the entire spine [63].

Axial psoriatic arthritis includes the following findings [57].

3.3.1. Vertebral Findings (Similar to Spa Findings)

1. Inflammation of the vertebral body superior or inferior corners, identified in the early stages as a low signal in T1w, a high signal in T2w and STIR, due to bone marrow edema and later, a high signal in T1w and T2w due to fatty bone marrow degeneration, known as the “shiny corner sign” or “Romanus lesions” (Figure 3). This is also found in ankylosing spondylitis.
2. Spondylodiscitis—inflammation of the whole vertebral end plate, involving the adjacent intervertebral disc (Andersson lesion) and soft tissue, identified on low signal T1w in both the intervertebral disc and adjacent end plates due to inflammation and bone marrow edema; on high signal T2w Fat-Sat and STIR in the disc space, adjacent endplates, and paravertebral soft tissue, also involving the psoas muscle, with the loss of the endplate cortex signal. Normally, muscle tissue has an intermediary signal in both T1w and T2w sequences, while inflammation leads to an increased signal in both

sequences. A T1w-sequence with gadolinium shows enhancement of the vertebral endplates and paravertebral soft tissue and peripheral enhancement in the case of collections. DWI sequences distinguish the acute and chronic stages (high signal vs. low signal). It is described also as a complication in the evolution of advanced SpA, mostly in the lumbar spine [64,65].

3. Facet joint inflammation—generated mostly by articular degeneration, usually associated with BME within the spinal pedicles, and common in most arthropathies.
4. Inflammation of the posterior and lateral elements—including the costovertebral joints, transverse and spinous processes, and the adjacent soft tissue
5. Bone erosion—best visualized on T1-weighted sequences as cortical (dark appearance) bone defects, contour irregularities, and the loss of the normal bright appearance of the adjacent bone marrow.
6. Bulky new bone formation—bone productions in high signal T1w; marginal and paramarginal vertical syndesmophytes distributed asymmetrically along the spine, with a peculiar feature of late PsA, while in SpA, syndesmophytes are typically bilateral and symmetrical, with only a marginal distribution and evolution from caudal to cranial [66]. In axial PsA, syndesmophytes extend towards the adjacent vertebra, while in AS, they are continuous from adjacent vertebra, with a tendency to the formation of osteophytic bridges and further evolution to the bamboo spine [67]. The vertebral joint spaces are preserved until the late stages of the disease.
7. Enthesitis—involving the supraspinal ligament, interspinal ligaments, and ligamentum flavum; normally, tendons have a homogenous low signal in all sequences [65].

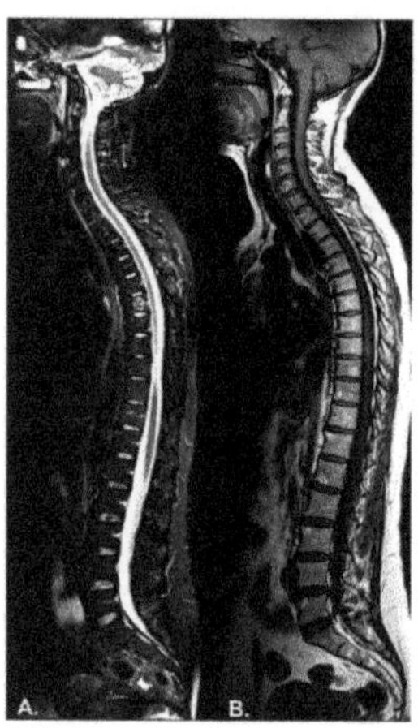

Figure 3. (**A**) Sagittal STIR, (**B**) T1-weighted sequence. Images show Romanus lesions as tiny subchondral bone marrow oedema (hyperintensities) in the antero-superior corners of lumbar vertebral bodies on STIR and, after the active phase, residual fatty transformation in the same regions on T1.

3.3.2. Sacroiliac Joints Findings

1. BME is evident in low signal T1w, hyper signal T2w, STIR images and +C T1w Fat-Sat, similar to blood vessels and spinal fluid; the signal intensity is directly proportional to the inflammation activity. It is usually located periarticular to or on the subchondral bone surfaces and it is an indicator of disease activity (Figure 4) [65].
2. Capsular inflammation—increased signal on T2w GRE and STIR sequences.
3. Enthesitis—entheses are normally seen as hypointense structures, whereas inflammation leads to an increase in signal; best visualized on T2w Fat-Sat and STIR sequences. There are similar findings in both PsA and SpA [65].
4. Joint space fluid.
5. Joint space enhancement.

6. Erosions–initially focal, later they will converge and will have a pseudo-enlargement aspect of the sacroiliac joint. In T1w images there is a loss of cortical bone signal (normally hypointense) and bone marrow fat (normally hyperintense).
7. Inflammation at the site of erosion.
8. Fat metaplasia in an erosion cavity or 'backfill' [65].
9. Sclerosis—better visualized on an X-ray or CT scan; a subchondral or periarticular area with a low signal compared to normal bone marrow on T1, T1FS (SPIR), and STIR sequences (Figure 5).
   Slight: <25% of the subcortical bone area.
   Moderate: 25% to <50% of the subcortical bone area.
   Severe: >50% of the subcortical bone area [29].
10. Ankylosis [65] Partial: Partial osseous bridging across the joint space. Total: Fusion of the joint facets [29].
11. Bone bud—new bone products that are not bridging the joint [68]. In axial PsA syndesmophytes extend into the SIJ space, while in AS, they are across the SIJ space [67].

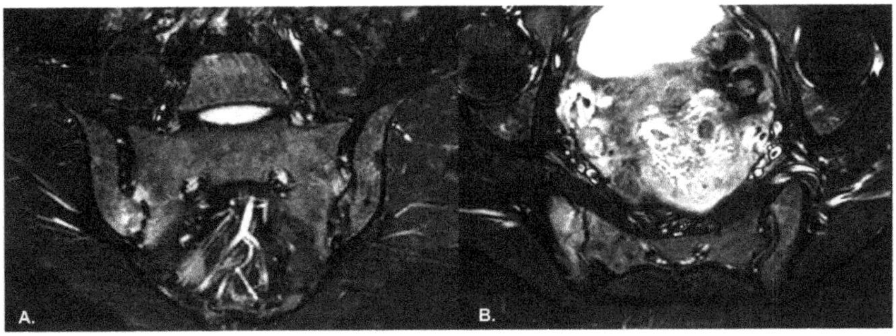

Figure 4. (**A**) Coronal and (**B**) axial STIR sequences show extensive subchondral oedema involving the sacroiliac joints, mainly on the right side and in the inferior regions, consistent with asymmetric active sacroiliitis in a patient with psoriasis.

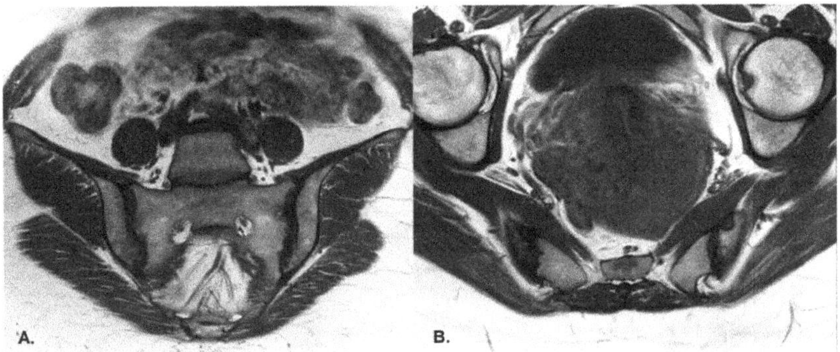

Figure 5. (**A**) Coronal and (**B**) axial T1-weighted sequences of the sacroiliac joints demonstrate subchondral sclerosis and erosions, predominantly on the right side, in the same patient as in Figure 4, consistent with sacroiliitis grade III.

The active inflammatory changes in PsA assessed by MRI include subchondral bone edema, enthesitis, capsulitis, synovitis, and enhancement inside or adjacent to the joint. Chronic inflammatory damage on MRI consists of periarticular erosions, fat metaplasia, subchondral bone sclerosis, joint space narrowing, syndesmophytes, and ankylosis [17,20,57].

Recent studies have shown conflicting results in terms of axial PsA assessment. MRI findings in PsA are less severe than in AxSpA, with lower scores for bone marrow edema at the vertebral and sacroiliac joints. Furthermore, there is a negative correlation with HLA-B27, which makes it challenging to develop a classification system [4]. Some studies have indicated that HLA-B27 positivity in axial forms of the disease is associated with high susceptibility and severity, including more extensive and severe inflammatory changes [7, 15,18,20].

Poddubnyy et al. observed that up to 44% of PsA patients were HLA-B27-positive, compared to 90% of axial SpA patients [38]. Meanwhile, in a study by Diaz et al. [21], only 10% of PsA patients were HLA-B27-positive.

PsA is a heterogeneous disease due to its frequent spine and sacroiliac joint involvement. The axial involvement can be extensive, with diffuse vertebral involvement associated with bulky vertical pseudosyndesmophytes that emerge from the spinal ligament [65,66], unlike in AS, where lesions are well defined and rarely associated with vertebral para marginal/"non-marginal bulky" syndesmophytes [5] (see Table 3).

Dorsal spine involvement in AS is characterized by higher vertebral osteoproliferative severity with symmetrical, marginal syndesmophytes [25], unlike in axPsA, where paravertebral ossification is uncommon, and bone productions are non-marginal and less symmetrical. Fusion facet joints were frequently found in AS lumbar spines, while PsA syndesmophytes may occur without sacroiliitis. Among patients with axPsA there was less severe asymmetrical sacroiliitis and less frequent IBP [5,18].

AS often involves important joint damage from the sacroiliac level evolving from caudal to cranial with the apophyseal joints' involvement and the formation of bulky syndesmophytes, while in axPsA inflammation was observed more frequently in the cervical spine (frequent fusion of facet joints), with minimal apophyseal joints involvement and less bone production [4,5], even in the absence of sacroiliitis or other significant spine damage in early disease [30,41].

On the other hand, Gazel et al. observed that active changes in the cervical spine are more frequent in AS patients than in PsA patients but with a similar rate of structural change. Compared to AS, PsA patients had numerically fewer changes in the cervical spine without reaching statistical significance [69].

Up to 75% of axial PsA patients with longstanding disease have cervical spine damage involving C1-C2, where odontoid erosions and atlantoaxial instability can be found, leading to further complications, such as spinal cord compression and neurological deficit [1,15]. The lower cervical spine might undergo bone production, inflammation of the inter-apophyseal joints, and ossification of the anterior longitudinal ligament [15].

According to Jadon et al., one out of four patients with PsA and AS fulfill the classification criteria for both diseases. The term psoriatic SpA is used for cases that share features of both entities [25]. Axial PsA patients seem to have similar sex predominance, disease activity, and functionality, but PsA patients are older at diagnosis time, with a median age of 49 years old, compared to 43 years old for AS patients, with much more smokers in the PSA group [27]. Moreover, SpA is genetically and immunologically linked to inflammatory bowel disease (IBD); thus, IBD may predict arthritis and two out of three SpA patients have subclinical IBD, with similar changes as those encountered in Crohn's disease [70]. IBD has a similar incidence in axial PsA and SpA and a higher incidence in pediatric patients with jPsA or psoriasis-associated conditions in JIA [11,27].

On the other hand, in a study that analyzed each vertebral level of structural damage, men with AS had higher score levels than women. The C5 and L4 vertebrae were more involved than others [71,72].

A study on 402 PsA and AS patients concluded that both have similar activity, metrology, and disability [25]. Similar findings were confirmed in another study published in 2021, which showed that both active and structural changes in PsA and AS patients had similar rates in the lumbar spine and SIJ and were less frequent in the cervical spine. In the thoracic spine more active and structural damage in PsA patients was recorded, but without reaching statistical significance [69] (Table 4).

The anterior chest wall (ACW) is usually overlooked while examining the extension of axial PsA. Research conducted between 2016 and 2019 using standard clinical practice methods examined 104 individuals with PsA, 45 of whom had the axial illness. Sacroiliitis affected 71% of the axial PsA patients. The most frequent bone marrow edema was identified in the sternoclavicular joints, followed by the manubriosternal and costosternal joints. MRI sequences showed bone marrow edema, synovial hyperemia, capsular structure thickness, erosions, bone irregularities, bone productions, and osteophytes [73].

Table 4. Comparison between the imaging features of axial PsA and AS.

	axPsA	AS
	44% HLA-B27-positive	90% HLA-B27-positive
Cervical spine	Facet joints fusion—more frequent	
Dorsal spine	Non-marginal syndesmophytes (spinal ligament origin) Asymmetrical Cranial to caudal evolution	Marginal syndesmophytes Symmetrical Caudal to cranial Paravertebral ossification
Lumbar spine		Facet joints fusion—more frequent
Sacroiliac	Less severe sacroiliitis Asymmetrical sacroiliitis Syndesmophytes extend into SIJ space Syndesmophytes may be observed in the absence of sacroiliitis	Severe sacroiliitis Symmetrical Syndesmophytes pass across the SIJ space
BME	Lower score in spine and SIJ	Higher score
Fat metaplasia	Less severe	More severe
Erosions	Less severe	More severe
Peripheral involvement **Enthesitis**	More often dactylitis, nail dystrophy Similar to AS	Less often Similar to PsA
Uveitis	Rare	Frequent
Inflammatory bowel disease	Similar prevalence	Similar prevalence

Recent studies have shown that many patients might have active spine changes without sacroiliitis [21,25,69]. In a study published in 2022, 91.7% of cases of MRI-spondylitis were in confirmed PsA patients, while 8.3% of cases were in patients with psoriasis and axial SpA [21].

Even though ALs are exceptionally identified at the disease onset, up to 6% of PsA and AS patients might encounter them. There were a few cases reported where symptomatic patients who were HLA-B27-negative did not meet the ASAS criteria for ax SpA but were identified on MRI scans with erosive disco-vertebral lesions without sacroiliitis modification. Nevertheless, these patients met the CASPAR and GRAPPA criteria for axial PsA [74].

Patients with PsA may have inflammatory or mechanical back pain, but may also have axial disease on imaging while experiencing no back pain or late pain onset [41], since over 30% of axPsA patients may not have IBP. Withal, IBP may be present in patients without axial PsA, so IBP presents a low correlation with imaging findings of sacroiliitis or spondylitis [30]. On the other hand, clinical complaints such as morning stiffness and pain during the night are highly associated with an ASAS MRI-positive and MRI-global impression [32]. MRI changes, such as subchondral BME (Figure 4), might be expected in both PsA patients and healthy people, especially recreational runners, professional ice hockey players, or women in the postpartum period (who present mechanic SIJ modification), leading to significant "false positive" diagnoses of PsA. However, sacroiliac joint inflammation does not exceed three quadrants in these cases, unlike in the SpA group [13,22,23,43,75,76].

In Gazel et al.'s study, only 16% of patients had active inflammation at both sites [69]. When small, well-defined, abnormal bone marrow signal intensities are seen on the SIJ, such as edema or sclerosis found on one, two, or, far less often, three and four consecutive slices, with no severe associated erosion, they are irrelevant. These MRI findings are better seen on coronal oblique images oriented on the S1–S3 axis, often in asymptomatic patients, or they might be associated with back pain. Diaz et al.'s study [21] showed that inflammatory changes in the spine and SIJ are less frequent compared to the high prevalence of back pain. At the same time, the axSpA and IBP criteria had poor sensitivity for MRI findings [1].

Mimics of sacroiliitis risk the over-diagnosis of inflammatory sacroiliitis with a subsequent negative impact on the patient's lifestyle. The most frequent mimickers are stress-related changes, infection, osteoarthritis, stress/insufficiency fractures, and osteitis condensans ilii [67].

Mechanical stress loading is concentrated in the anterior mid-third of the SIJ and is typically bilateral and triangular shaped. BME also occurs in healthy, physically active individuals at the antero-posterior inferior quadrants, probably due to the different distribution of mechanical forces [37,77], while degenerative changes, most commonly bone sclerosis, are found at the anterior and middle thirds of the joint and at the antero-superior quadrant. Table 5 summarizes the main pathologies of the differential diagnosis of BME, based on the typical locations. BME or sclerosis noted on the proximal/distal third of the posterior quadrant of the joint along with structural lesions, such as bone erosions or backfill, are strongly suggestive of sacroiliitis. [13,22,76,78]. The incidence of 30–41% of physically active individuals [78] that meet the ASAS criteria for axSpA advocates the importance of contextual MRI interpretation with a patient's clinical and serological profile.

Table 5. Differential diagnosis of inflammatory from non-inflammatory back pain based on the typical location of BME.

Condition	Characteristic Location of BME
axSpA	Posterior lower quadrant of the ilium Dorsal-cartilaginous segment of the SIJ
Healthy individuals	Lower iliac bone
Sports individuals	Overlaps with axSpA Posterior lower quadrant of the ilium Anterior upper quadrant of the sacrum
Postpartum	Overlaps with axSpA Typically, no structural changes
Degenerative	Anterior and middle thirds Ligamentous segment of the SIJ Associated with the degeneration of the pubic symphysis

Insufficiency fractures are typically seen in osteoporotic elderly people, more often post-menopausal women, but are also seen in cases of steroid-induced osteopenia, infiltrative disease, and a history of pelvic radiation, demonstrated by the sacral alae and bilateral predominance [78].

Stress fractures are typically more common in athletes, shown on MRI as a unilateral BME on the sacral side, with no involvement of the subchondral bone; a vertical fracture line should be seen. Young athletes may be affected by vertebral pedicles and pars interarticularis fractures, leading to spondylolysis [78,79]. *Osteitis Condensans Ilii* is another condition that needs to be excluded when SIJs are evaluated; it is characterized by extensive subchondral sclerosis, frequently involving the ilium and very little BME or fat replacement, without significant erosions of the articular surfaces, unlike sacroiliitis, with a similar location to mechanical etiology changes, in the anterior part of the mid-third of the SIJ, often bilateral and well-defined with a triangular shape, but more extensive sclerotic changes [22]. BME has a ventral-cartilaginous joint segment of the ilium pattern and extends beneath the arcuate line, unlike axSpA which is often located at the dorsal-cartilaginous segment of the SIJ and rarely exceeds the arcuate line [80].

Sibel Zehra Aydin et al. performed a study on 1195 patients with axial PsA where they analyzed the sensitivity of the ASAS IBP criteria. They observed that inflammatory back pain criteria are limited in axial PsA diagnosis, with moderate sensitivity and a lower diagnosis rate in up to half of women. In addition, they noted that the Calin criteria had better agreement with the imagistic investigation than other criteria [81].

A study published in 2013 [81] showed that the extent of lesions in axial PsA, non-radiologic axial SpA (nr-axSpA), and AS patients was similar. However, the scores for the number of severe lesions in SIJ and MRI were lower in axial PsA and nr-axSpA than AS, using the semi-quantitative Leeds Scoring System for BME lesions. In addition, they observed that most of the PsA patients who were HLA-B27-negative had normal MRI scans, while the HLA-B27-positive patients had a similar pattern of BME as AS patients. So, HLA-B27 positive defines a group with more severe lumbar spine and SIJ lesions [81].

Shan-Shan Li et al. selected 186 patients who met the CASPAR criteria for axPsA and associated dactylitis. They noticed that these patients had higher disease activity and more severe joint damage. On the other hand, dactylitic axPsA patients had inflammatory back pain in 50% of cases. Withal, dactylitis is more common in axial PsA, than in peripheral disease [76,82].

Conventional, segmental MRI investigations limit the ability to assess psoriatic arthritis or spondyloarthropathy evolution, as only up to 5% of patients have pure axial disease [1,5]. Therefore, a whole-body MRI has been introduced, especially for patients who have both spine and sacroiliac joint damage. It is essential to fully assess the disease extent and activity, since peripheral damage often precedes axial modifications and can also be evidence of silent skeletal lesions [7,68]. Thus, up to 70% of peripheral damage leads to axial changes during the evolution process. In the early stages, only up to 28% of patients are at risk of simultaneous axial damage [7].

Following standard protocols, advanced techniques have developed improved MRI imaging, which includes quantitative MRI sequences, such as dynamic contrast-enhanced MRI (DCE-MRI), diffusion-weighted imaging (DWI), and chemical shift-encoded MRI (CSE-MRI) [68].

- DWI has the advantage of not necessitating gadolinium administration and is helpful in the early detection of spinal cord damage. ADC discriminates between active-inactive juvenile inflammatory arthritis, a better diagnosis of SpA, and mechanical versus inflammatory back pain.
- CSE-MRI helps detect active inflammation and structural damage.
- DCE-MRI is useful in the evaluation of inflammation activity, detecting early disease from periarticular soft tissue inflammation, even in the absence of synovitis [81]. It is also helpful in distinguishing synovitis between RA and PsA.

New techniques have been developed for better cartilage assessment [68]:

- T2- and T1p-mapping—providing insight into early biochemical cartilage changes and evidence of atlantoaxial instability, and does not require contrast agent administration.
- DGEMRIC (delayed gadolinium-enhanced MRI of cartilage)—recommended for peripheral disease; useful in uncovering early biochemical cartilage disturbances; requires long acquisition protocols.

A 3T whole-body MRI study performed on 50 patients with psoriatic arthritis, spondylarthritis, or healthy subjects in 2021 revealed that 54 clinical enthesitis were observed from a total of 450 sites; the most common inflammation was detected in the anterior chest wall, involving the costo-sternal joints, sternal synchondrosis, and sternoclavicular joints in 44% of cases [83].

The IDEAL sequence provides excellent homogeneous fat suppression in areas of high magnetic susceptibility, delivering a better fat distribution map, helping in the quantitative assessment of the active sacroiliitis, and providing a more confident diagnosis than previous standard Fat-suppression sequences [84,85].

Even though axial PsA has some clinical and imaging particularities compared to AS, some studies still debate whether axial PsA is an independent disease or an axSpA with associated psoriasis [1].

A survey across 13 European countries between 2017 and 2018 found that patients visit 2.6 healthcare professionals before diagnosis, and there is a mean of 7.4 years' delay between the first visit to the doctor and the time of diagnosis [86]. Similar findings were noted in the Benavent et al. study [18].

Other studies have shown that an additional spine MRI scan should be recommended when axPsA is clinically suspected, and spine screening does not add value to SIJ MRI scans since there is a low level of confidence regarding active inflammation in the spine [21,69].

However, whole spine MRI allows for simultaneous axial and peripheral evaluation and should be considered for further studies to develop better diagnostic criteria and definition, a better understanding of treatment responses, and the timely use of biological agents [21,24,52,62].

The European League Against Rheumatism (EULAR) 2015 MRI recommendations include monitoring the disease activity, assessing structural changes, and predicting the outcome and treatment effects [87].

3.3.3. Specific Features of Juvenile Psoriatic Arthritis [88]

MRI findings in JIA include:

- Early disease—periarticular osteopenia, effusion, juxta-synovial soft-tissue swelling;
- Intermediate disease—narrowing of the spaces between joints, cortical erosions, epiphyseal overgrowth;
- Late disease—ankylosis, joint angular deformities, contractures, muscle atrophy.

However, it is questionable whether MRI-detected bony depressions may be regarded as an outcome metric to indicate structural damage in JIA. Children have incomplete ossification and the subchondral bony contour may appear uneven and fractured, so it can be difficult to differentiate a normal from a pathological appearance, but thinner cartilage may indicate JIA [89].

JPsA has two age-peak incidences with distinct features [34]:

1. The 2–5 years age group displays the phenotypic and pathophysiological features of common autoimmune diseases, including:
    - Female predominance, positive anti-nuclear antibodies (ANA), higher predisposition of chronic uveitis.
    - Similar features to oligo- and polyarticular JIA or early-onset ANA-positive JIA.
    - Arthritis, often involving the knee and ankle; the involvement of dactylitis and distal interphalangeal joints is highly suggestive of jPsA. Dactylitis is present in 20–40% of patients and is the first musculoskeletal finding at presentation in about 15% of cases; moreover, it has been observed a long time before skin psoriasis [70].
    - Higher incidence of small joint and wrist involvement.
    - Enthesitis in 22% of cases, typically at the Achilles tendon and plantar fascia insertions into the calcaneus, but many patients who encounter both arthritis and enthesitis are classified as ERA, according to the ILAR criteria [90,91]. It is less frequently encountered than in late-onset jPsA.
2. The 9–12 years age group have features of autoinflammation that emerge as enthesopathy, including:
    - Relative equal sex distribution, with a little male predisposition;
    - Resembles adult PsA features;

- Enthesitis in up to 60% of patients [91];
- Axial involvement, sacroiliitis;
- HLA-B27 positivity, but some of these cases will be diagnosed as ERA or UA by the ILAR criteria.

Psoriasis may occur in jPsA with a delay that can reach 10 years [34]; simultaneous sacroiliitis or enthesitis along with psoriatic or psoriatic-like skin lesions exclude patients from a diagnosis of jPsA and ERA, which represents a hindrance in the evaluation of patients. Psoriasis in young children is often more subtle and more similar to atopic eczema or erythema and scaling behind the ears [10,34].

Oligoarticular peripheral onset is seen in both juvenile and adult PsA, where juvenile evolution tends to be polyarticular and less severe in terms of bone erosions and deformities. In addition, psoriasis is associated with a more severe prognosis [10].

Altered gut and skin microbiomes seem to be associated with PsA, albeit gut dysbiosis (with the loss of microbial diversity) may or may not present gastrointestinal symptoms, hence colitis at baseline may predict a chronic course of arthritis. This association is more difficult to see in pediatric patients, since gut inflammation is best identified by endoscopy, but has limited indications in pediatrics due to the invasive procedure risks and the requirement for general anesthesia. Hence, fecal protectin (fCAL) showed a high correlation with endoscopic results and a further correlation with sacroiliac joint inflammation in both adults and children [35,70]. Strikingly, the ILAR classification subdivides patients with onset arthritis at pediatric age into groups with different names to those in the adult classification [92], even if some of them have counterparts in the adult categories, for example:

- Systemic JIA corresponds to adult-onset Still's disease;
- RF-positive polyarticular JIA is equivalent to RF-positive RA;
- ERA has adult-equivalent undifferentiated SpA.

Meanwhile, PsA and RF-negative polyarthritis are too heterogenous and do not have an adult correspondence or are conditions exclusively seen in the pediatric population, such as early-onset ANA-positive JIA, previously included in both PSA and RF-negative polyarthritis [93]. Olygoarticular JIA and RF-negative JIA seem to be specific entities relevant to the pediatric age group [94].

Once these patients reach adulthood, they will be reclassified into another disease category, based on adult classification criteria that may confuse the patient, and they will also be directed to different treatment protocols corresponding to the new disease entity [95,96]. DCE-MRI may be a high-predictive value tool in clinically inactive JIA patients for disease activity on MRI. It is useful to detect subclinical synovitis in JIA by synovial hypertrophy with hyper-vascularization and quick inflow-outflow that should differentiate active disease patients from inactive ones with "normal" residual synovial thickening. Therefore, synovial maximum enhancement at the baseline on MRI is an important discriminating feature [97]. However, the risk of brain gadolinium deposits should be carefully considered before administering gadolinium to pediatric patients [98].

3.4. Other Diagnostic Tools

The synovial biopsy may help to understand and compare PsA biology to that of other chronic inflammatory joint diseases, such as RA and SpA, and also contribute to the development of a biological foundation for PsA classification, since the synovial membrane is one of the main tissues affected by PsA. Still, up until now, there have only been a few studies [99]. The pathogenesis of psoriatic spondylitis is fueled by increased synovial vascularity with high levels of osteoproliferative cytokines, such as bone morphogenic protein, as well as inflammatory cytokines such as TNF α, compared to uSpA and RA; lower levels of T cells were found compared to RA, without reaching statistical significance [100,101]. Some studies [99,102,103] have shown that PsA has a similar synovial phenotype with AS and undifferentiated SpA, unlike RA; in addition, further studies have shown that PsA has a histological particularity involving the synovial vascularity pattern, but newer studies have observed no histological statistical significant differences between uSpA and PsA [101,103]. The synovial biopsy may be an additional tool in the diagnosis of atypical patients and may guide the treatment response follow-up, improve the treatment decision-making, and accelerate decisions in phase I–II clinical trials [104], although further studies are required.

PsA diagnostic criteria include the presence of psoriatic skin lesions, but other skin diseases may either clinically or histologically mimic psoriasis, therefore it is necessary to differentiate it for a correct diagnosis. This group of conditions includes psoriasiform/psoriatic-like lesions, such as psoriasiform dermatitis—lichenified dermatitis, seborrheic dermatitis, pityriasis rubra pilaris, allergic dermatitis, atopic dermatitis, nummular dermatitis, lichen simplex chronicus, dermatophytosis, dyshidrotic eczema, psoriasiform drug rash, parapsoriasis, and less often, mycosis fungoides and secondary syphilis [101,105]. In these cases, the best diagnostic tool is a skin biopsy; clinical-histological concordance in the diagnosis of psoriasis was found to be significantly influenced by clinical features such as the typical scale and Auspitz's sign, as well as histopathological evidence of suprapapillary thinning and granular layer absence. Lymphocytic exocytosis and the vertical orientation of collagen bundles were strongly linked with the diagnosis of psoriasiform dermatitis [106].

The lymphoid organs have not yet been investigated in PsA, despite the fact that activated immune cells can migrate from damaged psoriatic tissues to draining lymph nodes (LN), where they cause T helper cell differentiation and launch an inflammatory response. Since differentiated and activated T cells are essential for PsA, and LNs are the site of T-cell activation, this information will be extremely important for further studies [105,107–109].

3.5. *Pharmacological Therapy*

3.5.1. Treatment Recommendations and Responses

Due to the lack of evidence of clinical trials of drug classes specific to axial PsA patients, most of the treatment guidelines are based on axial SpA guidelines [110].

Two groups, GRAPPA (Group for Research and Assessment of Psoriasis and Psoriatic Arthritis) and EULAR provide evidence-based treatment recommendations to providers. In the first instance, the gold standard is non-biological treatment—NSAIDs—for mild symptoms, but their action is limited, so in case of ineffectiveness, conventional synthetic disease-modifying antirheumatic drugs (csDMARDs) such as Methotrexate or Leflunomide are recommended, especially for peripheric disease with PsO. In case of a lack of effectiveness, biological disease-modifying antirheumatic drugs (bDMARDs) are recommended. Recent studies have shown that psoriasis patients treated with bDMARDs had a lower incidence of PsA [111].

Other studies have demonstrated the effectiveness for the treatment of active PsA of Janus kinase (JAK) inhibitors, such as tofacitinib, which is in phase II trials; filgotinib; upadacitinib, which is in phase II/III trials [110]; monoclonal antibody-like TNF inhibitors (entanercept, infliximab, adalimumab, golimumab, certolizumab); and interleukin antagonist agents (IL-17A—secukinumab, ixekizumab; and IL-12/23—ustekinumab, which are more efficient than they are in SpA, since both IL-17A and IL-23 are involved in PsA immunopathogenesis; however, IL-17A seems to be more centered on axial SpA, while IL-23 is more specific for the pathogenesis of inflammatory bowel disease) [112].

The target of the management of chronic inflammatory disease is the absence of both clinical and paraclinical manifestations [113]. For many clinicians it is still a burden to decide whether or not to start biological therapy in very early-disease PsA patients, but prospective exploratory studies have indicated that very early bDMARDs showed an important decline in skin symptoms [114]. TNF-α drugs are preferred over other biologic drugs [115] and further studies have analyzed depression or anxiety levels among these patients and demonstrated that TNF-α antagonist-treated patients showed an improvement in mental state [116–118].

The treat-to-target strategy used in other rheumatoid diseases has also been found to be effective in PsA patients, but in clinical practice it has not yet been implemented, due to the large variability of symptoms and missing uniform agreement on which target should be followed in PsA treatment. The target of the therapy is to achieve remission/inactive disease (DAPSA-REM) or, at least, minimal disease activity (MDA) which is a primary outcome instrument for clinical studies in PsA that has validity and a good correlation with other multi-domain measures [119]. Nevertheless, studies have shown that over 48% of all PsA patient groups underwent orthopedic surgery during the disease evolution, with a median age of around 51 years old, much earlier than the median age in the healthy population. The most common joint-sacrificing surgeries involved in axial and peripheral PsA have been knee prosthesis and hip prosthesis. It is important to note that 36% of these surgeries had been performed before the diagnosis, so the early detection and vigorous treatment of PsA from the start is mandatory to limit disease development, as it may lower the need for surgical treatments and offer a chance to enhance patients' quality of life [120]. Inflammatory or mechanical lower back pain along with dactylitis is associated with a poor response to MTX, while TNF inhibitors and IL-17 inhibitor agents are strongly recommended as a first-line treatment as they both showed a good response in all six key domains of PsA (axial, peripheral, enthesitis, dactylitis, skin and nail lesions), as recommended at the last GRAPPA annual meeting of 2022 [118].

TNF inhibitors exceed the best treatment response in young patients with a short duration of disease, high C-reactive protein and normal body weight, and male sex is a good predictive factor. Meanwhile poor prognosis factors (dactylitis, high C-reactive protein, erosions) have better treatment responses to IL-23 inhibitor agents. Ustekinumab (IL12/23 inhibitor agent) has a similar efficiency in both normal and overweight patients, but it is indicated in mono/oligoarticular forms. Ustekinumab and apremilast (PDE inhibitor) obtained better responses in patients with low disease activity [118,121]. The treatment strategies for juvenile PsA are similar. JPsA patients more often present with peripheral arthritis and subsequently axial involvement, thus MTX is widely recommended as a second-line therapy, prior to Leflunomide or sulfasalazine, as it has better tolerability and fewer side effects [36].

Etanercept is the only TNF-inhibitor agent approved for pediatric use, recommended for children over 12 years old with a minimal response to conventional therapy. Ustekinumab (IL12/23) and scukinumab (IL-17A) are the subjects of ongoing trials. Tofacinib (JAK inhibitor) has been approved by the Food and Drug Administration and the European Medicines Agency for the treatment of early-onset jPsA in patients over 2 years old [10].

Methotrexate (MTX) is a widely used drug due to its effectiveness and the fact that it is less expensive than many of the alternatives recommended for the treatment of peripheral PSA and poly- and oligoarticular forms of JIA. Its long-term use has been related to severe liver toxicity, hepatic fibrosis, and eventually liver cirrhosis, although those findings only included old studies based on individuals who received a high dose administered daily. Newer studies have shown that MTX rarely induces liver fibrosis per se, but rather it acts synergistically with other hepatotoxic factors, such as diabetes mellitus or metabolic syndromes including obesity, increased body mass index, and dyslipidemia, which are strongly associated with PsA, or alcohol consumption [122].

Some studies have demonstrated the importance of the use of MRI imaging with low-frequency shear waves (range, 20–200 Hz) to create a visual elastogram that evaluates liver fibrosis (MRE). Most of the time, no data were available about the patients' condition before the start of MTX treatment, so future studies are needed for pre-therapeutic follow-up and during treatment to obtain statistically relevant results. A recent Indian study [123] showed that MRE and serum liver biomarkers might be useful as a substitute for a liver biopsy, which is associated with sampling errors, procedural risks, poor repeatability, and increased morbidity. Moreover, MRE showed high diagnostic accuracy, with higher results compared to the Fibroscan and acoustic radiation force impulse (ARFI) methods and a positive correlation with biomarkers (AST/ALT ratio). MRE is substantially more accurate at detecting fibrosis in obese patients and is unaffected by BMI, ascites, or body habitus, unlike ultrasound-based quantitative elastography techniques [123]. Moreover, the cumulative dose and treatment duration did not correlate with MRE values and mean liver stiffness values [123,124]. However, MRE has its limitations in patients with hemochromatosis or hemosiderosis due to the high hepatic iron overload that will shorten T1, T2, and T2* relaxation times and lower the signal intensity [124].

3.5.2. DMARDs Side Effects

As is the case for any medication, PsA drugs may have various side effects which can be exacerbated by each patient's comorbidities.

csDMARD medications may be associated most often with an increased risk of liver toxicity; gastrointestinal symptoms, such as nausea, vomiting, diarrhea, abdominal pain, and bloating; neurological/psychiatric symptoms, including headaches, dizziness, depression, anxiety, and sleep disturbance; and a cutaneous rash, ulceration, nodules, and itching or blistering. Less often, patients encountered blood abnormalities such as thrombocytopenia, transaminase increase, anemia, alkaline phosphatase elevation, respiratory symptoms, fatigue, fertility problems, or alopecia [36,125].

bDMARDs may increase the risk of respiratory or vaginal infections, liver toxicity, skin infections or cancer, tuberculosis reactivation, congestive cardiac failure, and demyelination syndromes [126], while a recent study showed that over one third of patients discontinued their treatment for fertility or family planning reasons [36]. The study in question was performed between 1994 and 2019 to evaluate real-world treatment tolerability, and showed that csDMARDs are less tolerated than bDMARDs, with a mean duration of treatment until discontinuation up to 10 months, while the duration of treatment with biologic medication reached 18 months [36].

4. Discussion

Psoriatic arthritis is a highly prevalent disease and a relevant health issue. PsA patients represent a heterogeneous group, with high variability of non-specific symptoms and high overlapping with other pathologies of the SpA group, especially AS, which requires a multidisciplinary approach.

The best tool for diagnosing PsA remains the CASPAR criteria. However, additional research is mandatory to create a worldwide accepted definition of axial involvement in PsA to develop diagnostic criteria that are easy to use, with high sensitivity and specificity. However, there are some psoriasiform skin lesions that may mimic psoriasis vulgaris which a skin biopsy might help to correctly diagnose. MRI has proved its essential role in early diagnosis, identifying pre-radiological modifications, such as bone marrow edema, fat metaplasia, and enthesitis. Asymmetrical sacroiliac erosions, cervical spine involvement, and skin psoriasis are more common characteristics of axial PsA than of axial SpA. MRI might give "false positive" results in cases of mechanical overload, such as for athletes or postpartum women, so clinical correlation remains essential for the patient's diagnosis.

The analyzed studies generally agreed that axial PsA and AS have similar spinal activity and disability, with the findings that AS may have numerically more vertebra involved, and that axial PsA has a slightly higher susceptibility in the cervical spine.

One-third of patients may have SIJ involvement preclinically [17]. However, whole-spine MRI is not recommended for screening in the absence of symptomatology due to low confidence in inflammatory changes in the spine. Furthermore, synovial biopsies may be useful in PsA diagnosis to further our understanding of the disease pathology, and may be a cornerstone in the future development of new targeted therapies.

Further studies are necessary since patients generally consult several doctors until they are correctly diagnosed. On average, more than 7 years pass between the first medical visit and the diagnosis. Universally established criteria will allow better study comparisons, understanding of the disease, and treatment management.

The high overlapping characteristics of axial PsA and SpA have allowed medical professionals to diagnose and manage these patients, even without specific studies in PsA. Nevertheless, there are still limited data referring to axial PsA, and further studies are necessary to better understand the disease pathway, early signs, definition, and management.

Juvenile psoriatic arthritis is a self-standing entity, different from adult PsA, that is characterized by dactylitis, with female predominance in the early-onset disease, while the late-onset disease has more similar features to SpA. However, further revisions of the actual classifications, with less restrictive criteria, will allow for easier resettlement from juvenile to adult disease and more homogeneous groups.

Newer techniques such as MRE may help with the further follow-up of patients undergoing biological and non-biological treatments to evaluate treatment hepatotoxicity, as it showed a higher success rate and diagnostic accuracy than Fibroscan, with comparable results to liver biomarkers [122–124]. Further studies on a larger population are necessary to validate the actual data.

Author Contributions: Conceptualization, L.S.P. and M.I.S.; methodology, N.S. and A.V.B.; software, D.J.; validation, L.S.P., N.S. and A.L.T.; formal analysis, M.I.S. and A.N.; investigation, M.I.S. and A.V.B.; resources, M.I.S., A.V.B. and A.L.T.; data curation, D.J. and D.C.V.; writing—original draft preparation, L.S.P.; writing—review and editing, N.S.; visualization, A.V.B. and M.N.M.; supervision, M.I.S.; project administration, A.L.T. and M.N.M.; funding acquisition, A.V.B. All authors have read and agreed to the published version of the manuscript.

Funding: The APC was paid by Dunarea de Jos University of Galati.

Institutional Review Board Statement: Not applicable.

Informed Consent Statement: Not applicable.

Data Availability Statement: Not applicable.

Acknowledgments: The paper was academically supported by Dunarea de Jos University of Galati through the Multidisciplinary Integrated Center of Dermatological Interface Research MIC-DIR (Centrul Integrat Multidisciplinar de Cercetare de Interfata Dermatologica—CIM-CID), "Dunărea de Jos" University, Galați, Romania.

Conflicts of Interest: The authors declare no conflict of interest.

References

1. Zeidler, H.K. Psoriatic Spondylitis: A Disease Manifestation in Debate: Evidences to Know for the Clinical Rheumatologist. *J. Clin. Rheumatol.* **2022**, *28*, 38–43. [CrossRef]
2. Crespo-Rodríguez, A.M.; Sanz Sanz, J.; Freites, D.; Rosales, Z.; Abasolo, L.; Arrazola, J. Role of Diagnostic Imaging in Psoriatic Arthritis: How, When, and Why. *Insights Imaging* **2021**, *12*, 121. [CrossRef]
3. Alinaghi, F.; Calov, M.; Kristensen, L.E.; Gladman, D.D.; Coates, L.C.; Jullien, D.; Gottlieb, A.B.; Gisondi, P.; Wu, J.J.; Thyssen, J.P.; et al. Prevalence of Psoriatic Arthritis in Patients with Psoriasis: A Systematic Review and Meta-Analysis of Observational and Clinical Studies. *J. Am. Acad. Dermatol.* **2019**, *80*, 251–265.e19. [CrossRef] [PubMed]
4. Lopez-Medina, C.; Ziade, N. Axial Disease in Psoriatic Arthritis: How Can We Define It, and Does It Have an Impact on Treatment? *Mediterr. J. Rheumatol.* **2022**, *33*, 142–149. [CrossRef] [PubMed]
5. Helliwell, P.S.; Helliwell, P.S. Axial Disease in Psoriatic Arthritis. *Rheumatology (UK)* **2020**, *59*, 1193–1195. [CrossRef]
6. Gubar, E.E.; Loginova, E.Y.; Korotaeva, T.V.; Glukhova, S.I.; Nasonov, E.L. Comparative Characteristics of Early Psoriatic Arthritis with and without Axial Skeleton Injury (a Subanalysis of the All-Russian Registry of Patients with Psoriatic Arthritis). *Rheumatol. Sci. Pract.* **2019**, *57*, 636–641. [CrossRef]
7. Feld, J.; Chandran, V.; Haroon, N.; Inman, R.; Gladman, D. Axial Disease in Psoriatic Arthritis and Ankylosing Spondylitis: A Critical Comparison. *Nat. Rev. Rheumatol.* **2018**, *14*, 363–371. [CrossRef]
8. Sankowski, A.J.; Łebkowska, U.M.; Ćwikła, J.; Walecka, I.; Walecki, J. Psoriatic Arthritis. *Pol. J. Radiol.* **2013**, *78*, 7–17. [CrossRef] [PubMed]
9. Christophers, E.; Barker, J.; Griffiths, C.; Daudén, E.; Milligan, G.; Molta, C.; Sato, R.; Boggs, R. The Risk of Psoriatic Arthritis Remains Constant Following Initial Diagnosis of Psoriasis among Patients Seen in European Dermatology Clinics. *J. Eur. Acad. Dermatol. Venereol.* **2010**, *24*, 548–554. [CrossRef] [PubMed]
10. Brunello, F.; Tirelli, F.; Pegoraro, L.; Dell'Apa, F.; Alfisi, A.; Calzamatta, G.; Folisi, C.; Zulian, F. New Insights on Juvenile Psoriatic Arthritis. *Front. Pediatr.* **2022**, *10*, 884727. [CrossRef]
11. Brandon, T.G.; Manos, C.K.; Xiao, R.; Ogdie, A.; Weiss, P.F. Pediatric Psoriatic Arthritis: A Population-Based Cohort Study of Risk Factors for Onset and Subsequent Risk of Inflammatory Comorbidities. *J. Psoriasis Psoriatic Arthritis* **2018**, *3*, 131–136. [CrossRef]
12. Carvalho, A.L.; Hedrich, C.M. The Molecular Pathophysiology of Psoriatic Arthritis—The Complex Interplay Between Genetic Predisposition, Epigenetics Factors, and the Microbiome. *Front. Mol. Biosci.* **2021**, *8*, 662047. [CrossRef] [PubMed]

13. Kiil, R.M.; Mistegaard, C.E.; Loft, A.G.; Zejden, A.; Hendricks, O.; Jurik, A.G. Differences in Topographical Location of Sacroiliac Joint MRI Lesions in Patients with Early Axial Spondyloarthritis and Mechanical Back Pain. *Arthritis Res. Ther.* **2022**, *24*, 75. [CrossRef]
14. Maksymowych, W.P.; Østergaard, M. Magnetic Resonance Imaging of Axial and Peripheral Disease in Psoriatic Arthritis: A Report From the 2020 GRAPPA Annual Meeting. *J. Rheumatol.* **2021**, *97*, 50–54. [CrossRef] [PubMed]
15. Giovannini, I.; Zabotti, A.; Cicciò, C.; Salgarello, M.; Cereser, L.; de Vita, S.; Tinazzi, I. Clinical Medicine Axial Psoriatic Disease: Clinical and Imaging Assessment of an Underdiagnosed Condition. *J. Clin. Med.* **2021**, *10*, 2845. [CrossRef] [PubMed]
16. Gottlieb, A.B.; Bakewell, C.; Merola, J.F. Musculoskeletal Imaging for Dermatologists: Techniques in the Diagnosis and Management of Psoriatic Arthritis. *Dermatol. Ther. (Heidelb.)* **2021**, *11*, 1199–1216. [CrossRef]
17. Braga, M.V.; de Oliveira, S.C.; Vasconcelos, A.H.C.; Lopes, J.R.; de Macedo Filho, C.L.; Ramos, L.M.A.; Rodrigues, C.E.M. Prevalence of Sacroiliitis and Acute and Structural Changes on MRI in Patients with Psoriatic Arthritis. *Sci. Rep.* **2020**, *10*, 11580. [CrossRef]
18. Benavent, D.; Plasencia, C.; Poddubnyy, D.; Kishimoto, M.; Proft, F.; Sawada, H.; López-Medina, C.; Dougados, M.; Navarro-Compán, V. Unveiling Axial Involvement in Psoriatic Arthritis: An Ancillary Analysis of the ASAS-PerSpA Study. *Semin. Arthritis. Rheum.* **2021**, *51*, 766–774. [CrossRef]
19. Feld, J.; Ye, J.Y.; Chandran, V.; Inman, R.D.; Haroon, N.; Cook, R.; Gladman, D.D. Axial Disease in Psoriatic Arthritis: The Presence and Progression of Unilateral Grade 2 Sacroiliitis in a Psoriatic Arthritis Cohort. *Semin. Arthritis. Rheum* **2021**, *51*, 464–468. [CrossRef]
20. Furer, V.; Levartovsky, D.; Wollman, J.; Wigler, I.; Paran, D.; Kaufman, I.; Elalouf, O.; Borok, S.; Anouk, M.; Maman, H.S.; et al. Prevalence of Nonradiographic Sacroiliitis in Patients with Psoriatic Arthritis: A Real-Life Observational Study. *J. Rheumatol.* **2021**, *48*, 1014–1021. [CrossRef]
21. Diaz, P.; Feld, J.; Eshed, I.; Eder, L. Characterising Axial Psoriatic Arthritis: Correlation between Whole Spine MRI Abnormalities and Clinical, Laboratory and Radiographic Findings. *RMD Open* **2022**, *8*, e002011. [CrossRef] [PubMed]
22. Castillo-Gallego, C.; Aydin, S.Z.; Emery, P.; McGonagle, D.G.; Marzo-Ortega, H. Magnetic Resonance Imaging Assessment of Axial Psoriatic Arthritis: Extent of Disease Relates to HLA-B27. *Arthritis Rheum.* **2013**, *65*, 2274–2278. [CrossRef] [PubMed]
23. Ye, L.; Liu, Y.; Xiao, Q.; Dong, L.; Wen, C.; Zhang, Z.; Jin, M.; Brown, M.A.; Chen, D. MRI Compared with Low-Dose CT Scanning in the Diagnosis of Axial Spondyloarthritis. *Clin. Rheumatol.* **2020**, *39*, 1295–1303. [CrossRef] [PubMed]
24. Poggenborg, R.P.; Pedersen, S.J.; Eshed, I.; Sørensen, I.J.; Møller, J.M.; Madsen, O.R.; Thomsen, H.S.; Østergaard, M. Head-to-Toe Whole-Body MRI in Psoriatic Arthritis, Axial Spondyloarthritis and Healthy Subjects: First Steps towards Global Inflammation and Damage Scores of Peripheral and Axial Joints. *Rheumatology (UK)* **2015**, *54*, 1039–1049. [CrossRef] [PubMed]
25. Jadon, D.R.; Sengupta, R.; Nightingale, A.; Lindsay, M.; Korendowych, E.; Robinson, G.; Jobling, A.; Shaddick, G.; Bi, J.; Winchester, R.; et al. Axial Disease in Psoriatic Arthritis Study: Defining the Clinical and Radiographic Phenotype of Psoriatic Spondyloarthritis. *Ann. Rheum. Dis.* **2017**, *76*, 701–707. [CrossRef] [PubMed]
26. Gensler, L.; Szumski, A.; Jones, H.; Baraliakos, X.; Gensler, L.S.; Szumski, A.; Heather Jones, M.E.; Xenofon Baraliakos, R. Does Psoriatic Axial Spondyloarthritis Phenotype Correlate with Imaging Morphotype? *Clin. Exp. Rheumatol.* **2020**, *38*, 329–332. [CrossRef]
27. Salinas, R.G.; Prado, E.S.; Ruta, S. Axial Involvement in Psoriatic Arthritis: Effect on Peripheral Arthritis and Differential Features with Axial Spondyloarthritis in South America. *J. Rheumatol.* **2021**, *48*, 1346–1348. [CrossRef]
28. Abdelaziz, M.M.; Ismail, N.; Gamal, A.M.; Lafy, R.; El-Adly, W. Comparative Analysis between Ankylosing Spondylitis and Axial Psoriatic Arthritis Patients. *Egypt. Rheumatol.* **2022**, *44*, 25–29. [CrossRef]
29. Arnbak, B.; Jensen, T.S.; Egund, N.; Zejden, A.; Hørslev-Petersen, K.; Manniche, C.; Jurik, A.G. Prevalence of Degenerative and Spondyloarthritis-Related Magnetic Resonance Imaging Findings in the Spine and Sacroiliac Joints in Patients with Persistent Low Back Pain. *Eur. Radiol.* **2016**, *26*, 1191–1203. [CrossRef]
30. Yap, K.S.; Ye, J.Y.; Li, S.; Gladman, D.D.; Chandran, V. Back Pain in Psoriatic Arthritis: Defining Prevalence, Characteristics and Performance of Inflammatory Back Pain Criteria in Psoriatic Arthritis. *Ann. Rheum. Dis.* **2018**, *77*, 1573–1577. [CrossRef]
31. Aydin, S.Z.; Kilic, L.; Kucuksahin, O.; Ureyen, S.B.; Kalyoncu, U. Performances of Inflammatory Back Pain Criteria in Axial Psoriatic Arthritis. *Rheumatology (Oxford)* **2017**, *56*, 2031–2032. [CrossRef] [PubMed]
32. Kivity, S.; Gofrit, S.G.; Baker, F.A.; Leibushor, N.; Tavor, S.; Lidar, M.; Eshed, I. Association between Inflammatory Back Pain Features, Acute and Structural Sacroiliitis on MRI, and the Diagnosis of Spondyloarthritis. *Clin. Rheumatol.* **2019**, *38*, 1579–1585. [CrossRef] [PubMed]
33. Zisman, D.; Gladman, D.D.; Stoll, M.L.; Strand, V.; Lavi, I.; Hsu, J.J.; Mellins, E.D. The Juvenile Psoriatic Arthritis Cohort in the CARRA Registry: Clinical Characteristics, Classification, and Outcomes. *J. Rheumatol.* **2017**, *44*, 342–351. [CrossRef]
34. Ekelund, M.; Aalto, K.; Fasth, A.; Herlin, T.; Nielsen, S.; Nordal, E.; Peltoniemi, S.; Rygg, M.; Zak, M.; Berntson, L. Psoriasis and Associated Variables in Classification and Outcome of Juvenile Idiopathic Arthritis—An Eight-Year Follow-up Study. *Pediatr. Rheumatol.* **2017**, *15*, 13. [CrossRef]
35. Lamot, L.; Miler, M.; Vukojević, R.; Vidović, M.; Lamot, M.; Trutin, I.; Gabaj, N.N.; Harjaček, M. The Increased Levels of Fecal Calprotectin in Children With Active Enthesitis Related Arthritis and MRI Signs of Sacroiliitis: The Results of a Single Center Cross-Sectional Exploratory Study in Juvenile Idiopathic Arthritis Patients. *Front. Med. (Lausanne)* **2021**, *8*, 650619. [CrossRef]

36. Jawad, I.I.; Nisar, M.K. How Well Are Biologic and Conventional DMARDs Tolerated in Psoriatic Arthritis: A Real World Study. *Eur. J. Rheumatol.* **2022**, *9*, 100–103. [CrossRef] [PubMed]
37. Grinnell-Merrick, L.L.; Lydon, E.J.; Mixon, A.M.; Saalfeld, W. Evaluating Inflammatory Versus Mechanical Back Pain in Individuals with Psoriatic Arthritis: A Review of the Literature. *Rheumatol. Ther.* **2020**, *7*, 667–684. [CrossRef]
38. Poddubnyy, D.; Jadon, D.R.; Van den Bosch, F.; Mease, P.J.; Gladman, D.D. Axial Involvement in Psoriatic Arthritis: An Update for Rheumatologists. *Semin. Arthritis Rheum* **2021**, *51*, 880–887. [CrossRef]
39. Akgul, O.; Ozgocmen, S. Classification Criteria for Spondyloarthropathies. *World J. Orthop.* **2011**, *2*, 107–115. [CrossRef]
40. D'Angelo, S.; Palazzi, C.; Gilio, M.; Leccese, P.; Padula, A.; Olivieri, I. Improvements in Diagnostic Tools for Early Detection of Psoriatic Arthritis. *Expert Rev. Clin. Immunol.* **2016**, *12*, 1209–1215. [CrossRef]
41. Poddubnyy, D.; Baraliakos, X.; van den Bosch, F.; Braun, J.; Coates, L.C.; Chandran, V.; Diekhoff, T.; van Gaalen, F.A.; Gensler, L.S.; Goel, N.; et al. Axial Involvement in Psoriatic Arthritis Cohort (AXIS): The Protocol of a Joint Project of the Assessment of SpondyloArthritis International Society (ASAS) and the Group for Research and Assessment of Psoriasis and Psoriatic Arthritis (GRAPPA). *Ther. Adv. Musculoskelet. Dis.* **2021**, *13*, 1759720X211057975. [CrossRef]
42. Gubar, E.; Korotaeva, T.; Loginova, E.; Korsakova, Y.; Glukhova, S. Pos0310 do axial psoriatic arthritis patients meet asas classification criteria for axial spondyloarthritis? *Ann. Rheum. Dis.* **2022**, *81*, 405. [CrossRef]
43. Poddubnyy, D. Classification vs Diagnostic Criteria: The Challenge of Diagnosing Axial Spondyloarthritis. *Rheumatology (UK)* **2020**, *59*, IV6–IV17. [CrossRef]
44. Wallman, J.; Alenius, G.-M.; Klingberg, E.; Sigurdardottir, V.; Wedrén, S.; Exarchou, S.; Lindström, U.; di Giuseppe, D.; Askling, J.; Jacobsson, L. Validity of Clinical Psoriatic Arthritis Diagnoses Made by Rheumatologists in the Swedish National Patient Register. *Scand. J. Rheumatol.* **2022**, 1–11. [CrossRef] [PubMed]
45. Nas, K.; Karkucak, M.; Durmuş, B.; Ulu, M.A.; Karatay, S.; Çapkin, E.; Ulusoy, H.; Gülkesen, A.; Sula, B.; Akgöl, G.; et al. The Performance of Psoriatic Arthritis Classification Criteria in Turkish Patients with Psoriatic Arthritis. *Int. J. Rheum. Dis.* **2017**, *20*, 985–989. [CrossRef] [PubMed]
46. Carneiro, S.; Palominos, P.E.; Anti, S.M.A.; Assad, R.L.; Gonçalves, R.S.G.; Chiereghin, A.; Lyrio, A.M.; Ximenes, A.C.; Saad, C.G.; Gonçalves, C.R.; et al. Brazilian Society of Rheumatology 2020 Guidelines for Psoriatic Arthritis. *Adv. Rheumatol.* **2021**, *61*, 69. [CrossRef]
47. Petty, R.E.; Southwood, T.R.; Manners, P.; Glass, D.N.; Goldenberg, J.; He, X.; Maldonado-Cocco, J.; Orozco-Alcala, J.; Prieur, A.-M. International League of Associations for Rheumatology Classification of Juvenile Idiopathic Arthritis: Second Revision, Edmonton, 2001. *J. Rheumatol.* **2004**, *31*, 390–392.
48. Gottlieb, A.B.; Wells, A.F.; Merola, J.F. Telemedicine and Psoriatic Arthritis: Best Practices and Considerations for Dermatologists and Rheumatologists. *Clin. Rheumatol.* **2022**, *41*, 1271–1283. [CrossRef] [PubMed]
49. Novelli, L.; Motta, F.; Ceribelli, A.; Guidelli, G.M.; Luciano, N.; Isailovic, N.; Vecellio, M.; Caprioli, M.; Clementi, N.; Clementi, M.; et al. A Case of Psoriatic Arthritis Triggered by SARS-CoV-2 Infection. *Rheumatology (UK)* **2021**, *60*, E21–E23. [CrossRef]
50. Colatutto, D.; Sonaglia, A.; Zabotti, A.; Cereser, L.; Girometti, R.; Quartuccio, L. Post-Covid-19 Arthritis and Sacroiliitis: Natural History with Longitudinal Magnetic Resonance Imaging Study in Two Cases and Review of the Literature. *Viruses* **2021**, *13*, 1558. [CrossRef]
51. Bardazzi, F.; Sacchelli, L.; Loi, C.; Filippi, F.; Guglielmo, A. Reply to "Dermatoses Caused by Face Mask Wearing during the COVID-19 Pandemic". *J. Eur. Acad. Dermatol. Venereol.* **2022**, *36*, e425–e427. [CrossRef]
52. Fassio, A.; Matzneller, P.; Idolazzi, L. Recent Advances in Imaging for Diagnosis, Monitoring, and Prognosis of Psoriatic Arthritis. *Front. Med. (Lausanne)* **2020**, *7*, 551684. [CrossRef]
53. Galluzzo, E.; Lischi, D.M.; Taglione, E.; Lombardini, F.; Pasero, G.; Perri, G.; Riente, L. Sonographic Analysis of the Ankle in Patients with Psoriatic Arthritis. *Scand J. Rheumatol.* **2000**, *29*, 52–55. [PubMed]
54. Wiell, C.; Szkudlarek, M.; Hasselquist, M.; Møller, J.M.; Vestergaard, A.; Nørregaard, J.; Terslev, L.; Østergaard, M. Ultrasonography, Magnetic Resonance Imaging, Radiography, and Clinical Assessment of Inflammatory and Destructive Changes in Fingers and Toes of Patients with Psoriatic Arthritis. *Arthritis Res. Ther.* **2007**, *9*, R119. [CrossRef] [PubMed]
55. Milosavljevic, J.; Lindqvist, U.; Elvin, A. Ultrasound and Power Doppler Evaluation of the Hand and Wrist in Patients with Psoriatic Arthritis. *Acta. Radiol.* **2005**, *46*, 374–385. [CrossRef] [PubMed]
56. Poggenborg, R.P.; Terslev, L.; Pedersen, S.J.; Østergaard, M. Recent Advances in Imaging in Psoriatic Arthritis. *Ther. Adv. Musculoskelet. Dis.* **2011**, *3*, 43–53. [CrossRef]
57. Mathew, A.J.; Østergaard, M.; Eder, L. Imaging in Psoriatic Arthritis: Status and Recent Advances. *Best Pract. Res. Clin. Rheumatol.* **2021**, *35*, 101690. [CrossRef] [PubMed]
58. Sieper, J.; Rudwaleit, M.; Baraliakos, X.; Brandt, J.; Braun, J.; Burgos-Vargas, R.; Dougados, M.; Hermann, K.G.; Landewe, R.; Maksymowych, W.; et al. The Assessment of SpondyloArthritis International Society (ASAS) Handbook: A Guide to Assess Spondyloarthritis. *Ann. Rheum. Dis.* **2009**, *68*, ii1–ii44. [CrossRef]
59. Sudoł-Szopinska, I.; Urbanik, A. Diagnostic Imaging of Sacroiliac Joints and the Spine in the Course of Spondyloarthropathies. *Pol. J. Radiol.* **2013**, *78*, 43–49. [CrossRef]
60. Baraliakos, X.; Hoffmann, F.; Deng, X.; Wang, Y.Y.; Huang, F.; Braun, J. Detection of Erosions in Sacroiliac Joints of Patients with Axial Spondyloarthritis Using the Magnetic Resonance Imaging Volumetric Interpolated Breath-Hold Examination. *J. Rheumatol.* **2019**, *46*, 1445–1449. [CrossRef]

61. Lukas, C.; Braun, J.; van der Heijde, D.; Hermann, K.-G.A.; Rudwaleit, M.; Østergaard, M.; Oostveen, A.; O'connor, P.; Maksymowych, W.P.; Lambert, R.G.W.; et al. Scoring Inflammatory Activity of the Spine ByMagnetic Resonance Imaging in AnkylosingSpondylitis: A Multireader Experiment. *J. Rheumatol.* **2007**, *34*, 862–870. [PubMed]
62. Her, M.; Kavanaugh, A. A Review of Disease Activity Measures for Psoriatic Arthritis: What Is the Best Approach? *Expert Rev. Clin. Immunol.* **2014**, *10*, 1241–1254. [CrossRef] [PubMed]
63. Krabbe, S.; Østergaard, M.; Pedersen, S.J.; Weber, U.; Kröber, G.; Makysmowych, W.; Lambert, R.G.W. Canada-Denmark MRI Scoring System of the Spine in Patients with Axial Spondyloarthritis: Updated Definitions, Scoring Rules and Inter-Reader Reliability in a Multiple Reader Setting. *RMD Open.* **2019**, *5*, e001057. [CrossRef] [PubMed]
64. Waldburger, M. Forme Particulière de Spondylarthrite Ankylosante Avec Spondylodiscite. *Rev. Med. Suisse.* **2005**, *9*, 1646.
65. Canella, C.; Schau, B.; Ribeiro, E.; Sbaffi, B.; Marchiori, E. MRI in Seronegative Spondyloarthritis: Imaging Features and Differential Diagnosis in the Spine and Sacroiliac Joints. *Am. J. Roentgenol.* **2013**, *200*, 149–157. [CrossRef] [PubMed]
66. Caso, F.; Costa, L.; Peluso, R.; Del Puente, A.; Scarpa, R. Psoriatic Arthritis. In *Mosaic of Autoimmunity: The Novel Factors of Autoimmune Diseases*; Elsevier: Amsterdam, The Netherlands, 2019; pp. 527–540. ISBN 9780128143087.
67. Poggenborg, R.P.; Glinatsi, D.; Østergaard, M. Magnetic Resonance Imaging in Psoriatic Arthritis. In *Psoriatic Arthritis and Psoriasis: Pathology and Clinical Aspects*; Springer International Publishing: Berlin/Heidelberg, Germany, 2016; pp. 199–208. ISBN 9783319195308.
68. Sudoł-Szopińska, I.; Giraudo, C.; Oei, E.H.G.; Jans, L. Imaging Update in Inflammatory Arthritis. *J. Clin. Orthop. Trauma.* **2021**, *20*, 101491. [CrossRef] [PubMed]
69. Gazel, U.; Ayan, G.; Solmaz, D.; Bakirci, S.; Karsh, J.; Sampaio, M.; Jibri, Z.; Aydin, S.Z. Limited Added Value of Whole Spine MRI in Spondyloarthritis for Disease Activity Assessment. *Clin. Rheumatol.* **2021**, *40*, 4909–4913. [CrossRef]
70. Stoll, M.L.; Punaro, M. Psoriatic Juvenile Idiopathic Arthritis: A Tale of Two Subgroups. *Curr. Opin. Rheumatol.* **2011**, *23*, 437–443. [CrossRef]
71. Garrido-Castro, J.L.; Calvo-Gutierrez, J.; Gonzalez-Navas, C.; Font-Ugalde, P.; Castro-Villegas, M.D.C.; Collantes-Estevez, E. FRI0218 Structural Damage Distribution at Vertebral Level in Patients with Axial Spondyloarthritis. *Ann. Rheum. Dis.* **2015**, *74*, 503. [CrossRef]
72. Chimenti, M.S.; Alten, R.; D'Agostino, M.A.; Gremese, E.; Kiltz, U.; Lubrano, E.; Moreno, M.; Pham, T.; Ramonda, R.; Spinelli, F.R.; et al. Sex-Associated and Gender-Associated Differences in the Diagnosis and Management of Axial Spondyloarthritis: Addressing the Unmet Needs of Female Patients. *RMD Open* **2021**, *7*, e001681. [CrossRef]
73. Fragoulis, G.E.; Evangelatos, G.; Konsta, M.; Iliopoulos, A. Anterior Chest Wall Involvement in Psoriatic Arthritis: A Forgotten Entity? *Rheumatology (UK)* **2020**, *59*, 3113–3114. [CrossRef] [PubMed]
74. Queiro, R.; Tejón, P.; Alonso, S.; Alperi, M.; Ballina, J. Erosive Discovertebral Lesion (Andersson Lesion) as the First Sign of Disease in Axial Psoriatic Arthritis. *Scand. J. Rheumatol.* **2013**, *42*, 220–225. [CrossRef]
75. Renson, T. Identifying Pitfalls and Opportunities of Magnetic Resonance Imaging in Spondyloarthritis. *Belg. J. Paediatr.* **2022**, *24*, 238–239.
76. Regierer, A.; Weiß, A.; Baraliakos, X.; Behrens, F.; Poddubnny, D.; Schett, G.; Lorenz, H.M.; Worsch, M.; Strangfeld, A. Comparison of patients with axial psa and patients with axspa and concomitant. *Ann. Rheum. Dis.* **2022**, *81*, 864–865. [CrossRef]
77. Badr, S.; Jacques, T.; Lefebvre, G.; Boulil, Y.; Diwan, R.A.; Cotten, A. Main Diagnostic Pitfalls in Reading the Sacroiliac Joints on Mri. *Diagnostics* **2021**, *11*, 2001. [CrossRef] [PubMed]
78. Caetano, A.P.; Mascarenhas, V.V.; Machado, P.M. Axial Spondyloarthritis: Mimics and Pitfalls of Imaging Assessment. *Front. Med. (Lausanne)* **2021**, *8*, 658538. [CrossRef] [PubMed]
79. Slobodin, G.; Lidar, M.; Eshed, I. Clinical and Imaging Mimickers of Axial Spondyloarthritis. *Semin. Arthritis Rheum* **2017**, *47*, 361–368. [CrossRef] [PubMed]
80. Ma, L.; Gao, Z.; Zhong, Y.; Meng, Q. Osteitis Condensans Ilii May Demonstrate Bone Marrow Edema on Sacroiliac Joint Magnetic Resonance Imaging. *Int. J. Rheum. Dis.* **2018**, *21*, 299–307. [CrossRef]
81. Chabros, P.; Pietrzak, A.; Gągała, J.; Kandzierski, G.; Krasowska, D. Psoriatic Arthritis—Classification, Diagnostic and Clinical Aspects. *Przegl. Dermatol.* **2020**, *107*, 32–43. [CrossRef]
82. Li, S.-S.; Du, N.; He, S.-H.; Liang, X.; Li, T.-F. Dactylitis Is Associated with More Severe Axial Joint Damage and Higher Disease Activity in Axial Psoriatic Arthritis. *J. Rheumatol.* **2022**, *49*, 1012–1019. [CrossRef]
83. Guo, Z.K.; Zhang, Y.M.; Kong, C.Y.; Liu, Y.; Ji, X.D.; Zhan, Y.; Qi, W.F.; Lei, X.W. Whole-Body Magnetic Resonance Imaging vs. Clinical Evaluation of Enthesitis in Patients with Spondyloarthritis. *Chin. Med. J. (Engl.)* **2021**, *134*, 2638–2640. [CrossRef]
84. Ren, C.; Zhu, Q.; Yuan, H. Mono-Exponential and Bi-Exponential Model-Based Diffusion-Weighted MR Imaging and IDEAL-IQ Sequence for Quantitative Evaluation of Sacroiliitis in Patients with Ankylosing Spondylitis. *Clin. Rheumatol.* **2018**, *37*, 3069–3076. [CrossRef]
85. Sarbu, M.I.; Sarbu, N.; Ene, D.C.; Corche, D.; Baz, R.; Negru, D.; Nechita, A.; Fotea, S.; Anghel, L.; Tatu, A.L. New Perspectives on Diagnosing Psoriatic Arthritis by Imaging Techniques. *Open Access Rheumatol.* **2021**, *13*, 343–352. [CrossRef]
86. Latest News and Advancements Within the Areas of Axial Spondyloarthritis and Psoriatic Arthritis Featured at EULAR 2021. *EMJ Rheumatol.* **2021**, *8*, 2–10.

87. Mandl, P.; Navarro-Compán, V.; Terslev, L.; Aegerter, P.; van der Heijde, D.; D'Agostino, M.A.; Baraliakos, X.; Pedersen, S.J.; Jurik, A.G.; Naredo, E.; et al. EULAR Recommendations for the Use of Imaging in the Diagnosis and Management of Spondyloarthritis in Clinical Practice. *Ann. Rheum. Dis.* **2015**, *74*, 1327–1339. [CrossRef]
88. Kan, J.H. Juvenile Idiopathic Arthritis and Enthesitis-Related Arthropathies. *Pediatr. Radiol.* **2013**, *43*, 172–180. [CrossRef]
89. Pradsgaard, D.Ø.; Spannow, A.H.; Heuck, C.; Herlin, T. Decreased Cartilage Thickness in Juvenile Idiopathic Arthritis Assessed by Ultrasonography. *J. Rheumatol.* **2013**, *40*, 1596–1603. [CrossRef] [PubMed]
90. Gelfand, J.M.; Weinstein, R.; Neimann, A.L.; Berlin, J.A.; Margolis, D.J. Prevalence and Treatment of Psoriasis in the United Kingdom A Population-Based Study. *Arch. Dermatol.* **2005**, *141*, 1537–1541. [CrossRef] [PubMed]
91. Tuttle, K.S.L.; Vargas, S.O.; Callahan, M.J.; Bae, D.S.; Nigrovic, P.A. Enthesitis as a Component of Dactylitis in Psoriatic Juvenile Idiopathic Arthritis: Histology of an Established Clinical Entity. *Pediatr. Rheumatol.* **2015**, *13*, 7. [CrossRef] [PubMed]
92. Ravelli, A.; Consolaro, A.; Schiappapietra, B.; Martini, A. The Conundrum of Juvenile Psoriatic Arthritis. *Clin. Exp. Rheumatol.* **2015**, *33* (Suppl. S93), S40–S43.
93. Martini, A.; Ravelli, A.; Avcin, T.; Beresford, M.W.; Burgos-Vargas, R.; Cuttica, R.; Ilowite, N.T.; Khubchandani, R.; Laxer, R.M.; Lovell, D.J.; et al. Toward New Classification Criteria for Juvenile Idiopathic Arthritis: First Steps, Pediatric Rheumatology International Trials Organization International Consensus. *J. Rheumatol.* **2019**, *46*, 190–197. [CrossRef] [PubMed]
94. Debrach, A.C.; Rougelot, A.; Beaumel, A.; Cabrera, N.; Belot, A.; Duquesne, A.; Aubry-Rozier, B.; Hofer, M.; Couret, M.; Larbre, J.P.; et al. Comparison of Paediatric and Adult Classification Criteria in Juvenile Idiopathic Arthritis during the Transition from Paediatric to Adult Care. *Jt. Bone Spine* **2021**, *88*, 105047. [CrossRef] [PubMed]
95. Guillaume-Czitrom, S.; Sibilia, J.; Nordal, E. Growing up with Chronic Arthritis: The Confusing Matter of Classification. *RMD Open* **2017**, *3*, e000417. [CrossRef] [PubMed]
96. Stoll, M.L.; Zurakowski, D.; Nigrovic, L.E.; Nichols, D.P.; Sundel, R.P.; Nigrovic, P.A. Patients with Juvenile Psoriatic Arthritis Comprise Two Distinct Populations. *Arthritis Rheum* **2006**, *54*, 3564–3572. [CrossRef] [PubMed]
97. Nusman, C.M.; Hemke, R.; Lavini, C.; Schonenberg-Meinema, D.; van Rossum, M.A.J.; Dolman, K.M.; van den Berg, J.M.; Maas, M.; Kuijpers, T.W. Dynamic Contrast-Enhanced Magnetic Resonance Imaging Can Play a Role in Predicting Flare in Juvenile Idiopathic Arthritis. *Eur. J. Radiol.* **2017**, *88*, 77–81. [CrossRef]
98. Murata, N.; Murata, K.; Gonzalez-Cuyar, L.F.; Maravilla, K.R. Gadolinium Tissue Deposition in Brain and Bone. *Magn. Reason. Imaging* **2016**, *34*, 1359–1365. [CrossRef]
99. Kruithof, E.; Baeten, D.; de Rycke, L.; Vandooren, B.; Foell, D.; Roth, J.; Cañete, J.D.; Boots, A.M.; Veys, E.M.; de Keyser, F. Open Access Synovial Histopathology of Psoriatic Arthritis, Both Oligo-and Polyarticular, Resembles Spondyloarthropathy More than It Does Rheumatoid Arthritis. *Arthritis Res. Ther.* **2005**, *7*, 569–580. [CrossRef]
100. Bolt, J.W.; van Ansenwoude, C.M.J.; Hammoura, I.; van de Sande, M.G.; van Baarsen, L.G.M. Translational Research Studies Unraveling the Origins of Psoriatic Arthritis: Moving Beyond Skin and Joints. *Front. Med. (Lausanne)* **2021**, *8*, 711823. [CrossRef]
101. van Kuijk, A.W.R.; Reinders-Blankert, P.; Smeets, T.J.M.; Dijkmans, B.A.C.; Tak, P.P. Detailed Analysis of the Cell Infiltrate and the Expression of Mediators of Synovial Inflammation and Joint Destruction in the Synovium of Patients with Psoriatic Arthritis: Implications for Treatment. *Ann. Rheum. Dis.* **2006**, *65*, 1551–1557. [CrossRef]
102. Reece, R.J.; Canete, J.D.; Parsons, W.J.; Emery, P.; Veale, D.J. Distinct vascular patterns of early synovitis in psoriatic, reactive, and rheumatoid arthritis. *Arthritis Rheum.* **1999**, *42*, 1481–1484. [CrossRef]
103. Tenazinha, C.; Barros, R.; Fonseca, J.E.; Vieira-Sousa, E. Histopathology of Psoriatic Arthritis Synovium—A Narrative Review. *Front. Med. (Lausanne)* **2022**, *9*, 860813. [CrossRef] [PubMed]
104. Ingegnoli, F.; Coletto, L.A.; Scotti, I.; Compagnoni, R.; Randelli, P.S.; Caporali, R. The Crucial Questions on Synovial Biopsy: When, Why, Who, What, Where, and How? *Front. Med. (Lausanne)* **2021**, *8*, 705382. [CrossRef] [PubMed]
105. Ramwadhdoebe, T.H.; Hahnlein, J.; van Kuijk, B.J.; Choi, I.Y.; van Boven, L.J.; Gerlag, D.M.; Tak, P.P.; van Baarsen, L.G. Human Lymph-Node CD8+ T Cells Display an Altered Phenotype during Systemic Autoimmunity. *Clin. Transl. Immunol.* **2016**, *5*, e67. [CrossRef] [PubMed]
106. Mehta, S.; Singal, A.; Singh, N.; Nath Bhattacharya, S.; Bhattacharya, N. A Study of Clinicohistopathological Correlation in A Study of Clinicohistopathological Correlation in Patients of Psoriasis and Psoriasiform Dermatitis Patients of Psoriasis and Psoriasiform Dermatitis. *Indian J. Dermatol. Venereol. Leprol.* **2009**, *75*, 100.
107. Karouzakis, E.; Hähnlein, J.; Grasso, C.; Semmelink, J.F.; Tak, P.P.; Gerlag, D.M.; Gay, S.; Ospelt, C.; van Baarsen, L.G.M. Molecular Characterization of Human Lymph Node Stromal Cells during the Earliest Phases of Rheumatoid Arthritis. *Front. Immunol.* **2019**, *10*, 1863. [CrossRef]
108. Ramwadhdoebe, T.H.; Hähnlein, J.; Maijer, K.I.; van Boven, L.J.; Gerlag, D.M.; Tak, P.P.; van Baarsen, L.G.M. Lymph Node Biopsy Analysis Reveals an Altered Immunoregulatory Balance Already during the At-Risk Phase of Autoantibody Positive Rheumatoid Arthritis. *Eur. J. Immunol.* **2016**, *46*, 2812–2821. [CrossRef]
109. van Baarsen, L.G.M.; de Hair, M.J.H.; Ramwadhdoebe, T.H.; Zijlstra, I.J.A.J.; Maas, M.; Gerlag, D.M.; Tak, P.P. The Cellular Composition of Lymph Nodes in the Earliest Phase of Inflammatory Arthritis. *Ann. Rheum. Dis.* **2013**, *72*, 1420–1424. [CrossRef]
110. Proft, F.; Torgutalp, M.; Muche, B.; Rios Rodriguez, V.; Verba, M.; Poddubnyy, D. Efficacy of Tofacitinib in Reduction of Inflammation Detected on MRI in Patients with Psoriatic ArthritiS PresenTing with Axial Involvement (PASTOR): Protocol of a Randomised, Double-Blind, Placebo-Controlled, Multicentre Trial. *BMJ Open.* **2021**, *11*, e048647. [CrossRef] [PubMed]

111. Acosta Felquer, M.L.; LoGiudice, L.; Galimberti, M.L.; Rosa, J.; Mazzuoccolo, L.; Soriano, E.R. Treating the Skin with Biologics in Patients with Psoriasis Decreases the Incidence of Psoriatic Arthritis. *Ann. Rheum. Dis.* **2022**, *81*, 74–79. [CrossRef]
112. Menter, A.; Krueger, G.G.; Paek, S.Y.; Kivelevitch, D.; Adamopoulos, I.E.; Langley, R.G. Interleukin-17 and Interleukin-23: A Narrative Review of Mechanisms of Action in Psoriasis and Associated Comorbidities. *Dermatol. Ther. (Heidelb.)* **2021**, *11*, 385–400. [CrossRef]
113. Aouad, K.; El-Zorkany, B. Treat-to-Target in Axial Spondyloarthritis: Are We There Yet? *Mediterr. J. Rheumatol.* **2022**, *33*, 137–141. [CrossRef]
114. Hioki, T.; Komine, M.; Ohtsuki, M. Diagnosis and Intervention in Early Psoriatic Arthritis. *J. Clin. Med.* **2022**, *11*, 2051. [CrossRef]
115. Ogdie, A.; Coates, L.C.; Gladman, D.D. Treatment Guidelines in Psoriatic Arthritis. *Rheumatology (UK)* **2021**, *59*, I37–I46. [CrossRef] [PubMed]
116. Milutinovic, S.; Veljkovic, K.; Zlatanovic, M.; Radunovic, G.; Damjanov, N. Depression/Anxiety Symptoms in Axial Spondyloarthritis and Psoriatic Arthritis Patients in Serbia: A Pilot Study. *Rheumatol. Int.* **2019**, *39*, 1595–1605. [CrossRef] [PubMed]
117. Singh, J.A.; Guyatt, G.; Ogdie, A.; Gladman, D.D.; Deal, C.; Deodhar, A.; Dubreuil, M.; Dunham, J.; Husni, M.E.; Kenny, S.; et al. Special Article: 2018 American College of Rheumatology/National Psoriasis Foundation Guideline for the Treatment of Psoriatic Arthritis. *Arthritis Rheumatol.* **2019**, *71*, 5–32. [CrossRef]
118. Advancements in Psoriatic Arthritis Management: Updated GRAPPA Treatment Recommendations. In Proceedings of the GRAPPA Annual Meeting and Trainee Symposium 2022, Brooklyn, NY, USA, 14–16 July 2022.
119. Wervers, K.; Vis, M.; Tchetveriko, I.; Gerards, A.H.; Kok, M.R.; Appels, C.W.Y.; van der Graaff, W.L.; van Groenendael, J.H.L.M.; Korswagen, L.A.; Veris-van Dieren, J.J.; et al. Burden of Psoriatic Arthritis According to Different Definitions of Disease Activity: Comparing Minimal Disease Activity and the Disease Activity Index for Psoriatic Arthritis. *Arthritis Care. Res. (Hoboken)* **2018**, *70*, 1764–1770. [CrossRef]
120. Haque, N.; Lories, R.J.; de Vlam, K. Orthopaedic Interventions in Patients with Psoriatic Arthritis: A Descriptive Report from the SPAR Cohort. *RMD Open.* **2016**, *2*, e000293. [CrossRef] [PubMed]
121. Mekhail, C.; Chouk, M.; Prati, C.; Wendling, D.; Verhoeven, F. Prognostic Factors of Good Response to DMARDs in Psoriatic Arthritis: A Narrative Review. *Expert Rev. Clin. Pharmacol.* **2020**, *13*, 505–519. [CrossRef] [PubMed]
122. Koutsompina, M.L.; Pappa, M.; Sakellariou, S.; Gialouri, C.G.; Fragoulis, G.E.; Androutsakos, T. Methotrexate-Related Liver Cirrhosis in Psoriatic Arthritis: A Case Report and Review of the Literature. *Mediterr. J. Rheumatol.* **2021**, *32*, 264–272. [CrossRef] [PubMed]
123. Roy, A.; Darapureddy, A.; Kumar, Y. Noninvasive Assessment of Liver Fibrosis by Magnetic Resonance Elastography in Patients with Rheumatic Disease on Long-Term Methotrexate Treatment. *Indian J. Rheumatol.* **2023**, *18*, 54–60. [CrossRef]
124. Hoganson, D.D.; Chen, J.; Ehman, R.L.; Talwalkar, J.A.; Michet, C.J.; Yin, M.; Crowson, C.S.; Matteson, E.L. Magnetic Resonance Elastography for Liver Fibrosis in Methotrexate Treatment. *Open J. Rheumatol. Autoimmune. Dis* **2012**, *02*, 6–13. [CrossRef] [PubMed]
125. Mease, P.J.; Armstrong, A.W. Managing Patients with Psoriatic Disease: The Diagnosis and Pharmacologic Treatment of Psoriatic Arthritis in Patients with Psoriasis. *Drugs* **2014**, *74*, 423–441. [CrossRef] [PubMed]
126. Mahmood, F.; Helliwell, P. Psoriatic Arthritis: A Review. *EMJ Reumatol.* **2016**, *3*, 114–117.

diagnostics

MDPI

Review

Challenging Patterns of Atypical Dermatofibromas and Promising Diagnostic Tools for Differential Diagnosis of Malignant Lesions

Olguța Anca Orzan [1,2], Alexandra Maria Dorobanțu [2,*], Cristian Dorin Gurău [3], Sibel Ali [2], Mara Mădălina Mihai [1,2], Liliana Gabriela Popa [1,2], Călin Giurcăneanu [1,2], Irina Tudose [4] and Beatrice Bălăceanu [2,*]

1 Dermatology Department, "Carol Davila" University of Medicine and Pharmacy, 020021 Bucharest, Romania
2 Dermatology Department, "Elias" University Emergency Hospital, 011461 Bucharest, Romania
3 Orthopedy and Traumatology Clinic, Clinical Emergency Hospital, 014451 Bucharest, Romania
4 Anatomic Pathology Laboratory, "Elias" University Emergency Hospital, 011461 Bucharest, Romania
* Correspondence: alexandramdorobantu@gmail.com (A.M.D.); balaceanubeatrice@yahoo.com (B.B.)

Abstract: Dermatofibroma (DF) or fibrous histiocytoma is one of the most frequent benign cutaneous soft-tissue lesions, characterized by a post-inflammatory tissue reaction associated with fibrosis of the dermis. Clinically DFs have a polymorphous clinical aspect from the solitary, firm, single nodules to multiple papules with a relatively smooth surface. However, multiple atypical clinico-pathological variants of DFs have been reported and, therefore, clinical recognition may become challenging, leading to a more burdensome identification and sometimes to misdiagnosis. Dermoscopy is considered an important tool in DFs diagnosis, as it improves diagnostic accuracy for clinically amelanotic nodules. Although typical dermoscopic patterns are most frequently seen in clinical practice, there have also been some atypical variants described, mimicking some underlying recurrent and sometimes harmful skin afflictions. Usually, no treatment is required, although an appropriate work-up may be necessary in specific cases, such as in the presence of atypical variants or a history of recent changes. This narrative review's aim is to summarize current evidence regarding clinical presentation, positive and differential diagnosis of atypical dermatofibromas and also to raise awareness about the importance of specific characteristics of atypical variants to better differentiate them from malignant conditions.

Keywords: dermatofibroma; atypical fibrous histiocytoma; dermoscopy; benign cutaneous tumours

1. Introduction

Dermatofibroma (DF), also known as fibrous histiocytoma, is a relatively common benign cutaneous tumour characterized by a post-inflammatory tissue reaction associated with fibrosis of the dermis [1–5]. It mostly occurs in young or middle-aged (20 to 40 years old) adults, generally in female patients, although there are histologic variants frequently encountered in males [1,3,6]. DFs with classical morphology have also been described in children aged less than 5 years old [7]. Although various locations have been noticed (head, face, auricle, neck, trunk, shoulder, pelvic girdles, and digits), DFs usually appear on the lower extremities [1,6–13]. DFs are generally asymptomatic but sometimes can become pruritic and tender [4,5]. On palpation, upon lateral compression of the skin, DFs characteristically sink below the level of the skin, a feature also known as the dimple sign [4,5,8,9,14].

The pathogenesis of DFs is unknown, although they usually arise as consequence of local trauma (tuberculin skin testing, skin tattooing, traumatism caused by razor, thorns or wood splinters etc), insect bites or an underlying condition (folliculitis) [2,4,15–18]. Even though local recurrence and rarely distant metastases have been mentioned in the scientific literature, DFs are considered benign lesions [19].

Yamamoto et al. Addressed the role of mast cells in the development of DFs, as they were found in solitary and multiple variants [20]. Mast cells could induce histopathologic changes, such as basal melanosis, acanthosis of the epidermis, and mononuclear cell recruitment [20].

Immunohistochemical testing identified the presence of factor XIIIa, which marks dermal dendritic cells [21–24]. MAC 387, which was labeled histiocytes, did not show relevant results, and the presence of CD68-positive histiocytes was not consistent [21–24]. One study analyzing 28 cases of dermatofibromas, showed that the majority of spindle-shaped cells, independently of the histological variant, stained positively for HSP47, a marker for skin fibroblasts [21]. Transforming growth factor-beta may also stimulate the fibrosis found in dermatofibromas [25,26].

Other studies suggested that the cell surface proteoglycan, fibroblast growth factor receptor 2, which plays a role in the epithelial–mesenchymal cross-talk, and syndecan-1, may also be involved in the pathogenesis of dermatofibromas [27,28]. Furthermore, CD14+ monocytes have been proposed as the original cells of dermatofibromas [22].

Regarding gene fusion, ALKgene rearrangement and overexpression has been found in both epithelioid and atypical dermatofibromas [29–32]. As such there have been reported rare autosomal dominant familial cases [2,4].

Reactive tissue alterations and neoplastic proliferation clinical clonality have been suggested as mechanisms involved in the pathogenesis of DFs [33,34]. Spontaneous development, lack of regression and the presence of clonal markers during the analysis of X-chromosome inactivation, may also support the clonal or neoplastic mechanism [19]. Mentzel et al. Investigated 7 cases of clinically aggressive dermatofibromas and underlined the malignant transformation of a cellular dermatofibroma into a spindle cell sarcoma [35]. Chromosomal aberrations by array-comparative genomic hybridization have been proposed as possible diagnostic tools for potentially metastatic dermatofibromas [36].

DFs usually have an excellent prognosis and do not require treatment unless the lesion is changing, bleeding, becomes symptomatic or suspicious, another diagnosis is more probable or the patient demands it clinical cosmetic reasons [4,19]. Complete surgical excision with clear margins for histopathologic examination is the most common therapy [4,19,37]. Atypical variants are more prone to recur and as a result, re-excision might be necessary [37]. Another alternative is liquid nitrogen cryotherapy [37].

2. Materials and Methods

A systematic literature search was done in the PubMed, Web of Science Core Collection, and Google Scholar databases, using the terms "atypical dermatofibroma", and "atypical fibrous histiocytoma". A total number of 1092 articles, 135 reviews, and 571 case reports were found. All the articles, reviews, and case reports included in the study were limited to English full text in humans. Finally, 134 studies were included in the review. The pictures are from the patients admitted in "Elias" University Emergency Hospital in the period 2021–2022. Written informed consent was obtained from all subjects involved in the study (Figure 1).

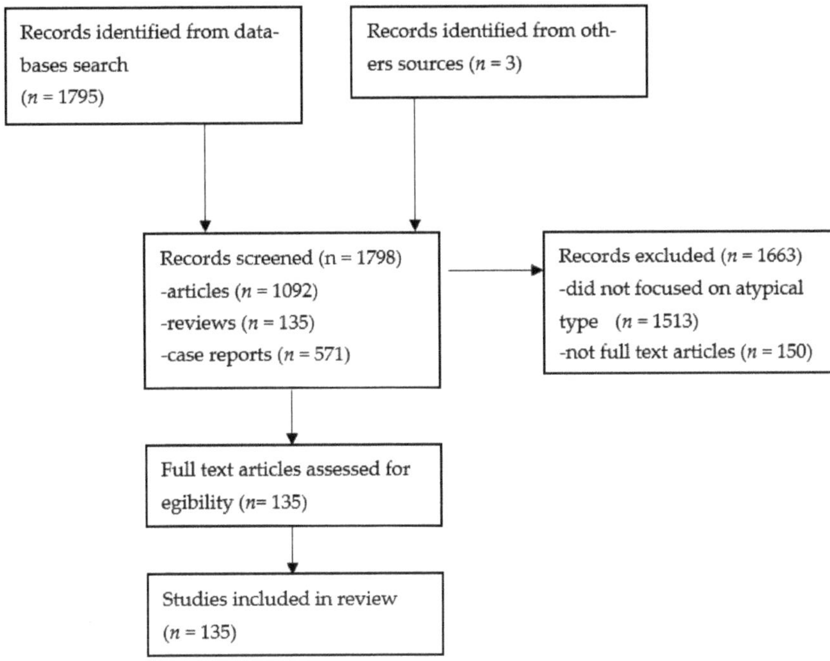

Figure 1. PRISMA flowchart [38].

3. Results

This narrative review aimed to reevaluate the clinical and land dermoscopic patterns of atypical dermatofibromas compared to the typical ones. Although the clinical diagnosis of DFs may be simple in daily practice, in the presence of various patterns, diagnosis of DFs can become challenging. Therefore, specific characterization of these atypical variants is essential in differentiating them from malignant conditions and assessing the risk of local recurrence.

3.1. Clinical Presentation

Clinically, DF usually presents as solitary, hyperkeratotic, small (0.3 to 1 cm) and slow-growing nodule with a red-brown surface [2,4,5,19] (Figure 2). The rate of recurrence seems to be higher in lesions initially greater than 1 cm [6]. Other clinical patterns include firm, flat, sometimes atrophic, single or multiple papules, plaques, with a variety of colors (light brown, dark brown, purple, red or yellow) [1,39] (Figure 3).

Colour may also vary depending on the Fitzpatrick fototype (Figure 4). The overlying skin can be pink, red, purple, gray, yellow, orange, blue, brown or black [40]. On palpation DFs have the consistency of a nodule, that moves freely over the subcutis [40]. The dimple sign is valuable in the diagnosis of DFs, although it may not always assure it [41].

Besides the classical clinical presentation, there have also been described some unusual atypical variants. Rare variants may include *metastasizing benign DFs*, which are usually larger than typical variants (more than 3 cm) [4,42,43]. Morphological features can be those of cellular, aneurysmal or atypical DFs and a greater number of mitosis has been noticed in this cases [4,42,43]. Extension into the subcutaneous layer and local recurrence has also been described [4,42,43]. Regarding metastatic sites, lymph nodes and lungs are the most frequent ones [4,42,43].

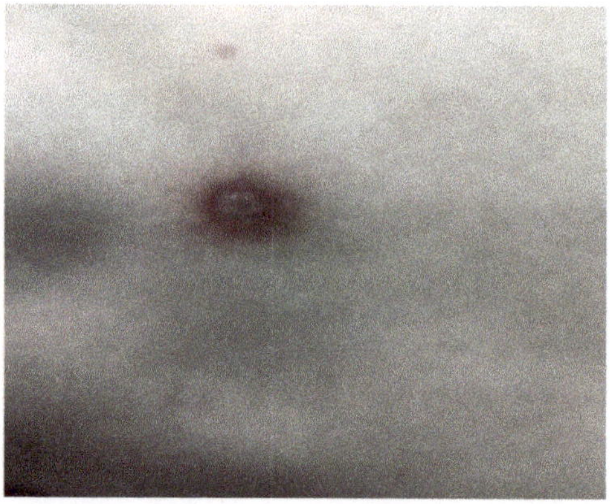

Figure 2. Clinical appearance of a DF in a young woman: a solitary, well-defined, hyperkeratotic nodule with a diametre of about 1 cm with a yellow-brown surface.

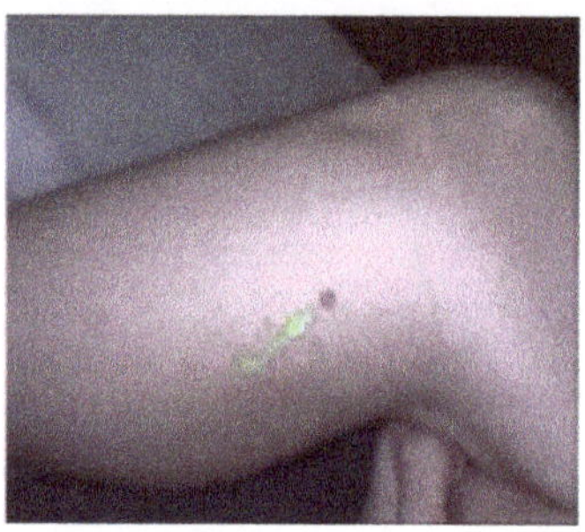

Figure 3. Another clinical pattern of a DF located on the leg: a flat, light brown, single papule.

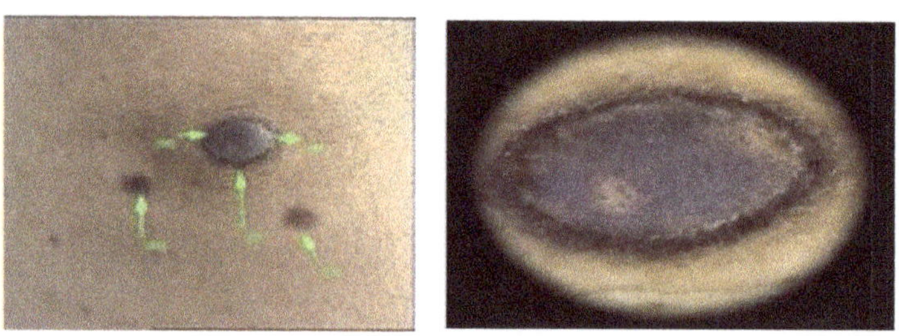

(**a**) (**b**)

Figure 4. (**a**) DF with a nodular, blue appearance in a 4th Fitzpatrick phototype patient. Differential diagnoses may include a blue nevus. (**b**) Dermoscopic image of a nodular, blue DF, with well-defined borders and some scales.

Giant lesions (larger than 5 cm) have also been described in the scientific literature [4,40,43] (Figure 5). The largest tumor reported measured 17 × 9 × 4 cm [44].

Multiple clustered DFs (more than 15) appear like a plaque with various single hyperpigmented papules [45,46]. They may also occur in children and can be either congenital or eruptive [45,46]. Atypical fibroxanthoma and dermatofibrosarcoma protuberans are differential dignosis that should be taken into consideration [45,46].

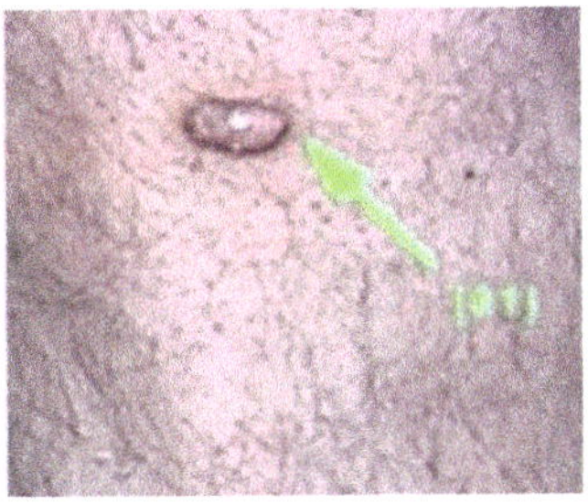

Figure 5. Giant DF in a young patient, with a diameter of about 5.5 cm.

Multiple eruptive, diffuse, and persistent DFs appear in less than 1% of cases, the majority of patients suffering from an underlying affliction, such as human immunodeficiency virus infection, autoimmune diseases (systemic lupus erythematosus, dermatomyositis, myasthenia gravis, pemphigus vulgaris), Graves disease, Hashimoto thyroiditis, chromosomal alterations (Down syndrome), hematologic malignancies (leukemia, cutaneous T-cell lymphoma, myelodysplastic syndrome, multiple myeloma) atopic dermatitis, metabolic disorders (hypercholesterolemia), glycosuria, hydronephrosis, diabetes mellitus, breast cancer, ulcerative colitis, Crohn's disease and sarcoidosis [47–61]. Moreover, multiple eruptive DFs have been linked to antiretroviral therapy (efalizumab and brentuximab vedotin), tyrosine kinase inhibitors (imatinib), and antitumor necrosis factor-alpha agents [62–65]. Some cases have also been described in pregnant women [4,66].

Other atypical presentations may include *polypoid*, *atrophic*, and DF with spreading *satelitosis* [67–69].

A Meyerson phenomenon adjacent to the DF has been seen [70] (Figure 6).

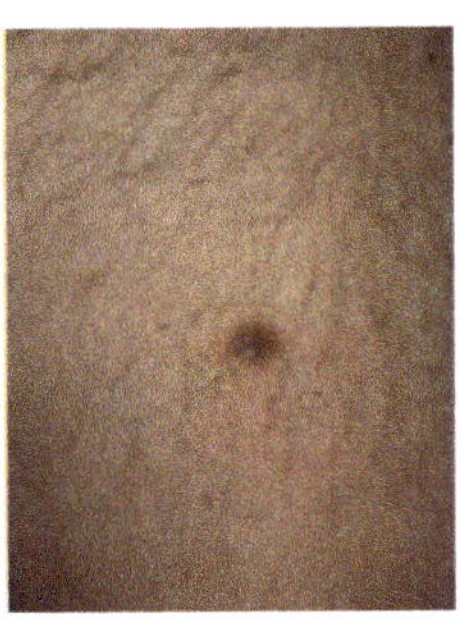

Figure 6. A Meyerson phenomenon is adjacent to the DF.

3.2. Diagnosis and Assessment

As multiple atypical variants of DF have been reported in the literature, clinical recognition may become challenging, leading to a more burdensome identification and sometimes to misdiagnosis [2,9]. Although diagnosis is commonly based on clinical presentation and history, further diagnostic tools such as dermoscopy, variable–frequency ultrasonography, fluorodeoxyglucose positron–emission tomography (FDG-PET) scans, and confocal laser scanning microscopy are necessary.

Dermoscopy is a non-invasive procedure useful for the diagnosis and management of pigmented tumours of the skin [9,71]. Among the scientific literature, various dermoscopic structures have been mentioned. DFs typical dermoscopic appearance includes the presence of a delicate, peripheral light-to-medium brown pigment network with a sharply demarcated central white scar-like area, white network and homogeneous pigmentation [1,8,9] (Figure 7).

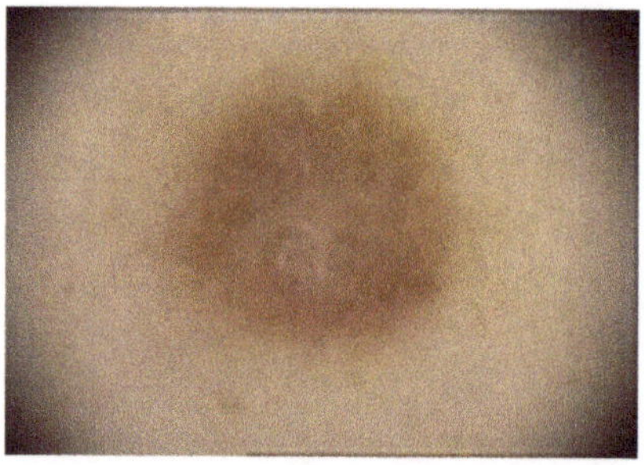

Figure 7. Typical dermoscopic appearance of a DF: delicate, peripheral light-to-medium brown pigment network with a sharply demarcated central white scar-like area, white network and homogeneous pigmentation.

For atypical variants, high definition optical coherence tomography can be useful as it correlates with histopathological types of DFs [72]. As DFs proved to have intense F-18 fluorodeoxyglucose uptake on positron emission tomography-computed tomography scan, the letter coud be a possible diagnostic alternative [73]. Although confocal laser scanning microsocopy is mainly used to evaluate melanocytic lesions, it is helpful in diagnosing DFs as their features (bright rings at the periphery, collagen structures at the center, central keratin) correlate with both dermoscopy and histopathology [74].

Although typical dermoscopic patterns are most frequently seen in clinical practice, some authors described atypical types of DFs (Table 1).

Table 1. Dermoscopic features of atypical dermatofibromas.

Study, Authors	Article Type	Year of Publication	Atypical Features with High Risk	Cases (*n*/%)
Aytekin S et al. [1]	Original article	2021	Irregular delicate/asymmetric pigment network Peripheral proeminent pigment network Irregular proeminent pigment network Atypical pigment network Irregular white network Irregular brown areas Dotted vessels Glomerular vessels Polymorphous/atypical vessels Ulceration White radial streaks	31 (21.8) 3 (2.1) 2 (1.4) 2 (1.4) 3 (2.1) 9 (6.3) 34 (23.9) 2 (1.4) 6 (4.2) 5 (3.5) 8 (5.6)
Genc Y et al. [75]	Report	2020	Melanoma-like BCC-like Keratoachantoma-like Spitzoid-like	11 (19.4) 3 (4.9) 2 (2.6) 26 (19.4)
Llambrich A et al. [3]	Research letter	2019	Dotted vessels Arborizing vessels Polymorphous/atypical vessels Shiny white streaks	18 (50) 7 (19.4) 18 (50) 16 (44.4)
Lin MJ et al. [76]	Original research	2018	Dotted/pinpoint vessels Sharply focused arborizing vessels Linear irregular vessels Glomerular vessels Polymorphous vascular pattern White lines Ulceration Blue/grey veil	2 (22) 0 1 (11) 1 (11) 1 (11) 0 0 0
Won KY et al. [77]	Original research	2017	Irregular shape Spiculated margins	8 (44) 12 (67)

Table 1. *Cont.*

Study, Authors	Article Type	Year of Publication	Atypical Features with High Risk	Cases (*n*/%)
Kelati A et al. [78]	Research article	2017	White streaks Ulceration Brown streks Negative-network-like appearance Dotted vessels Multicomponent melanoma-like Vascular tumor-like BCC-like Collision tumor-like Peripheral diffuse pink to red to reddish violet halo White ring around an ulceration Pink bluish pigmentation with vascularization Pigment network with a ring around follicular opening	18 (18) 6 (6) 6 (6) 3 (3) 23.3% 20 (20%) 0 0 0 7 (7%) 6 (6%) 7 (7%) 2 (2%)
Marinescu SA et al. [71]	Case report	2016	Pinky-milk areas Peripheral pigment network Polymorphous atypical vessels	
Roldán-Marín R et al. [79]	Case report	2014	Grey-green colour	
Ferrari A et al. [9]	Original article	2013	Melanoma-like Vascular tumour-like BCC-like Collision tumour-like	21 (16.2) 6 (4.6) 5 (3.8) 3 (2.3)
Zaballos et al. [8]	Prospective study	2008	Proeminent atypical pigment network Irregular pigment network Irregular white network Irregular brown areas Dotted vessels Glomerular vessels Polymorphous/atypical vessels Ulceration	13 (3.1) 8 (1.9) 3 (0.7) 125 (30.6) 3 (0.7) 10 (2.4) 18 (4.4)

Additional clinical features may include ring-like or donut-shaped globular structures, vascular structures and sometimes ulceration, comedo-like openings, scale, crusts, or peripheral collarette fissures [8,9].

The *pigment network* may vary from peripheral/total/irregular delicate to peripheral/total/irregular prominent and atypical appearance [1,3,9]. Aytekin et al. evaluated dermoscopically 142 DFs of 72 patients and concluded that pigment network was found in 57% of cases, the most common subtype being irregular delicate or asymmetric pigment network [1]. Delicate pigment network seen in DFs is commonly thin, varying from light to medium brown and it is considered that it results as hyperpigmentation of rete ridges rather than the proliferation melanocytes [1]. According to the study performed by Arpaia et al., the pigment network was darker in the center, becoming gradually pale towards the periphery with brownish thin streaks [80]. In 2000, Ferrari et al. noticed that the peripheral pigment network and central white scar-like patches are more prevalent in women and among the classical histopathologic type of DFs [36]. The results may alternate due to the quality of the dermoscopy (contact/non-contact, polarized/nonpolarized light) and the difference among genders [1]. Interestingly, Zaballos et al. also identified in some DFs a significant and/or atypical pigment network [8].

Central white scar-like patches are sharply demarcated with irregular white regions, histopathologically characterized by various grades of fibrosis in the dermis [1,9] (Figure 8).

Aytekin et al. have detected white scar-like patches in 37.3% of cases, the most common subtype being the central ones [1]. This structure is considered the most widespread feature of diffuse fibrous DF with a peripheral delicate pigment network [1]. It has also been noticed that the scar structure occasionally got a white radial streaks appearance, which gave the aspect of a spitzoid pattern [1]. Zaballos et al. evaluated 412 DFs and observed that white scar-like patches are mainly localized in the center part of the lesion [8]. Moreover, Arpaia et al. concluded that the central white patch was the most frequent dermoscopic feature, observed in 91.6% of cases [80].

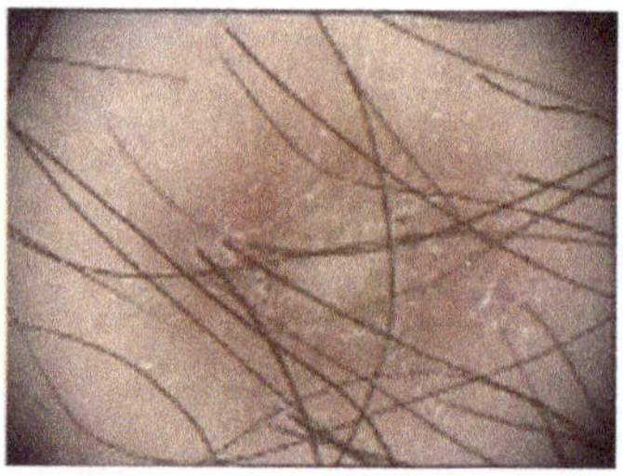

Figure 8. The dermoscopic appearance of central white scar-like patches sharply demarcated with irregular white regions.

The *white network* may be central, total, irregular, or crystalline-like [1,3,9]. Zaballos et al. identified a network of white lines and brown holes, which was later considered a variation of the white scar-like patch [8]. There has been raised awareness of the importance of distinguishing this structure from dysplastic nevi, Spitz nevi and the negative pigment network encountered in melanomas [1,8].

The *homogenous pigmentation* may include multiple colors (brown, yellow) or it may appear as hypopigmentation [1,3,9] (Figure 9).

Figure 9. The dermoscopic appearance of a homogenous yellow-brown pigmentation of a DF.

Ferrari et al. noticed that homogeneous pigmentation was most frequently in females and DFs with sebaceous hyperplasia, whereas peripheral homogeneous pigmentation was mostly encountered in men [36]. Karaarslan et al. observed a homogeneous blueish pigmentation that was associated with the hemosiderotic type of the DF [81]. Usually, hemosiderotic variants are indicated by the green color [79].

The *vascular pattern* has been widely discussed. Vascular structures are used in dermoscopy to diagnose melanoma and other pigmented or vascular tumors, which mimic melanoma [1,3]. Nevertheless, DFs may have peripheral, central, or total erythema, dotted, hairpin, glomerular, comma, or linear vessels, but also polymorphic and atypical ones [1,3,9] (Figure 10).

Figure 10. The dermoscopic appearance of a DF with peripheral erythema and dotted vessels.

Contrary to other studies, Genc et al. found vascular structures to be the most frequent dermatoscopic feature and described a red to brown halo phenomenon in 4.9% of DFs [75]. Another study performed by Agero et al. concluded that blood vessels were seen in 44% of DFs when using polarized light [82]. Ferrari et al. described 2 DFs with dotted vessel patterns, whereas Aytekin et al. stated that the most frequent vascular structures in their study were erythema and dotted vessels [1,9,36].

Other not so common dermatoscopic changes may involve ring-like structures, ulceration, scales, fissures, milia-like cysts, hemorrhage, crusts or white radial streaks [1,3,9,82]. Genc et al. have conducted a study which classified DFs depending on the dermatoscopic similarities to other lesions [75]:

- Melanoma-like: various colors and patterns, white structureless areas, polarizing-specific white lines, pink-red or blue-gray structureless areas, dark brown thick reticular lines, peripheral black clods and eccentric distribution of straight, curved, dotted and branched vessels [75].
- Basal cell carcinoma-like: arterial structures specifically in the papillary dermis mostly at the periphery of the lesion [75] (Figure 11).
- Keratoacanthoma-like: central keratin area with a surrounding radial arrangement of polymorphic vessels (curved, branched and dotted) [75].
- Seborrheic keratosis-like: thick curved lines, orange, brown or white clods, brown-black crusted structures, blue-gray structureless areas and loop, dotted or coiled vessels [75].
- Nevus-like: various hypopigmented structureless areas, having in between multifocal thin brown reticular lines [75] (Figures 4 and 13).
- Nevus sebaceous-like: white lines (associated with dermal fibrosis), peripheral thin brown reticular lines and central large yellow clods (associated with sebaceous hyperplasia) [75] (Figure 12).
- Xanthogranuloma-like: yellow structureless areas, coiled vessels with a peripheral reddish halo [75].
- Pyogenic granuloma-like: polymorphic vessels (curved, dotted, straight and branched) with irregular distribution, white lines and pink-red structureless areas [75].
- Spitzoid-like: pink-red structureless areas, shiny white lines, white structureless areas, light brown clods, halo phenomenon, dotted vessels [75].

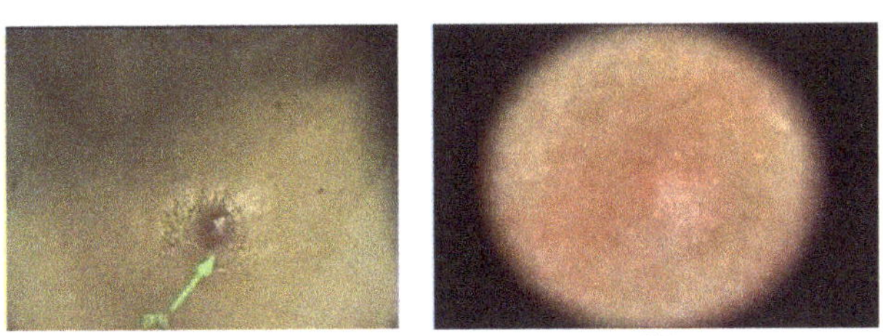

(a) (b)

Figure 11. (**a**) Clinical image of a DF with a solitary, well-defined, nodular, pink appearance in a female patient. Differential diagnoses may include basal cell carcinoma. (**b**) Dermoscopic image of a DF with arborizing vessels, along with the central white scar-like patch and fine delicate pigment network.

Llambrich et al. performed a retrospective review, analyzing clinical and dermoscopic features of 36 pink nodular DFs [3]. They underlined the importance of a correct differential diagnosis as pink nodular lesions with erythema, vascular structures, shiny white streaks and a central white patch may suggest malignancy, mainly amelanotic/hypomelanotic melanoma [3]. Moreover, regarding the dermopathological types, non-fibrocollagenous variants of DFs were proned to have atypical patterns [36,75].

Melanoma-like and pyogenic granuloma-like atypical patterns were seen mostly in the case of aneurysmal DFs [75]. Furthermore DFs may have a pinkish-red pigmentation, dotted vessels and superficial white scales resembling psoriasis [9]. A "collision tumour-like" pattern was also described having a white area with focal pigment network [9]. Particularly, collision-like patterns along with melanoma-like and vascular tumour-like patterns were most commonly noticed in men [9]. Aditionally there have been described palisading, granular cell, myxoid, lichenoid, balloon cell and signet-ring cell variants [2,6,9].

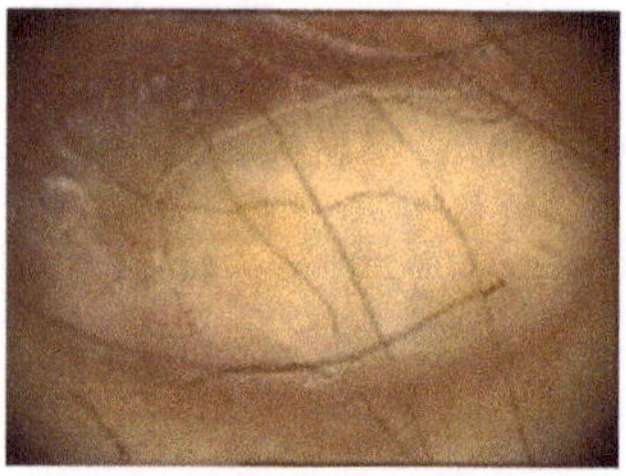

Figure 12. Dermoscopic image of a DF with a nevus sebaceous-like appearance: white lines (associated with dermal fibrosis), peripheral thin brown reticular lines and central yellow structures.

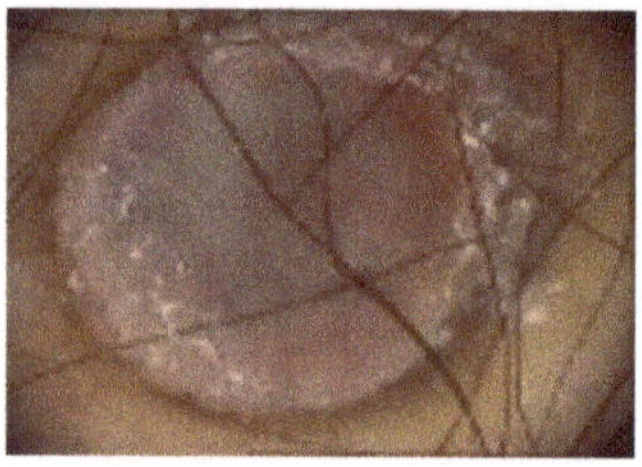

Figure 13. Dermoscopic image of a nevus-like DF, but also with coiled vessels and some scales.

As such, dermoscopy may be beneficial in increasing diagnosis and management accuracy, but since dermoscopic features may vary as well, it is certainly important to take into consideration other differential diagnoses [4,9,71,75].

3.3. Histologic Variants

Histologically, DFs contain uniform spindle cells organized in elongated fascicles [4]. Classical histopathological features of typical DFs include an overlying achantotic, hyperkeratotic and sometimes hyperpigmented epidermis [19,83–88] (Figure 14).

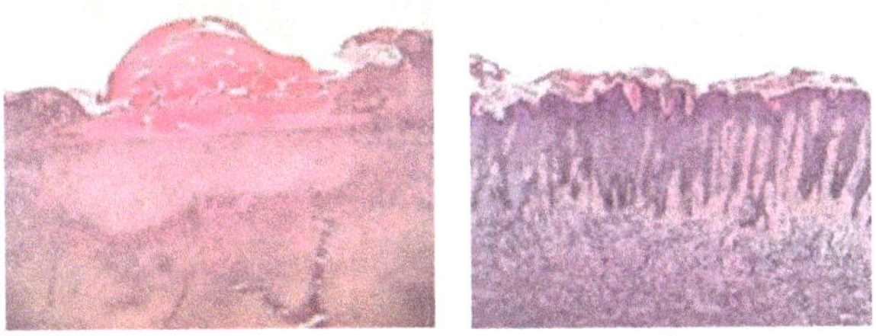

(a) (b)

Figure 14. (a) Histopathologic examination (hematoxylin and eosin, ×10): tumour proliferation localized in the papillary dermis and extending to the deep dermis, with interspersed collagen bundles, separated from the epidermis by a grenz zone. The overlying epidermis presents erosions centrally and collections in the keratin layer. (b) Histopathologic examination (hematoxylin and eosin, ×10): tumour proliferation composed of elongated and spindle-shape cells with elongated nuclei, in a fascicular-storiform configuration localized in the papillary dermis and extending to the deep dermis. The overlying epidermis has a hyperplastic appearance with hyperorthokeratosis, acanthosis, and elongation of the rate ridges. There is also follicular induction at the epidermis level.

Moreover, the epidermis usually exhibits elongated rete ridges containing hyperpigmented basal keratinocytes, aspect known as "dirty feet" sign [2]. There is also a proliferation of spindle-shaped fibrous cells mixed with histiocytoid cells at the level of the dermis [22,89,90]. Collagen bundles are commonly seen between the spindled fibrous cells along with the unaffected layeser, known as the "Grenz zone" [2,89–91].

The scientific literature has classified various types of DFs, regarding their histopathological characteristics [2]. Histological features may coexist in the same lesion [87]. As such there have been mentioned a lot of histopathological types: fibrocollagenous, cellular, keloidal, atrophic, aneurysmal, storiform, fibrocollagenous with sebaceous induction, lipidized, hemosiderotic, epithelioid, lichenoid, baloon cell, signet-ring myofibroblastic, clear cell, palisading, granular cell, myxoid and also the atypical type [2,6,71,83–85,87,88].

A histologic review performed by Alves et al. on 192 dermatofibromas stated that common fibrous hystiocitoma was the most frequent type, observed in 80% of cases [2]. Individual collagen bundles encompassed by lesional cells (fibroblasts, macrophages and blood vessels) and a predominantly lymphocytic inflammatory infiltrate may be seen [2].

Atypical DFs also known as DFs with monster cells, are poorly documented variants of typical fibrous histiocytomas [88]. Besides typical findings, atypical DF are comprised of pleomorphic spindle-like, hystiocite-like cells and multinucleate giant cells. [89].

LeBoit and Barr firstly described dermatofibroma with granular cells in 1991 [87,88]. This rare histologic variant can be confused with other malignant or benign cutaneous neoplasms such as: benign granular cell tumor, malignant granular cell tumor, primitive polypoid granular cell tumor, granular cell ameloblastoma, granular cell fibrous papule of the nose, granular cell basal cell carcinoma, granular cell schwannoma, granular cell leiomyoma, granular cell leiomyosarcoma or angiosarcoma and granular cell dermatofibrosarcoma protuberans [87]. Morfology of the lesion along with immunohistochemical evaluation might sometimes be decisive for the corect diagnosis [86]. For instance, benign granular cell tumors are positive for S-100 protein, CD63, CD68 and neuron-specific enolase whereas atypical fibroxanthoma stains negatively for S100 protein, Melan-A, human melanoma black (HMB)-45 pan-cytokeratin (CK) and actin and positively for CD68 and vimentin [87].

Immunohistochemistry can be useful to differentiate DFs from schwannomas, leiomyomas and leiomyosarcomas [87].

3.4. Differential Diagnosis

It is extremely important to recognize atypical DFs, as *cutaneous melanoma* is a vital clinical differential diagnosis and may display similar characteristics [9,19]. Nevertheless, other afflictions as well may be taken into accounts, such as intradermal nevi, basal cell carcinomas, keratoacanthomas, and dermatofibrosarcomas protuberans [4]. Differential diagnoses may also include angiokeratomas, Spitz-nevi, melanocytic nevi, blue nevi, granuloma annulare, supernumerary nipple, acrochordon, atypical fibroxanthoma, cutaneous metastasis, cutaneous T-cell lymphoma, cylindroma, pilomatrixoma or targetoid hemosiderotic hemangiomas [14,92,93].

Dermatofibrosarcoma protuberans (DFSP) appears as painless, slow-growing skin-colored nodule, with a finger-like projections pattern and should be distinguished from benign DFs as it is locally aggressive [4,19,94]. A delayed accurate diagnosis leads to clinical pitfalls [94]. A more cellular appearance and a "honeycomb" display of the subcutaneous fat is often seen in DFSP [19]. Immunohistochemical staining is also very useful, as there are various markers to differentiate the two **entities** [19,95–103]. Although DFSP stains positive for CD-34, nestin and collagen triple helix repeat containing-1 (Cthrc1) and negative for factor XIIIa, there has also been noticed an elevated expression of thrombospondin-1 (TSP-1) [19,98,100,101,103]. DFSP characteristically has a genomic reciprocal translocation in t (17;22) (q22;q13) that causes the fusion of the platelet-derived growth factor B-chain (PDGFB) and the promoter of the collagen type Iα1 (COL1A1) genes and might be detected by fluorescent in situ hybridisation or real-time PCR [94].

In comparison, DF stains positive for factor XIIIa, D2-40, insulin-like growth factor–binding protein 7 (IGFBP7), cathepsin K, CD99, leukocyte-specific protein 1 (LSP1) and 5-hydroxymethylcytosine (5-hmC) and negative for CD-34 [19,26,37,99,102,104–107]. Occasionally, the cellular type of DF may stain positive for CD34 [95,96]. Stromelysin-3 (ST-3) expression of DF shall also help to differentiate it from DFSP [97]. FGFR3/FOXN1 and FGF2/FGFR4 expression in the pathogenesis of DF is practical [104]. Fluorescence in situ hybridization (FISH) analysis is a valuable tool as well [108]. B-cell lymphoma 2 (Bcl-2) expression, autophagy marker Atg5, and phosphohistone-H3 can help to differentiate between DF and DFSP [109,110]. Moreover, Ki-67 staining shows a higher proliferation index in the case of DFSP [19].

Hemosiderotic dermatofibromas, dermatoscopically characterized by a blue/red center with white lines and maybe network and vessels at the periphery may lead to a dermoscopically differential diagnosis with *Kaposi sarcoma* due to the intense vascularity [28,29]. Nevertheless, Kaposi sarcoma stains are positive for CD31, CD34 and D2-40, and patients are also positive for HHV-8 [19].

CK20 positive Merkel cells, present in the follicular induction, crowding, no peripheral palisading, clear cell hyperplasia, and the absence of nuclear atypia are helpful pathologic features in differentiating DFs with follicular induction from *basal cell carcinomas* [19].

4. Discussion

This narrative review aimed to reevaluate the clinical and dermoscopic patterns of atypical dermatofibromas compared to the typical ones. Moreover, there have been mentioned some not-so-common etiopathogenic factors. Dermatofibromas are prevalent cutaneous benign tumours that most frequently affect young or middle-aged adults. Clinically, dermatofibromas appear as single or multiple firm papules or nodules with a smooth surface anywhere on the body, mainly on the lower extremities. They can vary in size and colour from light brown to dark brown, yellow, purple or red. Although the clinical diagnosis of DFs may be simple in daily practice, in the presence of various patterns, diagnosis of DFs can become challenging.

Therefore, it is essential to consider the possible links between dermoscopy and histology and complete surgical excision, especially in the presence of atypical variants or a history of recent changes. Thus, the precise definition of dermoscopic patterns for this frequent benign tumour is of major interest.

In a reverse manner, the possibility of a misdiagnosis of malignant skin disorders, inlcuding non-melanoma skin cancer, is a main challenge in terms of worldwide public health management. A lot of clinicians may face it and thus malignancy-related misdiagnosis remains one of the main issues in the dermatologic field. Similar to the case of melanomas or basal cell carcinomas being misdiagnosed as diabetic foot ulcer, they can be more easily considered, even by experts, as benign lesion, such as dermatofibromas [111]. As malignant tumors may sometimes mimic benign conditions, the main focus has been on finding a non-invasive, reliable, sensible and highly specific diagnostic methods to identify specific features which suggest malignancy.

Promising Differential Diagnostic Methods

Apart from dermoscopy and digital dermoscopy, not so common diagnostic tools may include variable–frequency ultrasonography, in which DFs appear as hypoechoic solid nodules and high-definition optical coherent tomography, in which DFs can resemble malignant conditions on FDG-PET scans [73,74]. Among other studied techniques there have been mentioned reflectance confocal microscopy (RCM), multiphoton microscopy, fluorescence evaluation, Raman spectroscopy and diffuse reflectance [112–115]. High-frequency ultrasonography has been a valuable diagnostic and prognostic tool in early detecting other types of malignancies, such as hepatocellular carcinoma, for a long time, but it recently has begun to represent a promising opportunity in dermatology by using deep learning-based algorithms to analyse automated images [77,116,117]. Complementary techniques, such as dermoscopy in conjunction with RCM, may also enhance diagnostic accuracy of melanocytic conditions [118–120]. Confocal laser scanning microscopy is also considered an alternative tool [118–120].

The use of fractal parameters and fractal analysis method in dermatology is promising in the evaluation of image parameters, independently of the adopted scale [121]. There are currently studies being carried regarding the usefulness of fractal parameters in building classes of disease units based upon pictures of cutaneous pigmented lesions [121,122]. It might provide fully automatic diagnostic systems able to determine the type of pigmented tumor and inform us regarding the most adequate management [121].

Taking into consideration that dermatology is a largely visual speciality, the high cost of travel expenses to urban centres, the long wait times to see a dermatologist and the shortage of dermatology services mainly in rural areas, several research studies analizying the role of an artificial intelligence (AI) system as a diagnostic tool for the management of skin conditions have been conducted, focusing mainly on the malignant ones [123–129]. Several studies proved that telehealth platforms, easily reachable through smartphone apps, could increase patients access to dermatological care, especially during COVID-19 pandemic [127,130–132]. Some of the apps can use AI to provide various differential diagnoses depending on the information provided: patient demographics, lesion type, location, symptoms and progression [127,129]. Artificial neural networks, such as convolutional neural networks (CNNs), can be used to analyse visual imagery, being very effective in recognizing automated images and equal or superior than dermatologist in recognizing skin cancers [123]. Implementing AI as a diagnostic aid in the clinical practice may be safe, useful and feasible for skin lesions accurate detection and for better differentiating malignant from benign ones [123–127].

In order to diminish the high degree of subjectivity and variability regarding specificity, sensitivity, and diagnostic accuracy when performing RCM, a lot of artificial intelligence algorithms were created to ensure alternatives, assistance and support od dermatologists on a daily basis [118]. AI in RCM has been used so far to point out the dermal–epidermal junction, evaluate the the quality of RCM mosaics and distinguish between different skin tumors [118].

A lot of other studies have analyzed the correlation between spectrophotometric parameters of skin color and behavioral/environmental factors to predict the risk of cutaneous malignancies [133]. They concluded that the measurment of skin melanin index measured on the arm or buttock is the simplest predictor and should be added in predictive models. Regarding the environmental/behavioral factors, the total number of sunburns appear to be the most important one. As such, spectrophotometric measurements may be considered a quick screening examination method of the skin [133].

In histopathology, AI is efficient in classifying and characterizing tissues, in detecting mitosis and segment histologic primitives as epithelium, nuclei and tubules [128].

As a consequence of complexity and intransparency of deep neural networks in classifying skin cancer, explainable artificial intelligence (XAI) has also been suggested as an alternative although further research studies are needed to evaluate the influence of XAI in detecting cutaneous cancer [134].

There are lot of opportunities that lie ahead, from automated classification of cutaneous cancer through convolutional neural networks, sequential digital dermoscopy and automated total body photography to AI and automated teledermoscopy [135]. However, the potential use of AI in clinical practice remains to be addressed due to their limitations and further studies need to be conducted in order to implement it every day medical practice [135].

5. Conclusions

Clinical diagnosis of typical dermatofibromas is easy, with a classic dermoscopic pattern of pigmented network and central white patch. However, in current clinical situations, dermatofibromas display a wide range of presentations and histological variants that make the differentiation from other tumours, such as malignant melanoma, very difficult. Specific characterization of these atypical variants is essential in differentiating them from possibly more aggressive lesions and assessing the risk of local recurrence. As a matter of fact, the definitive diagnosis of a skin condition, especially in a doubtful clinical diagnostic scenario, demands complete surgical resection and histopathological analysis. As artificial intelligence technologies had reached an impressive precision in identifying various skin lesions, along with other inovative diagnostic methods, we can emphasise that in the future it will lead to improved safety and patient care and maybe enhance dermatologists' productivity.

Author Contributions: Conceptualization, A.M.D. and O.A.O.; methodology, A.M.D. and O.A.O.; software, A.M.D., O.A.O., C.D.G., B.B., S.A., M.M.M., L.G.P., I.T. and C.G.; validation, A.M.D., O.A.O. and C.G.; formal analysis, A.M.D., O.A.O., C.D.G., B.B., S.A., M.M.M., L.G.P., I.T. and C.G; investigation, A.M.D., O.A.O., B.B., C.D.G. and I.T.; resources, A.M.D., O.A.O., C.D.G., B.B. and S.A.; M.M.M., L.G.P., I.T. and C.G.; data curation, A.M.D., O.A.O., B.B., C.D.G. and M.M.M.; writing—original draft preparation, A.M.D., O.A.O., B.B. and C.G.; writing—review and editing, A.M.D., O.A.O., C.D.G. and B.B.; visualization, A.M.D. and B.B.; supervision, A.M.D., O.A.O., B.B. and C.G.; project administration, A.M.D., O.A.O., B.B. and L.G.P. All authors have read and agreed to the published version of the manuscript.

Funding: This research received no external funding.

Institutional Review Board Statement: The study was approved by the local Ethical Committee of the "Elias" University Emergency Hospital. The research was conducted according to the Helsinki Declaration. The pictures are part of the authors' own archive and are from the patients admitted in "Elias" University Emergency Hospital in the last 2 years. They do not contain personal data.

Informed Consent Statement: Informed consent was obtained from all subjects involved in the study.

Data Availability Statement: This review summarizes data reported in the literature and it does not report primary data.

Conflicts of Interest: The authors declare no conflict of interest.

References

1. Aytekin, S.; Kaynak, E.; Ayhan, E. Dermoscopy of dermatofibromas: A new perspective. *Int. J. Clin. Pract.* **2021**, *75*, e14547. [CrossRef] [PubMed]
2. Alves, J.V.P.; Matos, D.M.; Barreiros, H.F.; Bártolo, E.A.F.L.F. Variants of dermatofibroma-a histopathological study. *An. Bras. Dermatol.* **2014**, *89*, 472–477. [CrossRef] [PubMed]
3. Llambrich, A.; Vila, A.; Terrasa, F.; Bañuls, J.; Nadal, C.; Zaballos, P. Dermoscopy of pink nodular dermatofibromas: A study of 36 cases. *Australas. J. Dermatol.* **2019**, *60*, e357–e360. [CrossRef]

4. uptodate.com. Available online: https://www.uptodate.com/contents/overview-of-benign-lesions-of-the-skin?search=dermatofibroma&source=search_result&selectedTitle=1~{}22&usage_type=default&display_rank=1#H1101419831 (accessed on 2 June 2022).
5. Kittler, H.; Rosendahl, C.; Cameron, A.; Tschandl, P. *Dermatoscopy: Pattern Analysis of Pigmented and Non-Pigmented Lesions*, 2nd ed.; Facultas Verlags und Buchhandels AG: Vienna, Austria, 2016; pp. 9–52.
6. Gaufin, M.; Michaelis, T.; Duffy, K. Cellular Dermatofibroma: Clinicopathologic Review of 218 Cases of Cellular Dermatofibroma to Determine the Clinical Recurrence Rate. *Dermatol. Surg.* **2019**, *45*, 1359–1364. [CrossRef]
7. Berklite, L.; Ranganathan, S.; John, I.; Picarsic, J.; Santoro, L.; Alaggio, R. Fibrous histiocytoma/dermatofibroma in children: The same as adults? *Hum. Pathol.* **2020**, *99*, 107–115. [CrossRef]
8. Zaballos, P.; Puig, S.; Llambrich, A.; Malvehy, J. Dermoscopy of Dermatofibromas: A prospective morphological study of 412 cases. *Arch. Dermatol.* **2008**, *144*, 75–83. [CrossRef] [PubMed]
9. Ferrari, A.; Argenziano, G.; Buccini, P.; Cota, C.; Sperduti, I.; De Simone, P.; Eibenschutz, L.; Silipo, V.; Zalaudek, I.; Catricalà, C. Typical and atypical dermoscopic presentations of dermatofibroma. *J. Eur. Acad. Dermatol. Venereol.* **2013**, *27*, 1375–1380. [CrossRef]
10. Bowling, J. Diagnostic dermatology. In *The Illustrated Guide*, 1st ed.; Wiley-Blackell: Oxford, UK, 2012; p. 77.
11. Lehmer, L.M.; Ragsdale, B.D. Digital dermatofibromas-Common lesion, uncommon location: A series of 26 cases and review of the literature. *Dermatol. Online J.* **2011**, *17*. [CrossRef]
12. Kadakia, S.; Chernobilsky, B.; Iacob, C. Dermatofibroma of the Auricle. *J. Drugs Dermatol.* **2016**, *15*, 1270–1272. [PubMed]
13. Kim, J.M.; Cho, H.J.; Moon, S.-H. Rare experience of keloidal dermatofibroma of forehead. *Arch. Craniofacial Surg.* **2018**, *19*, 72–74. [CrossRef]
14. emedicine.emedscape.com. Available online: https://emedicine.medscape.com/article/1056742-overview (accessed on 20 July 2022).
15. Evans, J.; Mattacks, C.; Clarke, T.; Pond, C.M. Dermatofibromas and Arthropod Bites: Is There Any Evidence to Link the Two? *Lancet* **1989**, *334*, 36–37. [CrossRef] [PubMed]
16. Lobato-Berezo, A.; Churruca-Grijelmo, M.; Martínez-Pérez, M.; Imbernón-Moya, A.; Vargas-Laguna, M.E.; Fernández-Cogolludo, E.; Aguilar-Martínez, A.; Gallego-Valdés, M. Dermatofibroma Arising within a Black Tattoo. *Case Rep. Dermatol. Med.* **2014**, *2014*, 745304. [CrossRef] [PubMed]
17. Watanabe, K.; Fukuda, H.; Niiyama, S.; Oharasaki, T.; Mukai, H. Multiple dermatofibromas subsequent to folliculitis. *Eur. J. Dermatol.* **2013**, *23*, 890–891. [CrossRef]
18. Nomura, E.; Yamamoto, T. Photoletter to the editor: Fibrous histiocytoma developing at the site of tuberculin skin test. *J. Dermatol. Case Rep.* **2012**, *6*, 130–131. [CrossRef] [PubMed]
19. Myers, D.J.; Fillman, E.P. Dermatofibroma. In *StatPearls [Internet]*; StatPearls Publishing LLC: Treasure Island, FL, USA, 2022.
20. Yamamoto, T.; Katayama, I.; Nishioka, K. Role of mast cells in dermatofibroma: Recent viewpoints into the pathogenesis. *Eur. J. Dermatol.* **2003**, *13*, 419–423. [PubMed]
21. Kuroda, K.; Tajima, S. Proliferation of HSP47-positive skin fibroblasts in dermatofibroma. *J. Cutan. Pathol.* **2008**, *35*, 21–26. [CrossRef]
22. Jin, S.Y.; Choi, J.S.; La Choi, Y.; Kim, D.H.; Lee, S.H. Identification of Leukocyte-Specific Protein 1-Positive Cells: A Clue to the Cell of Origin and a Marker for the Diagnosis of Dermatofibroma. *Ann. Dermatol.* **2015**, *27*, 157–162. [CrossRef]
23. Törőcsik, D.; Bárdos, H.; Nagy, L.; Ádány, R. Identification of factor XIII-A as a marker of alternative macrophage activation. *Cell Mol. Life Sci.* **2005**, *62*, 2132–2139. [CrossRef]
24. Cerio, R.; Spaull, J.; Jones, E.W. Histiocytoma cutis: A tumour of dermal dendrocytes (dermal dendrocytoma). *Br. J. Dermatol.* **2006**, *120*, 197–206. [CrossRef]
25. Kubo, M.; Ihn, H.; Yamane, K.; Tamaki, K. The expression levels and the differential expression of transforming growth factor-β receptors in dermatofibroma and dermatofibrosarcoma protuberans. *Br. J. Dermatol.* **2006**, *154*, 919–925. [CrossRef]
26. Yamamoto, T. Dermatofibroma: A possible model of local fibrosis with epithelial/mesenchymal cell interaction. *J. Eur. Acad. Dermatol. Venereol.* **2009**, *23*, 371–375. [CrossRef] [PubMed]
27. Sellheyer, K.; Smoller, B.R. Dermatofibroma: Upregulation of Syndecan-1 Expression in Mesenchymal Tissue. *Am. J. Dermatopathol.* **2003**, *25*, 392–398. [CrossRef] [PubMed]
28. Skroza, N.; Rotolo, S.; Ceccarelli, S.; Romano, F.; Innocenzi, D.; Frati, L.; Angeloni, A.; Marchese, C. Modulation of the expression of the FGFR2-IIIb and FGFR2-IIIc molecules in dermatofibroma. *J. Dermatol. Sci.* **2008**, *51*, 53–57. [CrossRef] [PubMed]
29. Walther, C.; Hofvander, J.; Nilsson, J.; Magnusson, L.; Domanski, H.A.; Gisselsson, D.; Tayebwa, J.; Doyle, L.A.; Fletcher, C.D.M.; Mertens, F. Gene fusion detection in formalin-fixed paraffin-embedded benign fibrous histiocytomas using fluorescence in situ hybridization and RNA sequencing. *Lab. Investig.* **2015**, *95*, 1071–1076. [CrossRef] [PubMed]
30. Panagopoulos, I.; Gorunova, L.; Bjerkehagen, B.; Lobmaier, I.; Heim, S. LAMTOR1-PRKCD and NUMA1-SFMBT1 fusion genes identified by RNA sequencing in aneurysmal benign fibrous histiocytoma with t(3;11)(p21;q13). *Cancer Genet.* **2015**, *208*, 545–551. [CrossRef]
31. Doyle, L.A.; Mariño-Enriquez, A.; Fletcher, C.D.M.; Hornick, J.L. ALK rearrangement and overexpression in epithelioid fibrous histiocytoma. *Mod. Pathol.* **2015**, *28*, 904–912. [CrossRef]

32. Szablewski, V.; Laurent-Roussel, S.; Rethers, L.; Rommel, A.; Vaneechout, P.; Camboni, A.; Willocz, P.; Copie-Bergman, C.; Ortonne, N. Atypical fibrous histiocytoma of the skin with CD30 and p80/ALK1 positivity and ALK gene rearrangement. *J. Cutan. Pathol.* **2014**, *41*, 715–719. [CrossRef]
33. Chen, T.-C.; Kuo, T.-T.; Chan, H.-L. Dermatofibroma is a clonal proliferative disease. *J. Cutan. Pathol.* **2000**, *27*, 36–39. [CrossRef]
34. Cerio, R.; Spaull, J.; Oliver, G.F.; Jones, E.W. A Study of Factor XHIa and MAC 387 Immunolabeling in Normal and Pathological Skin. *Am. J. Dermatopathol.* **1990**, *12*, 221–233. [CrossRef]
35. Mentzel, T.; Wiesner, T.; Cerroni, L.; Hantschke, M.; Kutzner, H.; Rütten, A.; Häberle, M.; Bisceglia, M.; Chibon, F.; Coindre, J.-M. Malignant dermatofibroma: Clinicopathological, immunohistochemical, and molecular analysis of seven cases. *Mod. Pathol.* **2013**, *26*, 256–267. [CrossRef]
36. Ferrari, A.; Soyer, H.P.; Peris, K.; Argenziano, G.; Mazzocchetti, G.; Piccolo, D.; De Giorgi, V.; Chimenti, S. Central white scarlike patch: A dermatoscopic clue for the diagnosis of dermatofibroma. *J. Am. Acad. Dermatol.* **2000**, *43*, 1123–1125. [CrossRef]
37. Lamgan, S.M.; Robinson, T.W.E. Cryotherapy for dermatofibromas. *Clin. Exp. Dermatol.* **1987**, *12*, 121–123. [CrossRef]
38. Page, M.J.; McKenzie, J.E.; Bossuyt, P.M.; Boutron, I.; Hoffmann, T.C.; Mulrow, C.D.; Shamseer, L.; Tetzlaff, J.M.; Akl, E.A.; Brennan, S.E.; et al. The PRISMA 2020 Statement: An Updated Guideline for Reporting Systematic Reviews. *BMJ* **2021**, *372*, n71. [CrossRef]
39. Mentzel, T. Cutaneous mesenchymal tumours: An update. *Pathology* **2014**, *46*, 149–159. [CrossRef] [PubMed]
40. Fitzpatrick, T.B.; Gilchrest, B.A. Dimple Sign to Differentiate Benign from Malignant Pigmented Cutaneous Lesions. *N. Engl. J. Med.* **1977**, *296*, 1518. [CrossRef] [PubMed]
41. Lookingbill, D.P. A Malignant Dimple. *N. Engl. J. Med.* **1977**, *297*, 841–842. [CrossRef] [PubMed]
42. Doyle, L.A.; Fletcher, C.D. Metastasizing "Benign" Cutaneous Fibrous Histiocytoma: A clinicopathologic analysis of 16 cases. *Am. J. Surg. Pathol.* **2013**, *37*, 484–495. [CrossRef] [PubMed]
43. Gershtenson, P.C.; Krunic, A.L.; Chen, H.M. Multiple clustered dermatofibroma: Case report and review of the literature. *J. Cutan. Pathol.* **2010**, *37*, e42–e45. [CrossRef] [PubMed]
44. Kalsi, H.; Rahman, A.; Harbol, T.; Sidhu, J. Giant Hemosiderotic Dermatofibroma: The Largest Giant Dermatofibroma Reported to Date. *Am. J. Dermatopathol.* **2015**, *37*, 778–782. [CrossRef] [PubMed]
45. Finch, J.; Berke, A.; McCusker, M.; Chang, M.W. Congenital Multiple Clustered Dermatofibroma in a 12-Year-Old Girl. *Pediatr. Dermatol.* **2014**, *31*, 105–106. [CrossRef]
46. Massone, C.; Parodi, A.; Virno, G.; Rebora, A. Multiple eruptive dermatofibromas in patients with systemic lupus erythematosus treated with prednisone. *Int. J. Dermatol.* **2002**, *41*, 279–281. [CrossRef] [PubMed]
47. García-Millán, C.; Aldanondo, I.; Fernández-Lorente, M.; Carrillo, R.; Jaén, P. Multiple Eruptive Dermatofibromas in 2 Patients Infected with the Human Immunodeficiency Virus. *Actas Dermosifiliogr.* **2007**, *98*, 702–706. [CrossRef] [PubMed]
48. Marque, M.; Pallure, V.; Huet, P.; Bessis, D.; Guillot, B. Multiple familial "eruptive" dermatofibromas. *Ann. Dermatol. Venereol.* **2013**, *140*, 452–454. [CrossRef] [PubMed]
49. Huang, P.-Y.; Chu, C.-Y.; Hsiao, C.-H. Multiple eruptive dermatofibromas in a patient with dermatomyositis taking prednisolone and methotrexate. *J. Am. Acad. Dermatol.* **2007**, *57*, S81–S84. [CrossRef] [PubMed]
50. López, N.; Fernández, A.; Bosch, R.J.; Herrera, E. Multiple eruptive dermatofibromas in a patient with Graves-Basedow disease. *J. Eur. Acad. Dermatol. Venereol.* **2008**, *22*, 402–403. [CrossRef]
51. Kimura, Y.; Kaneko, T.; Akasaka, E.; Nakajima, K.; Aizu, T.; Nakano, H.; Sawamura, D. Multiple eruptive dermatofibromas associated with Hashimoto's thyroiditis and myasthenia gravis. *Eur. J. Dermatol.* **2010**, *20*, 538–539. [CrossRef]
52. Monteagudo, B.; Suarez-Amor, O.; Cabanillas, M.; Leon-Mateos, A.; Perez-Valcarcel, J.; de las Heras, C. Down syndrome: Another cause of immunosuppression associated with multiple eruptive dermatofibromas? *Dermatol. Online J.* **2009**, *15*, 15. [CrossRef]
53. Alexandrescu, D.T.; Wiernik, P.H. Multiple Eruptive Dermatofibromas Occurring in a Patient with Chronic Myelogenous Leukemia. *Arch. Dermatol.* **2005**, *141*, 397–398. [CrossRef]
54. Bhattacharjee, P.; Umar, S.A.; Fatteh, S.M. Multiple Eruptive Dermatofibromas Occurring in a Patient with Myelodysplastic Syndrome. *Acta Derm. Venereol.* **2005**, *85*, 270–271. [CrossRef]
55. Zaccaria, E.; Rebora, A.; Rongioletti, F. Multiple eruptive dermatofibromas and immunosuppression: Report of two cases and review of the literature. *Int. J. Dermatol.* **2008**, *47*, 723–727. [CrossRef]
56. Yazici, A.C.; Baz, K.; Ikizoglu, G.; Koca, A.; Kokturk, A.; Apa, D.D. Familial eruptive dermatofibromas in atopic dermatitis. *J. Eur. Acad. Dermatol. Venereol.* **2006**, *20*, 90–92. [CrossRef] [PubMed]
57. Beyazit, Y.; Caner, S.; Kurt, M.; Kekilli, M.; Aydog, G.; Ibis, M. Dermatofibroma in a patient with Crohn's disease: A novel clinical manifestation. *J. Crohn's Colitis* **2010**, *4*, 490–491. [CrossRef] [PubMed]
58. Hiraiwa, T.; Hanami, Y.; Yamamoto, T. Hidradenitis suppurativa and multiple dermatofibromas in a patient with ulcerative colitis. *J. Dermatol.* **2013**, *40*, 1071–1072. [CrossRef] [PubMed]
59. Goldbach, H.; Wanat, K.; Rosenbach, M. Multiple eruptive dermatofibromas in a patient with sarcoidosis. *Cutis* **2016**, *98*, E15–E97. [PubMed]
60. Bachmeyer, C.; Cordier, F.; Blum, L.; Cazier, A.; Vérola, O.; Aractingi, S. Multiple eruptive dermatofibromas after highly active antiretroviral therapy. *Br. J. Dermatol.* **2000**, *143*, 1336–1337. [CrossRef] [PubMed]
61. Queirós, C.; Uva, L.; de Almeida, L.S.; Filipe, P. Multiple eruptive dermatofibromas associated with pregnancy—A case and literature review. *Dermatol. Online J.* **2019**, *25*, 12. [CrossRef]

62. Santos-Juanes, J.; Coto-Segura, P.; Mallo, S.; Galache, C.; Soto, J. Multiple Eruptive Dermatofibromas in a Patient Receiving Efalizumab. *Dermatology* **2008**, *216*, 363. [CrossRef]
63. Giavedoni, P.; Combalia, A.; Pigem, R.; Mascaró, J.M. Multiple Eruptive Dermatofibromas in a Patient Treated with Brentuximab Vedotin. *Actas. Dermosifiliogr.* **2019**, *110*, 419–420. [CrossRef]
64. Caldarola, G.; Bisceglia, M.; Pellicano, R. Multiple eruptive plaque-like dermatofibromas during anti-TNFα treatment. *Int. J. Dermatol.* **2013**, *52*, 638–641. [CrossRef]
65. Llamas-Velasco, M.; Fraga, J.; Solano-López, G.E.; Steegmann, J.L.; Diez, A.G.; Requena, L. Multiple eruptive dermatofibromas related to imatinib treatment. *J. Eur. Acad. Dermatol. Venereol.* **2014**, *28*, 979–981. [CrossRef]
66. Beatrous, S.V.; Riahi, R.R.; Grisoli, S.B.; Cohen, P.R. Associated conditions in patients with multiple dermatofibromas: Case reports and literature review. *Dermatol. Online J.* **2017**, *23*. [CrossRef]
67. Ohnishi, T.; Sasaki, M.; Nakai, K.; Watanabe, S. Atrophic dermatofibroma. *J. Eur. Acad. Dermatol. Venereol.* **2004**, *18*, 580–583. [CrossRef] [PubMed]
68. Kai, H.; Fujita, H.; Yamamoto, M.; Asahina, A. Polypoid dermatofibroma with a slim pedicle: A case report. *Dermatol. Online J.* **2012**, *18*. [CrossRef]
69. Curcó, N.; Jucgla, A.; Bordas, X.; Moreno, A. Dermatofibroma with spreading satellitosis. *J. Am. Acad. Dermatol.* **1992**, *27*, 1017–1019. [CrossRef]
70. Schofield, C.; Weedon, D.; Kumar, S. Dermatofibroma and halo dermatitis. *Australas. J. Dermatol.* **2012**, *53*, 145–147. [CrossRef] [PubMed]
71. Marinescu, S.A.; Tatu, A.L.; Mihai, I.R.; Giuglea, C. Correlations between clinics, dermoscopy and histopathology in a female with two dermatofibromas-a case report. *Rom. J. Morphol. Embryol.* **2016**, *57*, 323–326. [PubMed]
72. Picard, A.; Long-Mira, E.; Chuah, S.Y.; Passeron, T.; Lacour, J.-P.; Bahadoran, P. Interest of high-definition optical coherent tomography (HD-OCT) for non-invasive imaging of dermatofibroma: A pilot study. *J. Eur. Acad. Dermatol. Venereol.* **2016**, *30*, 485–487. [CrossRef]
73. Demir, M.K.; Ozdemir, H.; Gençhallaç, H.; Altaner, S.; Kartal, O. Dermatofibroma mimicking malignancy on integrated F-18 fluorodeoxyglucose PET-CT. *Diagn. Interv. Radiol.* **2009**, *15*, 61–63.
74. Guedes, R.V.D.A.P.; De Menezes, N.M.N.; Leite, I.B.; Baptista, M.A. Benign fibrous histiocytoma: Particular aspects on confocal laser scanning microscopy. *Eur. J. Dermatol.* **2012**, *22*, 288–289. [CrossRef]
75. Genc, Y.; Akay, B.N.; Heper, A.O.; Rosendahl, C.; Erdem, C. Dermatopathological characteristics of dermatofibromas from dermatoscopic clues. *Int. J. Dermatol.* **2020**, *59*, 66–75. [CrossRef]
76. Lin, M.J.; Xie, C.; Pan, Y.; Jalilian, C.; Kelly, J.W. Dermoscopy improves diagnostic accuracy for clinically amelanotic nodules. *Australas. J. Dermatol.* **2019**, *60*, 45–49. [CrossRef] [PubMed]
77. Won, K.Y.; Park, S.Y.; Jin, W.; Lew, B.-L. Dermatofibroma: Sonographic findings and pathologic correlation. *Acta Radiol.* **2018**, *59*, 454–459. [CrossRef] [PubMed]
78. Kelati, A.; Aqil, N.; Baybay, H.; Gallouj, S.; Mernissi, F.Z. Beyond classic dermoscopic patterns of dermatofibromas: A prospective research study. *J. Med. Case Rep.* **2017**, *11*, 266. [CrossRef]
79. Roldán-Marín, R.; Barreiro-Capurro, A.; García-Herrera, A.; Puig, S.; Alarcón-Salazar, I.; Carrera, C.; Malvehy, J. Green colour as a novel dermoscopic finding in the diagnosis of haemosiderotic dermatofibroma. *Australas. J. Dermatol.* **2014**, *55*, 196–197. [CrossRef] [PubMed]
80. Arpaia, N.; Cassano, N.; Vena, G.A. Dermoscopic Patterns of Dermatofibroma. *Dermatol. Surg.* **2005**, *31*, 1336–1339. [CrossRef]
81. Karaarslan, I.K.; Gencoglan, G.; Akalin, T.; Ozdemir, F. Different dermoscopic faces of dermatofibromas. *J. Am. Acad. Dermatol.* **2007**, *57*, 401–406. [CrossRef]
82. Agero, A.L.C.; Taliercio, S.; Dusza, S.W.; Salaro, C.; Chu, P.; Marghoob, A.A. Conventional and Polarized Dermoscopy Features of Dermatofibroma. *Arch. Dermatol.* **2006**, *142*, 1431–1437. [CrossRef]
83. Han, T.Y.; Chang, H.S.; Lee, J.H.K.; Lee, W.-M.; Son, S.-J. A Clinical and Histopathological Study of 122 Cases of Dermatofibroma (Benign Fibrous Histiocytoma). *Ann. Dermatol.* **2011**, *23*, 185–192. [CrossRef]
84. Zelger, B.G.; Sidoroff, A. Combined dermatofibroma: Co-existence of two or more variant patterns in a single lesion. *Histopathology* **2000**, *36*, 529–539. [CrossRef]
85. Calonje, E.; Brenn, T.; Lazar, A.; Mckee, P. *McKee's Pathology of the Skin with Clinical Correlations*, 4th ed.; Elsevier Saunders: Edinburgh, UK, 2012; pp. 1643–1662.
86. Zeidi, M.; North, J.P. Sebaceous induction in dermatofibroma: A common feature of dermatofibromas on the shoulder. *J. Cutan. Pathol.* **2015**, *42*, 400–405. [CrossRef]
87. Cazzato, G.; Colagrande, A.; Cimmino, A.; Marrone, M.; Stellacci, A.; Arezzo, F.; Lettini, T.; Resta, L.; Ingravallo, G. Granular Cell Dermatofibroma: When Morphology Still Matters. *Dermatopathology* **2021**, *8*, 371–375. [CrossRef] [PubMed]
88. LeBoit, P.E.; Barr, R.J.; Burall, S.; Metcalf, J.S.; Yen, T.S.B.; Wick, M.R. Primitive Polypoid Granular-Cell Tumor and Other Cutaneous Granular-Cell Neoplasms of Apparent Nonneural Origin. *Am. J. Surg. Pathol.* **1991**, *15*, 48–58. [CrossRef] [PubMed]
89. Lee, M.; Lee, W.J.; Jung, J.M.; Won, C.H.; Chang, S.E.; Choi, J.H.; Moon, K.C. Clinical and histological patterns of dermatofibroma without gross skin surface change: A comparative study with conventional dermatofibroma. *Indian J. Dermatol. Venereol. Leprol.* **2015**, *81*, 263–269. [CrossRef] [PubMed]

90. Victor, T.A. *Neoplasms with Follicular Differentiation*, 2nd ed.; Ackerman, B., Reddy, V.B., Soyer, H.P., Eds.; Blackwell Publishing Inc.: Hoboken, NJ, USA, 2002.
91. Kaddu, S.; McMenamin, M.E.; Fletcher, C.D.M. Atypical Fibrous Histiocytoma of the Skin: Clinicopathologic analysis of 59 cases with evidence of infrequent metastasis. *Am. J. Surg. Pathol.* **2002**, *26*, 35–46. [CrossRef] [PubMed]
92. Tatu, A.L. Black Nodule on the Forearm. *J. Cutan. Med. Surg.* **2017**, *21*, 157. [CrossRef]
93. Agarwal, A.; Gopinath, A.; Tetzlaff, M.T.; Prieto, V.G. Phosphohistone-H3 and Ki67: Useful Markers in Differentiating Dermatofibroma from Dermatofibrosarcoma Protuberans and Atypical Fibrohistiocytic Lesions. *Am. J. Dermatopathol.* **2017**, *39*, 504–507. [CrossRef]
94. Trøstrup, H.; Bigdeli, A.K.; Krogerus, C.; Kneser, U.; Schmidt, G.; Schmidt, V.J. A Multidisciplinary Approach to Complex Dermal Sarcomas Ensures an Optimal Clinical Outcome. *Cancers* **2022**, *14*, 1693. [CrossRef]
95. Volpicelli, E.R.; Fletcher, C.D.M. Desmin and CD34 positivity in cellular fibrous histiocytoma: An immunohistochemical analysis of 100 cases. *J. Cutan. Pathol.* **2012**, *39*, 747–752. [CrossRef]
96. John, A.M.; Holahan, H.H.; Singh, P.; Handler, M.Z.; Lambert, W.C. When Immunohistochemistry Deceives Us: The Pitfalls of CD34 and Factor XIIIa Stains in Dermatofibroma and Dermatofibrosarcoma Protuberans. *Skinmed* **2017**, *15*, 53–55.
97. Kim, H.J.; Lee, J.Y.; Kim, S.H.; Seo, Y.J.; Lee, J.H.; Park, J.K.; Kim, M.H.; Cinn, Y.W.; Cho, K.H.; Yoon, T.Y. Stromelysin-3 expression in the differential diagnosis of dermatofibroma and dermatofibrosarcoma protuberans: Comparison with factor XIIIa and CD34. *Br. J. Dermatol.* **2007**, *157*, 319–324. [CrossRef]
98. Maekawa, T.; Jinnin, M.; Ihn, H. The expression levels of thrombospondin-1 in dermatofibroma and dermatofibrosarcoma protuberans. *Eur. J. Dermatol.* **2011**, *21*, 534–538. [CrossRef] [PubMed]
99. Li, J.; Yu, Y.; Yang, Y.; Wang, L.; Cao, J.; Liang, X.; Xiao, X.; Tu, Y.; Chen, H. IGFBP7, a novel immunohistochemical marker in differentiating dermatofibroma from dermatofibrosarcoma protuberans. *J. Eur. Acad. Dermatol. Venereol.* **2012**, *26*, 382–385. [CrossRef] [PubMed]
100. Mori, T.; Misago, N.; Yamamoto, O.; Toda, S.; Narisawa, Y. Expression of nestin in dermatofibrosarcoma protuberans in comparison to dermatofibroma. *J. Dermatol.* **2008**, *35*, 419–425. [CrossRef] [PubMed]
101. Wang, L.; Xiang, Y.N.; Zhang, Y.H.; Tu, Y.T.; Chen, H.X. Collagen triple helix repeat containing-1 in the differential diagnosis of dermatofibrosarcoma protuberans and dermatofibroma. *Br. J. Dermatol.* **2011**, *164*, 135–140. [CrossRef]
102. Yan, X.; Takahara, M.; Xie, L.; Tu, Y.; Furue, M. Cathepsin K expression: A useful marker for the differential diagnosis of dermatofibroma and dermatofibrosarcoma protuberans. *Histopathology* **2010**, *57*, 486–488. [CrossRef] [PubMed]
103. Toyozawa, S.; Yamamoto, Y.; Ishida, Y.; Kondo, T.; Nakamura, Y.; Furukawa, F. Immunohistochemical Analysis of CXCR4 Expression in Fibrohistiocytic Tumors. *Acta Histochem. Cytochem.* **2010**, *43*, 45–50. [CrossRef] [PubMed]
104. Ishigami, T.; Hida, Y.; Matsudate, Y.; Murao, K.; Kubo, Y. The involvement of fibroblast growth factor receptor signaling pathways in dermatofibroma and dermatofibrosarcoma protuberans. *J. Med. Investig.* **2013**, *60*, 106–113. [CrossRef]
105. Kazlouskaya, V.; Malhotra, S.; Kabigting, F.D.; Lal, K.; Elston, D.M. CD99 Expression in Dermatofibrosarcoma Protuberans and Dermatofibroma. *Am. J. Dermatopathol.* **2014**, *36*, 392–396. [CrossRef]
106. West, K.L.; Cardona, D.M.; Su, Z.; Puri, P.K. Immunohistochemical Markers in Fibrohistiocytic Lesions: Factor XIIIa, CD34, S-100 and p75. *Am. J. Dermatopathol.* **2014**, *36*, 414–419. [CrossRef]
107. Mikoshiba, Y.; Ogawa, E.; Uchiyama, R.; Uchiyama, A.; Uhara, H.; Okuyama, R. 5-Hydroxymethylcytosine is a useful marker to differentiate between dermatofibrosarcoma protuberans and dermatofibroma. *J. Eur. Acad. Dermatol. Venereol.* **2016**, *30*, 130–131. [CrossRef]
108. Karanian, M.; Pérot, G.; Coindre, J.-M.; Chibon, F.; Pedeutour, F.; Neuville, A. Fluorescence in situ hybridization analysis is a helpful test for the diagnosis of dermatofibrosarcoma protuberans. *Mod. Pathol.* **2015**, *28*, 230–237. [CrossRef] [PubMed]
109. Kaur, H.; Kaur, J.; Gill, K.S.; Mannan, R.; Arora, S. Subcutaneous Dermatofibroma: A Rare Case Report with Review of Literature. *J. Clin. Diagn. Res.* **2014**, *8*, FD01–FD02. [CrossRef] [PubMed]
110. Motegi, S.-I.; Fujiwara, C.; Sekiguchi, A.; Yamazaki, S.; Yokoyama, Y.; Yasuda, M.; Ishikawa, O. Possible contribution of PDGF-BB-induced autophagy in dermatofibrosarcoma protuberans: Autophagy marker Atg5 could be a differential marker between dermatofibrosarcoma protuberans and dermatofibroma. *J. Dermatol. Sci.* **2019**, *93*, 139–141. [CrossRef] [PubMed]
111. Lyundup, A.V.; Balyasin, M.V.; Maksimova, N.V.; Kovina, M.V.; Krasheninnikov, M.E.; Dyuzheva, T.G.; Yakovenko, S.A.; Appolonova, S.A.; Schiöth, H.B.; Klabukov, I.D. Misdiagnosis of diabetic foot ulcer in patients with undiagnosed skin malignancies. *Int. Wound J.* **2022**, *19*, 871–887. [CrossRef] [PubMed]
112. Giovannacci, I.; Meleti, M.; Garbarino, F.; Cesinaro, A.M.; Mataca, E.; Pedrazzi, G.; Reggiani, C.; Paganelli, A.; Truzzi, A.; Elia, F.; et al. Correlation between Autofluorescence Intensity and Histopathological Features in Non-Melanoma Skin Cancer: An Ex Vivo Study. *Cancers* **2021**, *13*, 3974. [CrossRef] [PubMed]
113. Dessinioti, C.; Antoniou, C.; Stratigos, A.J. New targeted approaches for the treatment and prevention of nonmelanoma skin cancer. *Expert Rev. Dermatol.* **2011**, *6*, 625–634. [CrossRef]
114. Drakaki, E.; Vergou, T.; Dessinioti, C.; Stratigos, A.J.; Salavastru, C.; Antoniou, C. Spectroscopic methods for the photodiagnosis of nonmelanoma skin cancer. *J. Biomed. Opt.* **2012**, *18*, 061221. [CrossRef]
115. Seidenari, S.; Arginelli, F.; Bassoli, S.; Cautela, J.; French, P.M.W.; Guanti, M.; Guardoli, D.; König, K.; Talbot, C.; Dunsby, C. Multiphoton Laser Microscopy and Fluorescence Lifetime Imaging for the Evaluation of the Skin. *Dermatol. Res. Pract.* **2011**, *2012*, 810749. [CrossRef]

116. Balaceanu, L.A. Biomarkers vs. imaging in the early detection of hepatocellular carcinoma and prognosis. *World J. Clin. Cases* **2019**, *7*, 1367–1382. [CrossRef]
117. Czajkowska, J.; Juszczyk, J.; Piejko, L.; Glenc-Ambroży, M. High-Frequency Ultrasound Dataset for Deep Learning-Based Image Quality Assessment. *Sensors* **2022**, *22*, 1478. [CrossRef]
118. Malciu, A.M.; Lupu, M.; Voiculescu, V.M. Artificial Intelligence-Based Approaches to Reflectance Confocal Microscopy Image Analysis in Dermatology. *J. Clin. Med.* **2022**, *11*, 429. [CrossRef] [PubMed]
119. Maron, R.C.; Utikal, J.S.; Hekler, A.; Hauschild, A.; Sattler, E.; Sondermann, W.; Haferkamp, S.; Schilling, B.; Heppt, M.V.; Jansen, P.; et al. Artificial Intelligence and Its Effect on Dermatologists' Accuracy in Dermoscopic Melanoma Image Classification: Web-Based Survey Study. *J. Med. Internet Res.* **2020**, *22*, e18091. [CrossRef]
120. Tschandl, P.; Wiesner, T. Advances in the diagnosis of pigmented skin lesions. *Br. J. Dermatol.* **2018**, *178*, 9–11. [CrossRef] [PubMed]
121. Styła, M.; Giżewski, T. The Study of Usefulness of a Set of Fractal Parameters to Build Classes of Disease Units Based on Images of Pigmented Skin Lesions. *Diagnostics* **2021**, *11*, 1773. [CrossRef] [PubMed]
122. Tschandl, P.; Rosendahl, C.; Kittler, H. The HAM10000 dataset, a large collection of multi-source dermatoscopic images of common pigmented skin lesions. *Sci. Data* **2018**, *5*, 180161. [CrossRef]
123. Felmingham, C.; MacNamara, S.; Cranwell, W.; Williams, N.; Wada, M.; Adler, N.R.; Ge, Z.; Sharfe, A.; Bowling, A.; Haskett, M.; et al. Improving Skin Cancer Management with Artificial Intelligence (SMARTI): Protocol for a preintervention/postintervention trial of an artificial intelligence system used as a diagnostic aid for skin cancer management in a specialist dermatology setting. *BMJ Open* **2022**, *12*, e050203. [CrossRef]
124. Tschandl, P.; Rinner, C.; Apalla, Z.; Argenziano, G.; Codella, N.; Halpern, A.; Janda, M.; Lallas, A.; Longo, C.; Malvehy, J.; et al. Human–computer collaboration for skin cancer recognition. *Nat. Med.* **2020**, *26*, 1229–1234. [CrossRef]
125. Esteva, A.; Kuprel, B.; Novoa, R.A.; Ko, J.; Swetter, S.M.; Blau, H.M.; Thrun, S. Dermatologist-level classification of skin cancer with deep neural networks. *Nature* **2017**, *542*, 115–118. [CrossRef] [PubMed]
126. Nagendran, M.; Chen, Y.; Lovejoy, C.A.; Gordon, A.C.; Komorowski, M.; Harvey, H.; Topol, E.J.; Ioannidis, J.P.A.; Collins, G.S.; Maruthappu, M. Artificial intelligence versus clinicians: Systematic review of design, reporting standards, and claims of deep learning studies. *BMJ* **2020**, *368*, m689. [CrossRef]
127. Ouellette, S.; Rao, B.K. Usefulness of Smartphones in Dermatology: A US-Based Review. *Int. J. Environ. Res. Public Health* **2022**, *19*, 3553. [CrossRef]
128. Patrzyk, S.; Bielecki, W.; Woźniacka, A. A study of attitudes among Polish dermatologists and dermatology trainees regarding modern technologies in medicine. *Adv. Dermatol. Allergol.* **2022**, *39*, 531–537. [CrossRef] [PubMed]
129. Breslavets, M.; Shear, N.H.; Lapa, T.; Breslavets, D.; Breslavets, A. Validation of artificial intelligence application in clinical dermatology. *J. Am. Acad. Dermatol.* **2022**, *86*, 201–203. [CrossRef] [PubMed]
130. Koh, U.; Horsham, C.; Soyer, H.P.; Loescher, L.J.; Gillespie, N.; Vagenas, D.; Janda, M. Consumer Acceptance and Expectations of a Mobile Health Application to Photograph Skin Lesions for Early Detection of Melanoma. *Dermatology* **2019**, *235*, 4–10. [CrossRef] [PubMed]
131. Fluhr, J.W.; Gueguen, A.; Legoupil, D.; Brenaut, E.; Abasq, C.; Araújo, H.; Misery, L. Teledermatology in Times of COVID-19 Confinement: Comparing Patients' and Physicians' Satisfaction by the Standardized Brest Teledermatology Questionnaire. *Dermatology* **2021**, *237*, 191–196. [CrossRef]
132. Asabor, E.N.; Bunick, C.G.; Cohen, J.M.; Perkins, S.H. Patient and physician perspectives on teledermatology at an academic dermatology department amid the COVID-19 pandemic. *J. Am. Acad. Dermatol.* **2021**, *84*, 158–161. [CrossRef]
133. Fijałkowska, M.; Koziej, M.; Żądzińska, E.; Antoszewski, B.; Sitek, A. Assessment of the Predictive Value of Spectrophotometric Skin Color Parameters and Environmental and Behavioral Factors in Estimating the Risk of Skin Cancer: A Case–Control Study. *J. Clin. Med.* **2022**, *11*, 2969. [CrossRef]
134. Hauser, K.; Kurz, A.; Haggenmüller, S.; Maron, R.C.; von Kalle, C.; Utikal, J.S.; Meier, F.; Hobelsberger, S.; Gellrich, F.F.; Sergon, M.; et al. Explainable artificial intelligence in skin cancer recognition: A systematic review. *Eur. J. Cancer* **2022**, *167*, 54–69. [CrossRef]
135. Liopyris, K.; Gregoriou, S.; Dias, J.; Stratigos, A.J. Artificial Intelligence in Dermatology: Challenges and Perspectives. *Dermatol. Ther.* **2022**, *12*, 2637–2651. [CrossRef]

diagnostics

MDPI

Review

Skin Dialogues in Atopic Dermatitis

Elena Porumb-Andrese [1,†], Claudia Florida Costea [2,*], Andrei Cucu [3,*], Gabriela Rusu-Zota [4,†], Daciana Elena Braisteanu [1,†], Vlad Porumb [5,†], Mihaela Monica Scutariu [6], Alexandra Maria Dorobanțu [7] and Ramona Gabriela Ursu [8]

1 Department Dermatology, Faculty of Medicine, University of Medicine and Pharmacy "Grigore T. Popa", Universitatii St. No. 16, 700115 Iasi, Romania
2 Department of Surgery II, Discipline of Ophthalmology, Faculty of Medicine, University of Medicine and Pharmacy "Grigore T. Popa", Universitatii St. No. 16, 700115 Iasi, Romania
3 2nd Neurosurgery Clinic, Professor Doctor Nicolae Oblu Emergency Clinical Hospital, 700309 Iasi, Romania
4 Department of Pharmacology, Clinical Pharmacology and Algesiology, Faculty of Medicine, University of Medicine and Pharmacy "Grigore T. Popa", Universitatii St. No. 16, 700115 Iasi, Romania
5 Department of Surgery I, Faculty of Medicine, University of Medicine and Pharmacy "Grigore T. Popa", Universitatii St. No. 16, 700115 Iasi, Romania
6 Department of Dental Technology, Faculty of Dental Medicine, University of Medicine and Pharmacy "Grigore T. Popa", Universitatii St. No. 16, 700115 Iasi, Romania
7 Dermatology Department, Elias Emergency University Hospital, 011461 Bucharest, Romania
8 Department of Microbiology, Faculty of Medicine, University of Medicine and Pharmacy "Grigore T. Popa", Universitatii St. No. 16, 700115 Iasi, Romania
* Correspondence: costea10@yahoo.com (C.F.C.); andreiucucu@yahoo.com (A.C.)
† These authors contributed equally to this work.

Abstract: Atopic dermatitis (AD) is a chronic skin disorder associated with significant quality-of-life impairment and increased risk for allergic and non-allergic comorbidities. The aim of this review is to elucidate the connection between AD and most common comorbidities, as this requires a holistic and multidisciplinary approach. Advances in understanding these associations could lead to the development of highly effective and targeted treatments.

Keywords: atopic dermatitis; comorbidities; cardiometabolic risk; neoplasia risk

1. Introduction

In the past two decades, allergic disorders have become a major global health problem that is currently estimated to afflict between 25% and 40% of the population in the developed world. These diseases, such as atopic dermatitis, asthma, or allergic rhinitis, are associated with significant morbidity and quality-of-life impairment, in addition to presenting a high socioeconomic burden [1].

What does atopy mean? The word is derived from the Greek atopia, meaning "different". The term "atopy" was originally proposed in 1923 by Coca and Cooke in order to describe asthma and allergic rhinitis, and the concept of atopic dermatitis was introduced 10 years later in 1933 by Wise and Sulzberger [2].

Atopic dermatitis (AD) is one of the most common and complex chronic skin disorders characterized by pruritus and inflammation, with most patients having a personal or family history of "atopic march" [3]. The World Allergy Organization estimates that the disease affects 10–30% of the pediatric population and 2–10% of adults in developed countries, with onset occurring in the first 6 months of life in 45% of affected children and during the first year of life in 60% cases. In terms of prevalence, AD in industrialized countries has tripled in the past 30 years, and although it is more common in young people, AD in most adults continues for many years [4,5].

The most important function of the skin is to provide a first-line defense between the body and the environment. The epidermal barrier serves three primary functions: limiting passive water loss, restricting environmental chemical absorption, and preventing microbial infection [6]. Although the pathophysiology of AD is not completely understood, numerous studies have demonstrated that skin barrier dysfunction is a central feature of AD, and with immune dysregulation as the second contributing factor to the pathobiology. In the acute phase, T helper 2 and T helper 22 responses are increased in the skin, with some implications from T helper 17. The disease continues its progression with an increased role played by T helper 1 and T helper 2 cells. The disease is considered a product of the interplay between environmental factors and genetic susceptibility [7].

AD is clinically diagnosed through extensive history and physical examination of the morphology, distribution, and clinical course of lesions. It generally presents three clinical phases: acute (with vesicles, erythema, oedema, and crusts), subacute (scaly and dry skin, papules), and chronic (dyspigmentation, thickening, xerosis and lichenification), although phases may coexist. Pruritus is a hallmark of AD, and excoriations secondary to scratching are often present. Generalized xerosis is a ubiquitous clinical feature that is often associated with palmoplantar hyperlinearity (Figure 1). In 1979, Hanifin and Rajka (H-R) advanced major criteria for this diagnosis; the criteria are comprehensive and considered the "gold standard". The United Kingdom Working Party criteria are an abridged version of the H-R criteria and tend to work better in children than adults. The American Academy of Dermatology has also developed a streamlined version of the H-R criteria [8].

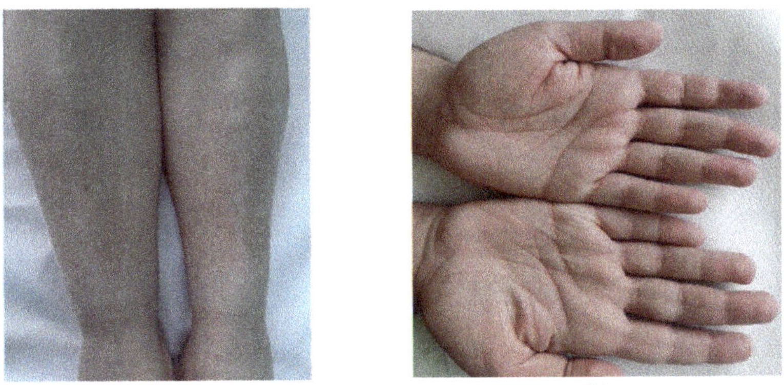

(a) (b)

Figure 1. Cutaneous xerosis (**a**); palmar hyperlinearity (**b**).

Given its tremendous burden on physical, mental, and social health, numerous scoring systems have been described to evaluate AD. The scores can be classified into two categories: those aimed at evaluating skin involvement, such as the Eczema Area and Severity Index (EASI) for mild-to-severe AD in both pediatric and adult patients; the Self-Administered Eczema Area and Severity Index (SA-EASI); SCROAD (SCORing Atopic Dermatitis); Patient-Oriented SCOring Atopic Dermatitis (PO-SCORAD); and those aimed at evaluating quality of life, such as the Dermatology Life Quality Index (DLQI), and Patient-Oriented Eczema Measures (POEM). QoL measurement has become an essential aspect of monitoring disease symptoms in AD to assess the safety and efficacy of therapies in clinical trials [9]. Other alternatives are the Hospital Anxiety and Depression Scale (HADS), the Numerical Rating Scale (NRS), and the Visual Analogue Scale (VAS); however, there are no specific tests to define and evaluate AD, and further validation of these methods in AD is needed.

Patients with AD are at high risk of developing allergic and non-allergic comorbidities, and early onset is considered a risk factor in the development of such associations. The relationship between comorbidities and AD is likely bidirectional and multifactorial, reflecting the systemic nature of the disease (Figure 2). Allergic conditions include asthma, allergic rhinitis, atopic keratoconjunctivitis, blepharitis, eosinophilic esophagitis, and food allergy.

Non-allergic comorbidities include cutaneous bacterial infections (Staphylococcus aureus colonization and Streptococcus pyogenes), cutaneous viral infections (eczema herpeticum, eczema coxsackium, ezcema vaccinatum, and molluscum contagiosum), or extracutaneous infections, as well as mental health comorbidities (attention-deficit hyperactivity, autistic spectrum disorder, anxiety, and depression), cardiovascular diseases, obesity, autoimmune diseases, or gastrointestinal immune-mediated disorders [7,10]. The mechanisms underlying these associations are mostly unknown, and the absolute magnitude of the risks has not been well defined to date.

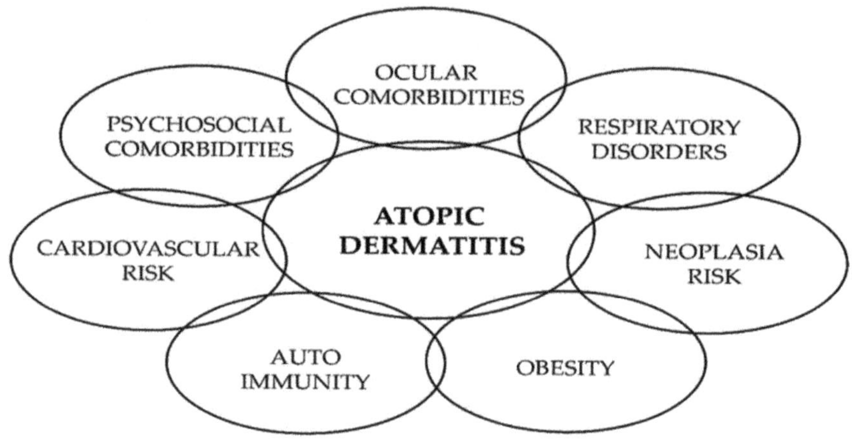

Figure 2. Common comorbidities of atopic dermatitis.

2. Respiratory Disorders

2.1. Allergic Rhinitis

Chronic rhinitis is divided into allergic rhinitis (AR) and non-allergic rhinitis. In a case of AR, patients test positive in a skin prick test or for the presence of IgE in serum. Patients with non-AR test negative for both. Among the atopic disorders, AR is the most prevalent, affecting up to 40% of the population [11]. The disease is defined as chronic inflammation of nasal mucosa driven by type-2 helper T (Th2) cells with symptoms of sneezing, airflow obstruction, nasal itching, and clear nasal discharge [12]. Numerous predisposing risk factors have been described for AR, such as history of allergic diseases, early-life antibiotic use, male sex, maternal smoking, elevated serum IgE levels, and indoor allergen exposure [13].

Epidemiological studies have shown that the presence of AR is an increased risk factor for development of asthma; up to 40% of people with AR have or will have asthma [14].

In terms of therapeutic options, various therapeutic agents are used, such as mast cell stabilizers, chemical mediator receptor antagonists (antihistamines, antileukotrienes, prostaglandin D2, and thromboxane A2 receptor antagonists), Th2 cytokine inhibitors, steroids, and alpha-sympathomimetics [15].

2.2. Allergic Bronchial Asthma

AR and bronchial asthma (BA) can be considered as different aspects of one systemic allergic disorder. Generally, the history begins with AD and progresses, in an additive manner, to include asthma and AR, which is defined as a chronic inflammatory disorder of the airways that translates clinically in recurrent episodes of wheezing, chest tightness, or coughing [16].

A positive correlation has been demonstrated between the severity of AD and the risk of developing bronchial asthma (BA). In a large study of 3225 AR patients, Valero et al. examined this relationship and concluded that respiratory allergic disease is a systemic condition and that some characteristics of AR may affect the development of asthma [17].

On the other hand, in middle-income countries, studies have suggested that atopic sensitization it is not strongly correlated with sequential development of AD, BA, and AR. A possible explanation for this relationship is that environmental exposures in late childhood may dissociate atopic conditions from each other [18,19].

3. Psychological Distress in Atopic Dermatitis

Recent studies have also shown that an important pathogenic role is played in illness development by emotional stress and various personality traits. Both acute and chronic stress exacerbate disease symptoms by altered innervation of the skin and elevated tonus of the sympathetic nervous system [20,21].

Associations have also been recognized between allergic diseases and psychiatric disorders, such as depression, anxiety, and, perhaps the best-studied mental health comorbidity associated with ADHD, attention deficit/hyperactivity disorder [22]. The association between AD and ADHD is unsurprising, given the Diagnostic and Statistical Manual of Mental Disorders, 5th edition, diagnostic criteria for ADHD, with characteristic features that overlap with those of an itchy child. The chronic use of antihistamines, an often-used therapy in AD, was significantly also associated with increased ADHD symptoms [23]. In a German study, it was shown that the risk of depression for these patients is comparable with the risk of depression in patients suffering from cancers [24]. In a Swedish study with 34,313 participants, adults with AD were found to have an increased risk of severe depression, anxiety, and reduced quality of life [25]. The specificity of the relationship remains unclear, but a recent meta-analysis highlights a two-way direction between mental health and atopic disorder [26].

A recent study from Korea that included 72,435 adolescent middle and high school students with AD revealed that AD patients are prone to depressive feelings, as well as ideation and planning attempts at suicidal behaviors [27].

4. Ocular Comorbidities

Ocular complications of atopic dermatitis include blepharitis, conjunctivitis, modifications of the lacrimal canaliculi, cataract, uveitis, keratoconus, and retinal detachments [28,29].

Keratoconus was described by Hilgatner in 1937 as atopic dermatitis complication; patients diagnosed with keratoconus have symptoms of eye rubbing [30]. Bawazeer et al. mentioned that eye rubbing occurs in the cases of progressive keratoconus and atopy may contribute to keratoconus probably by eye rubbing [31]. Allergic disorders (atopic keratoconjunctivitis, asthma, and atopic dermatitis) are also associated with corneal hydrops in keratoconus [32].

Bacteria that colonize eyelid skin and conjunctival fornixes are more numerous in patients with atopic dermatitis than in healthy individuals. The most frequently identified pathogen in patients with atopic dermatitis is Staphylococcus aureus (67%) [33].

Other ocular complications during atopic dermatitis include blepharitis, retinal detachment, subcapsular cataract, keratoconjunctivitis, uveitis, and ocular herpes simplex. Retinal detachment is a serious complication of atopic dermatitis and is often caused by eye rubbing [34–38].

Cataract in atopic dermatitis is usually bilateral, with lens opacities in the anterior and posterior subcapsular region [39]. Sasabe et al. proved the association between atopic dermatitis in patients with cataract and high levels of IgE [40].

Patients with severe atopic dermatitis were associated with advanced glaucoma induced by glucocorticosteroids, and eye rubbing in these patients may be a cause of ocular trauma, possibly causing globular rupture [39,41].

In the case of children, in a study of 59 children with atopic dermatitis, Carmi et al. observed several associated eye diseases, including chronic blepharitis (1 case), papillofollicular conjunctivitis (11 cases), purulent bacterial conjunctivitis (1 case), nuclear cataract (1 case), and amblyopia (1 case) [35].

4.1. *Atopic Keratoconjunctivitis (AKC)*

Atopic keratoconjunctivitis (AKC) is another comorbidity of AD characterized by redness, ocular pruritus, burning discomfort, pain in the eye, or photophobia. AKC is a noninfectious, inflammatory condition that affects the cornea and conjunctiva and occurs in an estimated 25% to 42% of patients with AD [31]. AKC can progress to recurrent corneal erosions, corneal ulcers, keratoconus, cataracts, blepharitis, tear dysfunctions, or infectious keratitis. It is also a potentially blinding condition if not treated early [41–45].

Atopic keratoconjunctivitis was the most common ocular manifestation identified in children diagnosed with atopic dermatitis [45].

The diagnosis of atopic dermatitis in adult and pediatric patients requires ophthalmological check-ups in order to identify possible associated eye diseases (Figure 3).

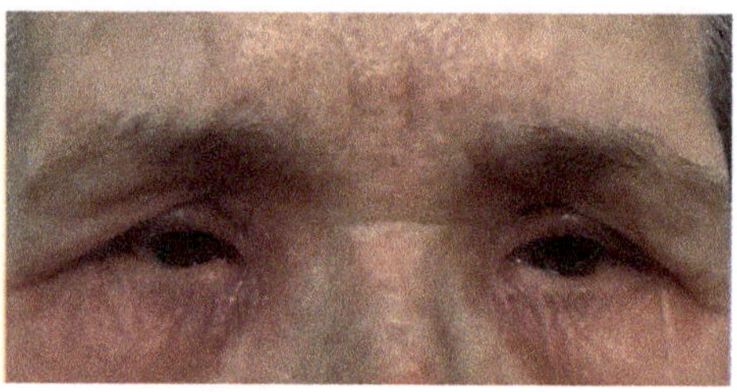

Figure 3. Periocular atopic dermatitis (PD).

Treatment of atopic dermatitis is essential in the prevention of eye rubbing, which can cause severe eye complications, such as retinal detachment or worsening of the keratoconus, causing permanent impairment of visual acuity. A recent study showed that patients with PD treated with oral antihistamines and topical steroids were approximately four times more likely to experience cure than those treated with topical steroids alone [46]. Topical calcineurin inhibitors are another highly effective approach for treatment of conjunctival involvement [47]. This could underline the importance of allergy testing for this category of patients.

4.2. *Blepharitis*

Blepharitis is an inflammatory condition of the eyelids that is estimated to affect more than 6% of patients with AD. Clinically, patients may experience pruritus and local irritation, tearing, burning and foreign sensation, photophobia, or crusting of the eyelids [48].

5. Immune Dialogue between the Gastrointestinal Tract and Skin

The role of gut microbiota in the development and severity of symptoms of AD remains unclear. However, it has been proposed that probiotics may lead to an induction of regulatory of T cells, along with suppression of regulatory cytokines Il-10 and TGF-β [49]. Systemic antibiotic treatment has been reported to increase the risk of AD, possibly due to changes in intestinal microbiota [50].

Other studies revealed a significant reduction in the microbial diversity of the oral cavity in AD patients, as well as an apparent alteration in genera [51].

More than 90% of all microorganisms in humans are found in the large intestine. Previous studies have revealed that changes in the gut microbiome in early life are closely correlated with age of onset and severity of AD; abundant evidence suggests that host-microbiome interactions can determine health status. The development of skin changes may be promoted not only by gut microbiome diversity but also by specific interactions between the gut microbiome, the immune system, and the host [52]. In patients with AD, *Clostridium difficile*, *Escherichia coli*, and *Staphylococcus aureus* levels in the gut microbiome are higher than in healthy controls [53]. Transepidermal water loss decreased in human subjects after oral supplementation of *Lactobacillus brevis* SBC8803 [54]. Transepidermal water loss is an indicator of skin barrier function. In animal models of AD, infection with *Helicobacter pylori* exerts a protective effect on the severity of skin lesions, such as erythema, swelling, and hyperemia [55]. The potential for probiotics to treat inflammatory skin diseases has been a focal point of research. For example, a number of randomized, controlled trials have reported a reduction in disease severity following administration of oral probiotics to children with AD [56,57]. Conversely, other studies have noted no differences between oral probiotics and placebo in preventing or improving AD symptoms [58].

Another study using *Lactobacillus acidophilus* L-92 in children between 10 months and 3 years old who had undergone conventional treatment and with concomitant food allergies (FA) showed a decrease in allergic sensitization, with decreased SCORAD index and decreased total IgE levels after only 24 weeks. These results suggest that L-92 works as an adjunctive treatment in patients with AD and FA [59].

Irritable bowel syndrome, a functional gastrointestinal disorder characterized by chronic or recurrent abdominal pain, cramping, gas, bloating, and diarrhea/constipation, was found to be more common in AD patients that in healthy controls in a recent study conducted by Kaya İslamoğlu et al. The relationship could be explained by inflammation, an important factor in the pathogenesis of both diseases, with elevated serum inflammatory cytokines and decreased serum anti-inflammatory cytokines for irritable bowel syndrome. In adults, risk of celiac disease was also increased but not in children [60].

6. Cardiometabolic Risk

Until now, AD has been linked to comorbidities such as allergies, asthma, and psychiatric disorders; however, recent studies have suggested an association between AD and cardiovascular disease in both adults and pediatric patients. Cardiovascular comorbidities for AD include hypertension, myocardial infarction, stroke, angina, coronary heart disease, heart failure, peripheral artery disease, thrombosis, and atrial fibrillation. The link between atopy and cardiovascular risk is based on the theory of a shared inflammatory pathway [61].

A meta-analysis of 30 studies showed higher odds of overweight or/and obesity in children and adults with AD in studies from North America and Asia [62]. A Taiwanese cohort study showed a significantly increased risk of risk of ischemic stroke but not of myocardial infarction (MI) in patients with AD [63]. In Europe, a recent large German study revealed a modestly increased risk of hypertension, angina, and peripheral artery disease but not of MI and stroke [64].

Compared to other chronic inflammatory skin conditions, such as psoriasis, the risk for cardiovascular diseases in AD is much lower, although it remains controversial.

A possible link between AD and type 2 diabetes is likely multifactorial and possibly explained by genetic predisposition; AD-associated therapies, including corticosteroids; lifestyle; or systemic low-grade inflammation [65,66]. In contrast, children diagnosed with type 1 diabetes mellitus seem to be less likely to develop AD compared with controls; however, evidence of this association is limited, and large control studies are needed to elucidate the exact link between AD and diabetes [67]. Nevertheless, it is difficult to determine whether the potential association of AD with these types of comorbidities is due to pathophysiological effects, adverse events associated with medications used to treat AD, or a combination of the two factors.

Obesity

A recently meta-analysis of epidemiological studies showed that obesity is associated with an increased prevalence of AD in both children and adults. Such an association can be interpreted in three ways: obesity causes AD by providing chronic inflammatory status; obesity can influence sex hormones by increasing 17-β estradiol plasma concentrations, and estrogens have proinflammatory properties; or poor diet habits or sedentary lifestyle could lead to both illness [61].

The association between obesity and atopy may be linked to genetics or shared risk factors, such as gut microbiome. However, the reported results are conflicting. For example, a large study with more than 15,000 children showed that excessive gestational weight gain was associated with a 10% increase in risk of AD [68], whereas another study revealed that obese children younger than 2 years were 15 times more likely to have the disease [69].

Obesity is known to modify cutaneous and systemic inflammation via increased cytokines secretion (tumor necrosis factor alpha and interleukin 6) [70].

These molecules may predispose patients to hypersensitivity reactions. Another explanation could be the predisposition to a more sedentary lifestyle, which may increase the risk of obesity [71].

A cross-sectional analysis of a large population-based cohort (441,746 adults with AD) revealed a small association between AD and overweight and obese patients but no association with severe forms of obesity [72]. An interesting aspect is that these studies used body mass index to determine obesity rather than measurements of abdominal obesity/waist circumference, possibly resulting in a stronger association between AD and obesity. However, the authors drew the common conclusion that maintaining ideal body mass index during early childhood important aspect for patients with atopic march. A cross-sectional study of 116,816 patients from Israel revealed that metabolic syndrome to was less prevalent in people with AD, whereas another study from Korea revealed this association only in female subjects. In general, females are more prone to severe allergies because estrogen is generally known to enhance the activity of eosinophil and inhibit the production of cortisol [73].

7. Cancer and Atopic Dermatitis

The association between cancer and atopic dermatitis remains controversial. A recent meta-analysis showed that there was a small but significantly increased risk of lymphoma in the AD cohort and that AD severity is a significant risk factor [74]. On the other hand, a meta-analysis found no association with acute myelogenous leukemia in two studies and an inverse association between AD and acute lymphoblastic leukemia in six studies [75]. In a prospective adult Dutch cohort study, Hajdarbegovic et al. found an inverse association between AD and actinic keratosis but no association with non-melanoma skin cancer (NMSC) [76]. In a recent systematic review, AD was found to be associated with a significantly elevated risk of NMSC but with a non-significantly decreased risk of melanoma. However, Karim et al. showed that allergic diseases appeared to reduce the risk of developing skin cancers [77,78]. No association was found between AD and brain, pancreatic, or bladder cancer [79].

The relationship between allergies and cancer is based on four hypotheses: chronic inflammation, inappropriate T-helper 2, immune-skewing immunosurveillance, and prophylaxis [80].

8. Cutaneous Infections

AD is associated with cutaneous bacterial infections and more severe forms of cutaneous viral infections. A recent large study concluded that pediatric and adult AD patients were associated with significantly higher odds of erysipelas, impetigo, carbuncle/furuncle, cellulitis, methicillin-resistant and methicillin-sensitive Staphylococcus aureus infections, herpes simplex, molluscum contagiosum, human papillomavirus, zoster virus, dermatophytosis, and candidiasis of the skin/nails [81].

8.1. Bacterial Infections

Approximately 80–90% of patents with AD are colonized with *S aureus*, which produces enterotoxins that are known to break down the skin barrier and enhance type 2 inflammation [82]. Clinically, features that suggest secondary bacterial skin infection are extensive disease, and oozing, crusting, or asymmetrical lesions [83]. Doctors should also be aware of the existence of a rare syndrome that associates dermatitis with recurrent skin infections. Job syndrome is a rare, primary immunodeficiency characterized by the clinical presentation of AD, staphylococcal infections, and recurrent pulmonary infections. The disease is distinguished by elevated IgE levels with an early onset in primary childhood and is classified into two types: type I (autosomal dominant hyper IgE syndrome), with abnormalities in multiple systems, the including skeletal, dental, and immune systems; and type II (autosomal recessive hyper IgE syndrome), with no musculoskeletal alterations [84].

Recent studies have shown that local treatment with topical corticosteroids, emollients, and bleach baths resulted in decreased *S. aureus* and increased microbial diversity, including *Corynebacterium*, *Propionibacterium*, or *Streptoccocus*. Fundamental knowledge about RNA, in association with the considerable technological advances in recent years provided by cutting-edge technologies in molecular biology, has led to biotechnological and medical applications of RNA, including in AD. We identified some interesting studies (see Table 1) which using mRNA analysis for different bacterial species to optimize AD therapy [85–89].

Table 1. Associations between AD and bacterial mRNA analysis.

Authors, Year, Country	Sample Type/Objective of the Study	mRNA Analysis Assay	Results	Novelty
Nakatsuji, T. et al., 2021, CA, USA	The safety and mechanisms of action of *Staphylococcus hominis* A9 (ShA9), a bacterium isolated from healthy human skin, as a topical therapy for AD.	Expression of mRNA for psmα	Improvement in local eczema severity was suggested by post hoc analysis of participants with S. aureus directly killed by ShA9.	The safety and potential benefits of bacteriotherapy for AD.
Jeong, D.Y. et al., 2020, Sunchang, South Korea	The efficacy of *Pediococcus acidilactici* SRCM102024 (PA) for treatment of AD in HaCaT cells and NC/Nga mice.	mRNA expression of PAR-2, TNF-α, IL-4, and IL-13	Oral treatment of PA relieved AD symptoms in a dose-dependent manner, preventing overactivation of the immune response.	Oral PA intake may be a safe and effective alternative therapyfor AD.
Moran, M.C. et al., 2019, New York, USA	Virulence factors can be directly detected at the protein level from human samples.	mRNA analysis	Staphylococcal virulence factors can be quantified at the protein level directly from skin swabs of atopic dermatitis patients.	Secreted virulence factors are present on the skin of atopic patients, as well as the physiological impact of *S. aureus* in inflammatory diseases, such as atopic dermatitis.

8.2. Viral Infections

Approximately 20% of patients with AD develop eczema herpeticum, an acute, life-threatening viral infection caused by herpes simplex virus [90]. Recurrence was associated with an earlier onset of AD but not with total IgE serum levels. This form of Herpes *simplex* infection is positively associated with onset of AD at an early age, more severe disease, greater T-helper 2 polarity, and other skin infections [91]. Molluscum contagiosum is another viral infection frequently associated with higher recurrence of AD flares is some pediatric patients, who are more likely to have a family history of atopy [92].

8.3. Fungal Infection

Malassezia spp. is common in AD patients, with sensitization rates particularly higher in cases with head and neck types of AD. Some recent studies investigated a possible correlation between IgE-mediated sensitization to *Malassesia* spp. and disease severity. This correlation seems to be more frequent in adults than in children [93].

Filaggrin mutations could increase the risk of *human papilloma virus* in the cervical mucosa. Immunosuppressive drugs used to treat AD were investigated with respect to the associated risk of this type of infection. It seems that AD patients treated with oral immunosuppressive therapy have no additional risk of developing cervical neoplasia [94].

Children with AD have a higher prevalence of extracutaneous infections, such as streptococcal pharyngitis, recurrent otitis, urinary tract infections, and infective endocarditis [95]. Most recently, the impact of COVID-19 on patients with AD was established by specific questionnaires that highlighted the heavy burden placed on patients during such periods. Predictive factors were identified, such as sex (females tend to be more vulnerable) and age (individuals under 40 years old). Reports also suggest that lower education level, unemployment, and poor economic status were significant risk factors for developing depressive symptoms during the pandemic period in the general population [96].

In terms of dermatologic conditions, AD has been associated with alopecia areata, vitiligo, dermato/polymyositis, cellulitis, actinic keratoses, systemic lupus erythematosus, connective tissue disease, Sjogren syndrome, and seborrheic dermatitis [97].

In a recent study that included 70,584 individuals with AD and 270,783 controls, eating disorders were positively associated with AD, with the highest association with bulimia nervosa, followed by binge eating disorders [98].

Comorbidities were shown to differ according to geographical region, for example, cardiovascular comorbidities were less registered in Europe (large German, Danish, and Canadian investigations) that in the USA. This could be explained by different therapeutic options available in different countries, divergent diagnostic criteria, age of patients, or lifestyle factors (e.g., smoking and diet) [99]. Other studies have shown that children had a higher prevalence of infections, whereas adults had a higher prevalence of mental disorders [100]. Large variations in the prevalence of comorbidities in countries with ethnically homogenous populations could be explained by the fact that, alongside genetic predisposition, environmental factors play an important role in the pathogenesis of the disease.

9. Conclusions

Recent studies have redefined the concept of AD, which is no longer considered a disease that affects only the skin but a systemic illness; therefore, the concept of classic atopic may be further expanded. The causative mechanisms underlying AD and its associated comorbidities are largely unknown, but many of these conditions are directly connection with patient age, chronicity, or inadequate disease control. There is a growing list of potential comorbidities that suggests that patients should be treated with a multidisciplinary approach. Understanding these comorbidity patterns can help us to ensure better care of patients with AD. AD could be considered a systemic inflammatory disease with multiorgan involvement beyond the skin. Recent advancements in the understanding of the molecular pathogenesis are closely related to the development of therapeutic agents specifically targeting type 2 inflammatory pathways. Important is to mention the role of biologic drug dupilumab; its approval for the treatment of AD and of different type 2 immune mediated conditions led to the concept of treating skin manifestations together with comorbidities in AD. The introduction of dupilumab and of other recently approved targeted drugs or those in advanced stages of development has been made possible by advances in understanding of molecular mechanisms.

Author Contributions: Conceptualization: R.G.U. and E.P.-A.; methodology, C.F.C., A.C. and V.P.; resources, G.R.-Z., D.E.B., M.M.S. and A.M.D.; writing—original draft preparation E.P-A. and R.G.U. All authors have read and agreed to the published version of the manuscript.

Funding: This research received no external funding.

Institutional Review Board Statement: Not applicable.

Informed Consent Statement: Not applicable.

Data Availability Statement: Not applicable.

Conflicts of Interest: The authors declare no conflict of interest.

References

1. Asher, M.I.; Montefort, S.; Bjorksten, B.; Lai, C.K.; Strachan, D.P.; Weiland, S.K.; Williams, H.; The ISAAC Phase Three Study Group. Worldwide time trends in the prevalence of symptoms of asthma, allergic rhinoconjunctivitis, and eczema in childhood: Isaac phases one and three repeat multicountry cross-sectional surveys. *Lancet* **2006**, *368*, 733–743. [CrossRef]
2. Dhar, S.; Banerjee, R. Atopic dermatitis in infants and children in India. *Indian J. Dermatol. Venereol. Leprol.* **2010**, *76*, 504–513. [CrossRef]

3. Kramer, O.N.; Strom, M.A.; Ladizinski, B.; Lio, P.A. The history of atopic dermatitis. *Clin. Dermatol.* **2017**, *35*, 344–348. [CrossRef]
4. Henriksen, L.; Simonsen, J.; Haerskjold, A.; Linder, M.; Kieler, H.; Thomsen, S.F.; Stensballe, L.G. Incidence rates of atopic dermatitis, asthma, and allergic rhinoconjunctivitis in Danish and Swedish children. *J. Allergy Clin. Immunol.* **2015**, *136*, 360–366.e2. [CrossRef] [PubMed]
5. Chello, C.; Carnicelli, G.; Sernicola, A.; Gagliostro, N.; Paolino, G.; Di Fraia, M.; Faina, V.; Muharremi, R.; Grieco, T. Atopic dermatitis in the elderly Caucasian population: Diagnostic clinical criteria and review of the literature. *Int. J. Dermatol.* **2020**, *59*, 716–721. [CrossRef]
6. Wickett, R.R.; Visscher, M.O. Structure and function of the epidermal barrier. *Am. J. Infect. Control.* **2006**, *34*, S98–S110. [CrossRef]
7. Lyons, J.J.; Milner, J.D.; Stone, K.D. Atopic Dermatitis in Children: Clinical features, pathophysiology, and treatment. *Immunol. Allergy Clin. N. Am.* **2014**, *35*, 161–183. [CrossRef] [PubMed]
8. Maliyar, K.; Sibbald, C.; Pope, E.; Sibbald, R.G. Diagnosis and Management of Atopic Dermatitis: A Review. *Adv. Skin Wound Care* **2018**, *31*, 538–550. [CrossRef]
9. Ali, F.; Vyas, J.; Finlay, A. Counting the Burden: Atopic Dermatitis and Health-Related Quality of Life. *Acta Derm. Venereol.* **2020**, *100*, adv00161. [CrossRef]
10. Paller, A.; Jaworski, J.C.; Simpson, E.L.; Boguniewicz, M.; Russell, J.J.; Block, J.K.; Tofte, S.; Dunn, J.D.; Feldman, S.R.; Clark, A.R.; et al. Major Comorbidities of Atopic Dermatitis: Beyond Allergic Disorders. *Am. J. Clin. Dermatol.* **2018**, *19*, 821–838. [CrossRef]
11. Kakli, H.A.; Riley, T.D. Allergic Rhinitis. *Prim. Care Clin. Off. Pract* **2016**, *43*, 465–475. [CrossRef]
12. Wheatley, L.M.; Togias, A. Clinical practice. Allergic Rhinitis. *N. Engl. J. Med.* **2015**, *372*, 456–463. [CrossRef]
13. Saulyte, J.; Regueira, C.; Montes-Martínez, A.; Khudyakov, P.; Takkouche, B. Active or Passive Exposure to Tobacco Smoking and Allergic Rhinitis, Allergic Dermatitis, and Food Allergy in Adults and Children: A Systematic Review and Meta-Analysis. *PLoS Med.* **2014**, *11*, e1001611. [CrossRef] [PubMed]
14. Shaaban, R.; Zureik, M.; Soussan, D.; Neukirch, C.; Heinrich, J.; Sunyer, J.; Wjst, M.; Cerveri, I.; Pin, I.; Bousquet, J.; et al. Rhinitis and onset of asthma: A longitudinal population-based study. *Lancet* **2008**, *372*, 1049–1057. [CrossRef]
15. Okubo, K.; Kurono, Y.; Ichimura, K.; Enomoto, T.; Okamoto, Y.; Kawauchi, H.; Suzaki, H.; Fujieda, S.; Masuyama, K. Japanese Society of Allergology. Japanese guidelines for allergic rhinitis 2020. *Allergol. Int.* **2020**, *69*, 331–345. [CrossRef]
16. Padem, N.; Saltoun, C. Classification of asthma. *Allergy Asthma Proc.* **2019**, *40*, 385–388. [CrossRef]
17. Valero, A.; Pereira, C.; Loureiro, C.; Martínez-Cócera, C.; Murio, C.; Rico, P.; Palomino, R.; Dávila, I. Interrelationship between skin sensitization, rhinitis, and asthma in patients with allergic rhinitis: A study of Spain and Portugal. *J. Investig. Allergy Clin. Immunol.* **2009**, *19*, 167–172.
18. Hill, D.A.; Spergel, J.M. The atopic march: Critical evidence and clinical relevance. *Ann. Allergy Asthma Immunol.* **2018**, *120*, 131–137. [CrossRef] [PubMed]
19. Soto, M.T.S.; Patiño, A.; Nowak, D.; Radon, K. Prevalence of asthma, rhinitis and eczema symptoms in rural and urban school-aged children from Oropeza Province-Bolivia: A cross-sectional study. *BMC Pulm. Med.* **2014**, *14*, 40. [CrossRef]
20. Liezmann, C.; Klapp, B.; Peters, E. Stress, atopy and allergy. A re-evaluation from a psychoneuroimmunologic perspective. *Dermato-Endocrinology* **2011**, *3*, 37–40. [CrossRef]
21. Schut, C.; Weik, U.; Tews, N.; Gieler, U.; Deinzer, R.; Kupfer, J. Psychophysiological Effects of Stress Management in Patients with Atopic Dermatitis: A Randomized Controlled Trial. *Acta Derm. Venereol.* **2013**, *93*, 57–61. [CrossRef]
22. Slattery, M.J.; Essex, M.J. Specificity in the association of anxiety, depression, and atopic disorders in a community sample of adolescents. *J. Psychiatr. Res.* **2011**, *45*, 788–795. [CrossRef]
23. Schmitt, J.; Buske-Kirschbaum, A.; Tesch, F.; Trikojat, K.; Stephan, V.; Abraham, S.; Bauer, A.; Nemat, K.; Plessow, F.; Roessner, V. Increased attention-deficit/hyperactivity symptoms in atopic dermatitis are associated with history of antihistamine use. *Allergy* **2017**, *73*, 615–626. [CrossRef] [PubMed]
24. Treudler, R.; Zeynalova, S.; Riedel-Heller, S.; Zuelke, A.; Roehr, S.; Hinz, A.; Glaesmer, H.; Kage, P.; Loeffler, M.; Simon, J.C. Depression, anxiety and quality of life in subjects with atopic eczema in a population-based cross-sectional study in Germany. *J. Eur. Acad. Dermatol. Venereol.* **2019**, *34*, 810–816. [CrossRef] [PubMed]
25. Theodosiou, G.; Montgomery, S.; Metsini, A.; Dalgard, F.; Kobyletzki, L. Burden of Atopic Dermatitis in Swedish Adults: A Population-based Study. *Acta Derm. Venereol.* **2019**, *99*, 964–970. [CrossRef]
26. Chida, Y.; Hamer, M.; Steptoe, A. A Bidirectional Relationship between Psychosocial Factors and Atopic Disorders: A Systematic Review and Meta-Analysis. *Psychosom. Med.* **2008**, *70*, 102–116. [CrossRef] [PubMed]
27. Lee, S.; Shin, A. Association of atopic dermatitis with depressive symptoms and suicidal behaviors among adolescents in Korea: The 2013 Korean Youth Risk Behavior Survey. *BMC Psychiatry* **2017**, *17*, 3. [CrossRef]
28. Brandonisio, T.M.; Bachman, J.A.; Sears, J.M. Atopic dermatitis: A case report and current clinical review of systemic and ocular manifestations. *Optometry* **2001**, *72*, 94–102.
29. Handa, S.; Jain, A.; Kanwar, A.J.; Kaujalgi, R. Ocular abnormalities in atopic dermatitis in Indian patients. *Indian J. Dermatol. Venereol. Leprol.* **2009**, *75*, 148–151. [CrossRef]
30. Hilgartner, H.; Gilbert, J.T. A Preliminary Report of a Case of Keratoconus, Successfully Treated with Organotherapy, Radium, and Short-Wave Diathermy. *Am. J. Ophthalmol.* **1937**, *20*, 1032–1039. [CrossRef]
31. Bawazeer, A.M.; Hodge, W.G.; Lorimer, B. Atopy and keratoconus: A multivariate analysis. *Br. J. Ophthalmol.* **2000**, *84*, 834–836. [CrossRef] [PubMed]

32. Barsam, A.; Brennan, N.; Petrushkin, H.; Xing, W.; Quartilho, A.; Bunce, C.; Foot, B.; Cartwright, N.K.; Haridas, A.; Agrawal, P.; et al. Case-control study of risk factors for acute corneal hydrops in keratoconus. *Br. J. Ophthalmol.* **2016**, *101*, 499–502. [CrossRef]
33. Brémond-Gignac, D.; Nischal, K.K.; Mortemousque, B.; Gajdosova, E.; Granet, D.B.; Chiambaretta, F. Atopic Keratoconjunctivitis in Children: Clinical Features and Diagnosis. *Ophthalmology* **2015**, *123*, 435–437. [CrossRef] [PubMed]
34. Hida, T.; Tano, Y.; Okinami, S.; Ogino, N.; Inoue, M. Multicenter retrospective study of retinal detachment associated with atopic dermatitis. *Jpn. J. Ophthalmol.* **2000**, *44*, 407–418. [CrossRef]
35. Carmi, E.; Defossez-Tribout, C.; Ganry, O.; Cene, S.; Tramier, B.; Milazzo, S.; Lok, C. Ocular Complications of Atopic Dermatitis in Children. *Acta Derm. Venereol.* **2006**, *86*, 515–517. [CrossRef] [PubMed]
36. Geyer, O.; Pianka, P.; Goldstein, M.; Ben-Nun, Y. Case report: Retinal detachment associated with atopic dermatitis. *Can. Fam. Physician* **1998**, *44*, 576–577.
37. Taniguchi, H.; Ohki, O.; Yokozeki, H.; Katayama, I.; Tanaka, A.; Kiyosawa, M.; Nishioka, K. Cataract and Retinal Detachment in Patients with Severe Atopic Dermatitis Who Were Withdrawn from the Use of Topical Corticosteroid. *J. Dermatol.* **1999**, *26*, 658–665. [CrossRef]
38. Orihara, T.; Hirota, K.; Yokota, R.; Kunita, D.; Itoh, Y.; Rii, T.; Koto, T.; Hiraoka, T.; Inoue, M.; Hirakata, A. Comparison of Retinal Detachment Associated with Atopic Dermatitis with that of a Blunt Trauma. *Nippon Ganka Gakkai Zasshi* **2015**, *119*, 457–463.
39. Lee, S.H.; Lee, S.-H.; Lee, S.H.; Park, Y.L. Cataract, Glaucoma, and Dry Eye Disease in Adults with Atopic Dermatitis: A Nationwide Cross-Sectional Study from the Republic of Korea. *Ann. Dermatol.* **2019**, *31*, 37–43. [CrossRef]
40. Sasabe, T.; Suwa, Y.; Kawamura, T.; Aoki, T. Cataracts occur in patients with atopic dermatitis when the serum IgE increases. *Nippon Ganka Gakkai Zasshi* **1997**, *101*, 389–392. [CrossRef]
41. Govind, K.; Whang, K.; Khanna, R.; Scott, A.W.; Kwatra, S.G. Atopic dermatitis is associated with increased prevalence of multiple ocular comorbidities. *J. Allergy Clin. Immunol. Pract* **2018**, *7*, 298–299. [CrossRef] [PubMed]
42. Jan, R.-L.; Weng, S.-F.; Wang, J.-J.; Tseng, S.-H.; Chang, Y.-S. Association between atopic keratoconjunctivitis and the risk of corneal ulcer. *Br. J. Ophthalmol.* **2020**, *105*, 1632–1637. [CrossRef] [PubMed]
43. Weng, S.; Jan, R.; Wang, J.; Tseng, S.; Chang, Y. Association between atopic keratoconjunctivitis and the risk of keratoconus. *Acta Ophthalmol.* **2020**, *99*, e54–e61. [CrossRef]
44. Chen, J.J.; Applebaum, D.S.; Sun, G.S.; Pflugfelder, S.C. Atopic keratoconjunctivitis: A review. *J. Am. Acad. Dermatol.* **2014**, *70*, 569–575. [CrossRef] [PubMed]
45. Guglielmetti, S.; Dart, J.K.; Calder, V. Atopic keratoconjunctivitis and atopic dermatitis. *Curr. Opin. Allergy Clin. Immunol.* **2010**, *10*, 478–485. [CrossRef]
46. Pandit, S.A.; Glass, L.R.D. Non-glaucoma periocular allergic, atopic, and irritant dermatitis at an academic institution: A retrospective review. *Orbit* **2018**, *38*, 112–118. [CrossRef] [PubMed]
47. Sernicola, A.; Gattazzo, I.; Di Staso, F.; Giordano, D.; Capalbo, A.; Persechino, F.; Scuderi, G.; Persechino, S. Treatment of refractory conjunctivitis associated to dupilumab with topical pimecrolimus applied to the eyelid skin. *Dermatol. Ther.* **2019**, *32*, e13134. [CrossRef]
48. Hsu, J.I.; Pflugfelder, S.C.; Kim, S.J. Ocular complications of atopic dermatitis. *Cutis* **2019**, *104*, 189–193.
49. Kwon, H.-K.; Lee, C.-G.; So, J.-S.; Chae, C.-S.; Hwang, J.-S.; Sahoo, A.; Nam, J.H.; Rhee, J.H.; Hwang, K.-C.; Im, S.-H. Generation of regulatory dendritic cells and $CD4^+$ $Foxp3^+$ T cells by probiotics administration suppresses immune disorders. *Proc. Natl. Acad. Sci. USA* **2010**, *107*, 2159–2164. [CrossRef] [PubMed]
50. Tsakok, T.; McKeever, T.; Yeo, L.; Flohr, C. Does early life exposure to antibiotics increase the risk of eczema? A systematic review. *Br. J. Dermatol.* **2013**, *169*, 983–991. [CrossRef]
51. Li, W.; Xu, X.; Wen, H.; Wang, Z.; Ding, C.; Liu, X.; Gao, Y.; Su, H.; Zhang, J.; Han, Y.; et al. Inverse Association between the Skin and Oral Microbiota in Atopic Dermatitis. *J. Investig. Dermatol.* **2019**, *139*, 1779–1787.e12. [CrossRef] [PubMed]
52. Lee, S.-Y.; Lee, E.; Park, Y.M.; Hong, S.-J. Microbiome in the Gut-Skin Axis in Atopic Dermatitis. *Allergy Asthma Immunol. Res.* **2018**, *10*, 354–362. [CrossRef] [PubMed]
53. Lee, E.; Lee, S.-Y.; Kang, M.-J.; Kim, K.J.; Won, S.; Kim, B.-J.; Choi, K.Y.; Kim, B.-S.; Cho, H.-J.; Kim, Y.; et al. Clostridia in the gut and onset of atopic dermatitis via eosinophilic inflammation. *Ann. Allergy Asthma Immunol.* **2016**, *117*, 91–92.e1. [CrossRef]
54. Salem, I.; Ramser, A.; Isham, N.; Ghannoum, M.A. The Gut Microbiome as a Major Regulator of the Gut-Skin Axis. *Front. Microbiol.* **2018**, *9*, 1459. [CrossRef]
55. Xue, Q.; Li, X.; Li, Y.; Xu, J.; Wu, Z.; Wang, J. Dialogue between gastrointestinal tract and skin: New insights into the Helicobacter pylori and atopic dermatitis. *Helicobacter* **2020**, *26*, e12771. [CrossRef]
56. Navarro-Lopez, V.; Ramirez-Bosca, A.; Ramon-Vidal, D.; Ruzafa-Costas, B.; Genoves-Martinez, S.; Chenoll-Cuadros, E.; Carrión-Gutiérrez, M.; Horga de la Parte, J.; Prieto-Merino, D.; Codoñer-Cortés, F.M. Effect of oral administration of a mixture of probiotic strains on SCORAD index and use of topical steroids in young patients with moderate atopic dermatitis a randomized clinical trial. *JAMA Dermatol.* **2018**, *154*, 37–43. [CrossRef] [PubMed]
57. Wu, Y.J.; Wu, W.F.; Hung, C.W.; Ku, M.S.; Liao, P.F.; Sun, H.L.; Lu, K.H.; Sheu, J.N.; Lue, K.H. Evaluation of efficacy and safety of Lactobacillus rhamnosus in children aged 4–48 months with atopic dermatitis: An 8-week, double-blind, randomized, placebo-controlled study. *J. Microbiol. Immunol. Infect.* **2017**, *50*, 684–692. [CrossRef]
58. Cabana, M.D.; McKean, M.; Caughey, A.B.; Fong, L.; Lynch, S.; Wong, A.; Leong, R.; Boushey, H.A.; Hilton, J.F. Early Probiotic Supplementation for Eczema and Asthma Prevention: A Randomized Controlled Trial. *Pediatrics* **2017**, *140*, e20163000. [CrossRef]

59. Nakata, J.; Hirota, T.; Umemura, H.; Nakagawa, T.; Kando, N.; Futamura, M.; Nakamura, Y.; Ito, K. Additive effect of *Lactobacillus acidophilus* L-92 on children with atopic dermatitis concomitant with food allergy. *Asia Pac. Allergy* **2019**, *9*, e18. [CrossRef]
60. Islamoğlu, Z.G.K.; Unal, M.; Küçük, A. Atopic dermatitis in adults and irritable bowel syndrome: A cross-sectional study. *Indian J. Dermatol.* **2019**, *64*, 355–359. [CrossRef]
61. Zhang, A.; Silverberg, J.I. Association of atopic dermatitis with being overweight and obese: A systematic review and metaanalysis. *J. Am. Acad. Dermatol.* **2015**, *72*, 606–616.e4. [CrossRef] [PubMed]
62. Hjuler, K.F.; Böttcher, M.; Vestergaard, C.; Deleuran, M.; Raaby, L.; Bøtker, H.E.; Iversen, L.; Kragballe, K. Increased Prevalence of Coronary Artery Disease in Severe Psoriasis and Severe Atopic Dermatitis. *Am. J. Med.* **2015**, *128*, 1325–1334.e2. [CrossRef] [PubMed]
63. Su, V.Y.-F.; Chen, T.-J.; Yeh, C.-M.; Chou, K.-T.; Hung, M.-H.; Chu, S.-Y.; Su, K.-C.; Chang, Y.-S.; Lin, Y.-H.; Liu, C.-J. Atopic dermatitis and risk of ischemic stroke: A nationwide population-based study. *Ann. Med.* **2014**, *46*, 84–89. [CrossRef]
64. Standl, M.; Tesch, F.; Baurecht, H.; Rodríguez, E.; Müller-Nurasyid, M.; Gieger, C.; Peters, A.; Wang-Sattler, R.; Prehn, C.; Adamski, J.; et al. Association of Atopic Dermatitis with Cardiovascular Risk Factors and Diseases. *J. Investig. Dermatol.* **2016**, *137*, 1074–1081. [CrossRef]
65. Andersen, Y.M.; Egeberg, A.; Gislason, G.H.; Skov, L.; Knop, F.K.; Thyssen, J.P. Adult atopic dermatitis and the risk of type 2 diabetes. *J. Allergy Clin. Immunol.* **2017**, *139*, 1057–1059. [CrossRef]
66. Donath, M.Y.; Shoelson, S.E. Type 2 diabetes as an inflammatory disease. *Nat. Rev. Immunol.* **2011**, *11*, 98–107. [CrossRef]
67. Kusari, A.; Han, A.M.; Schairer, D.; Eichenfield, L.F. Atopic Dermatitis: New Developments. *Dermatol. Clin.* **2018**, *37*, 11–20. [CrossRef]
68. Chen, Y.; Zhu, J.; Lyu, J.; Xia, Y.; Ying, Y.; Hu, Y.; Qu, J.; Tong, S.; Li, S. Association of Maternal Prepregnancy Weight and Gestational Weight Gain with Children's Allergic Diseases. *JAMA Netw. Open* **2020**, *3*, e2015643. [CrossRef]
69. Silverberg, J.I.; Kleiman, E.; Lev-Tov, H.; Silverberg, N.B.; Durkin, H.G.; Joks, R.; Smith-Norowitz, T.A. Association between obesity and atopic dermatitis in childhood: A case-control study. *J. Allergy Clin. Immunol.* **2011**, *127*, 1180–1186.e1. [CrossRef] [PubMed]
70. Grieco, T.; Dies, L.; Sernicola, A.; Chello, C.; Gagliostro, N.; Carnicelli, G.; Paolino, G. Potential clinical and serological predictors of chronic spontaneous urticaria relapse in patients under omalizumab treatment. *Immunotherapy* **2020**, *12*, 1173–1181. [CrossRef]
71. Silverberg, J.I.; Greenland, P. Eczema and cardiovascular risk factors in 2 US adult population studies. *J. Allergy Clin. Immunol.* **2015**, *135*, 721–728.e6. [CrossRef]
72. Ascott, A.; Mansfield, K.; Schonmann, Y.; Mulick, A.; Abuabara, K.; Roberts, A.; Smeeth, L.; Langan, S. Atopic eczema and obesity: A population-based study. *Br. J. Dermatol.* **2020**, *184*, 871–879. [CrossRef]
73. Shalom, G.; Dreiher, J.; Kridin, K.; Horev, A.; Khoury, R.; Battat, E.; Freud, T.; Comaneshter, D.; Cohen, A. Atopic dermatitis and the metabolic syndrome: A cross-sectional study of 116 816 patients. *J. Eur. Acad. Dermatol. Venereol.* **2019**, *33*, 1762–1767. [CrossRef]
74. Legendre, L.; Barnetche, T.; Mazereeuw-Hautier, J.; Meyer, N.; Murrell, D.; Paul, C. Risk of lymphoma in patients with atopic dermatitis and the role of topical treatment: A systematic review and meta-analysis. *J. Am. Acad. Dermatol.* **2015**, *72*, 992–1002. [CrossRef]
75. Linabery, A.M.; Jurek, A.M.; Duval, S.; Ross, J.A. The Association between Atopy and Childhood/Adolescent Leukemia: A Meta-Analysis. *Am. J. Epidemiol.* **2010**, *171*, 749–764. [CrossRef] [PubMed]
76. Hajdarbegovic, E.; Blom, H.; Verkouteren, J.; Hofman, A.; Hollestein, L.; Nijsten, T. Atopic dermatitis is not associated with actinic keratosis: Cross-sectional results from the Rotterdam study. *Br. J. Dermatol.* **2016**, *175*, 89–94. [CrossRef]
77. Zhu, Y.; Wang, H.; He, J.; Yang, L.; Zhou, X.; Li, Z.; Zhou, H.; Zhao, H.; Li, Y. Atopic Dermatitis and Skin Cancer Risk: A Systematic Review. *Dermatol. Ther.* **2022**, *12*, 1167–1179. [CrossRef]
78. Karim, A.F.; Westenberg, L.E.H.; Eurelings, L.E.M.; Otten, R.; Gerth van Wijk, R. The association between allergic diseases and cancer: A systematic review of the literature. *Neth. J. Med.* **2019**, *77*, 42–66.
79. Deckert, S.; Kopkow, C.; Schmitt, J. Nonallergic comorbidities of atopic eczema: An overview of systematic reviews. *Allergy* **2013**, *69*, 37–45. [CrossRef]
80. Josephs, D.H.; Spicer, J.F.; Corrigan, C.; Gould, H.; Karagiannis, S.N. Epidemiological associations of allergy, IgE and cancer. *Clin. Exp. Allergy* **2013**, *43*, 1110–1123. [CrossRef]
81. Ren, Z.; Silverberg, J.I. Association of Atopic Dermatitis With Bacterial, Fungal, Viral, and Sexually Transmitted Skin Infections. *Dermatitis* **2020**, *31*, 157–164. [CrossRef] [PubMed]
82. Wang, V.; Boguniewicz, J.; Boguniewicz, M.; Ong, P.Y. The infectious complications of atopic dermatitis. *Ann. Allergy Asthma Immunol.* **2020**, *126*, 3–12. [CrossRef] [PubMed]
83. Girolomoni, G.; Busà, V.M. Flare management in atopic dermatitis: From definition to treatment. *Ther. Adv. Chronic Dis.* **2022**, *13*, 20406223211066728. [CrossRef] [PubMed]
84. Hafsi, W.; Yarrarapu, S.N.S. Job Syndrome. 31 August 2021. In *StatPearls [Internet]*; StatPearls Publishing: Treasure Island, Finland, 2022.
85. Nakatsuji, T.; Hata, T.R.; Tong, Y.; Cheng, J.Y.; Shafiq, F.; Butcher, A.M.; Salem, S.S.; Brinton, S.L.; Spergel, A.K.R.; Johnson, K.; et al. Development of a human skin commensal microbe for bacteriotherapy of atopic dermatitis and use in a phase 1 randomized clinical trial. *Nat. Med.* **2021**, *27*, 700–709. [CrossRef]

86. Jeong, D.-Y.; Ryu, M.-S.; Yang, H.-J.; Jeong, S.-Y.; Zhang, T.; Yang, H.J.; Kim, M.J.; Park, S. Pediococcus acidilactici intake decreases the clinical severity of atopic dermatitis along with increasing mucin production and improving the gut microbiome in Nc/Nga mice. *Biomed. Pharmacother.* **2020**, *129*, 110488. [CrossRef]
87. Moran, M.C.; Cahill, M.P.; Brewer, M.G.; Yoshida, T.; Knowlden, S.; Perez-Nazario, N.; Schlievert, P.M.; Beck, L.A. Staphylococcal Virulence Factors on the Skin of Atopic Dermatitis Patients. *mSphere* **2019**, *4*, e00616-19. [CrossRef]
88. Porumb-Andrese, E.; Ursu, R.G.; Ivanov, I.; Caruntu, I.-D.; Porumb, V.; Ferariu, D.; Damian, C.; Ciobanu, D.; Terinte, C.; Iancu, L.S. The BRAF V600E Mutation Detection by quasa Sensitive Real-Time PCR Assay in Northeast Romania Melanoma Patients. *Appl. Sci.* **2021**, *11*, 9511. [CrossRef]
89. Porumb-Andrese, E.; Scutariu, M.M.; Luchian, I.; Schreiner, T.G.; Mârţu, I.; Porumb, V.; Popa, C.G.; Sandu, D.; Ursu, R.G. Molecular Profile of Skin Cancer. *Appl. Sci.* **2021**, *11*, 9142. [CrossRef]
90. Traidl, S.; Roesner, L.; Zeitvogel, J.; Werfel, T. Eczema herpeticum in atopic dermatitis. *Allergy* **2021**, *76*, 3017–3027. [CrossRef]
91. Silverberg, J.I. Comorbidities and the impact of atopic dermatitis. *Ann. Allergy Asthma Immunol.* **2019**, *123*, 144–151. [CrossRef]
92. Silverberg, N.B. Molluscum contagiosum virus infection can trigger atopic dermatitis disease onset or flare. *Cutis* **2018**, *102*, 191–194. [PubMed]
93. Glatz, M.; Bosshard, P.P.; Hoetzenecker, W.; Schmid-Grendelmeier, P. The Role of *Malassezia* spp. in Atopic Dermatitis. *J. Clin. Med.* **2015**, *4*, 1217–1228. [CrossRef]
94. Garritsen, F.; Verheijen, R.; Gerestein, C.; Van Zuilen, A.; Oosterhaven, J.; Van Dijk, M.; Bruijnzeel-Koomen, C.; Schuttelaar, M.; De Bruin-Weller, M. Is there an increased risk of cervical neoplasia in atopic dermatitis patients treated with oral immunosuppressive drugs? *J. Eur. Acad. Dermatol. Venereol.* **2017**, *32*, 271–275. [CrossRef] [PubMed]
95. Brunner, P.M.; Silverberg, J.I.; Guttman-Yassky, E.; Paller, A.S.; Kabashima, K.; Amagai, M.; Luger, T.A.; Deleuran, M.; Werfel, T.; Eyerich, K.; et al. Increasing Comorbidities Suggest that Atopic Dermatitis Is a Systemic Disorder. *J. Investig. Dermatol.* **2017**, *137*, 18–25. [CrossRef] [PubMed]
96. Xiong, J.; Lipsitz, O.; Nasri, F.; Lui, L.M.W.; Gill, H.; Phan, L.; Chen-Li, D.; Iacobucci, M.; Ho, R.; Majeed, A.; et al. Impact of COVID-19 pandemic on mental health in the general population: A systematic review. *J. Affect. Disord.* **2020**, *277*, 55–64. [CrossRef]
97. Roh, Y.S.; Huang, A.H.; Sutaria, N.; Choi, U.; Wongvibulsin, S.; Choi, J.; Bordeaux, Z.A.; Parthasarathy, V.; Deng, J.; Patel, D.P.; et al. Real-world comorbidities of atopic dermatitis in the US adult ambulatory population. *J. Am. Acad. Dermatol.* **2021**, *86*, 835–845. [CrossRef]
98. Kauppi, S.; Jokelainen, J.; Timonen, M.; Tasanen, K.; Huilaja, L. Atopic dermatitis and the risk of eating disorders: A population-based cohort study. *J. Am. Acad. Dermatol.* **2021**, *87*, 474–476. [CrossRef]
99. Zeynalova, N.; Schimpf, S.; Setter, C.; Yahiaoui-Doktor, M.; Zeynalova, S.; Lordick, F.; Loeffler, M.; Hinz, A. The association between an anxiety disorder and cancer in medical history. *J. Affect. Disord.* **2019**, *246*, 640–642. [CrossRef] [PubMed]
100. Bekić, S.; Martinek, V.; Talapko, J.; Majnarić, L.; Vasilj Mihaljević, M.; Škrlec, I. Atopic Dermatitis and Comorbidity. *Healthcare* **2020**, *8*, 70. [CrossRef] [PubMed]

diagnostics

MDPI

Case Report

The Benefits and Challenges of the Multimodal Treatment in Advanced/Metastatic Malignant Melanoma

Alexandru-Rares Stoian [1,2,†], Gabriela Rahnea-Nita [1,3,*], Anda-Natalia Ciuhu [3,†], Laurentia Gales [1,4,†], Rodica-Maricela Anghel [1,4], Laura-Florentina Rebegea [5,6,7], Roxana-Andreea Rahnea-Nita [1,3,†], Liliana-Florina Andronache [1,†], Ioana Soare [8,†] and Gabriela Stoleriu [6]

1 Clinical Department, Faculty of Medicine, The University of Medicine and Pharmacy "Carol Davila", 8 Eroii Sanitari Street, 050474 Bucharest, Romania
2 "Bagdasar-Arseni" Emergency Clinical Hospital, 041915 Bucharest, Romania
3 "Sf. Luca" Chronic Disease Hospital, 041915 Bucharest, Romania
4 The Oncological Institute "Prof. Dr. Alexandru Trestioreanu", 022328 Bucharest, Romania
5 Radiotherapy Department, "Sf. Ap. Andrei" County Emergency Clinical Hospital, 800579 Galati, Romania
6 Clinical Department, Faculty of Medicine and Pharmacy, "Dunarea de Jos" University of Galati, 800008 Galati, Romania
7 Research Center in the Field of Medical and Pharmaceutical Sciences, ReFORM-UDJ, 800010 Galati, Romania
8 Clinical Department, The Faculty of Medicine, "Titu Maiorescu" University, 040051 Bucharest, Romania
* Correspondence: gabriela_rahnea@yahoo.com; Tel.: +40-213343010
† These authors contributed equally to this work.

Abstract: Currently, the treatment of malignant melanoma offers the longest and the most studied experience of innovative treatments in malignant pathology. The algorithm of the therapeutic decision in advanced or metastatic melanoma must comprise: the timing of the therapeutic initiation, the sequencing of the specific oncological treatment (radiotherapy and chemotherapy still being therapeutic alternatives in selected cases), the diagnosis and the management of adverse reactions. We present the case of a patient diagnosed with metastatic malignant melanoma in November 2019, who progressed successively under new systemic treatment throughout the 3 years of treatment and experienced skin reactions of various degrees of severity. The comprehensive response to secondary hilar pulmonary lymphatic determinations under subsequent chemotherapy was specific to the presented case. The occurrence of vitiligo secondary to immunotherapy is a favorable prognostic factor, but the occurrence of secondary cerebral determinations is an extremely severe prognostic factor in malignant melanoma and a challenge in making the therapeutic decision. Previous treatment with immune checkpoint inhibitors may trigger a favorable response to systemic chemotherapy. The early and accurate diagnosis of the adverse events of the new therapies requires a multidisciplinary approach, because it can radically change the therapeutic decision.

Keywords: malignant melanoma; new therapeutic approach; chemotherapy; adverse events

1. Introduction

Malignant melanoma (MM) has shown an increase in incidence in the last decade, with exposure to ultraviolet radiation as the main triggering factor [1–3]. In a study conducted in England, an increase in the incidence of the disease from an average of 837 cases/year between 1981–1985 to 6963 cases/year between 2016–2018 was highlighted [1,2]. On the other hand, in a study carried out in the United States, a rapid decrease in mortality was highlighted through the efficiency of immune and targeted therapies in malignant melanoma [4].

Described since the 14th century B.C.E. (before the Common Era) in the mummies of Peru, malignant melanoma started to be recognized, named and subsequently studied beginning in the 1800s when Rene Laennec introduced the term "melanosis" and Sir

Robert Carswell named the malignant pigment lesions of the skin "melanoma" [5,6]. Since 1975, specific systemic therapies for malignant melanoma have been developed, including the approval of Dacarbazine and, subsequently, treatments with high-dose alpha interferon (IFN-α) and high-dose interleukin-2 (IL-2) [6]. Since the 2010s, progress in systemic treatment has become exponential, in response to decades of molecular and immunologic studies of cancers [7,8].

The mechanism of action of Dacarbazine, a standard treatment for malignant melanoma for more than 30 years, has not been elucidated yet [9]. It is considered to have a cytotoxic effect due to its structure as an alkylating agent and because it inhibits deoxyribonucleic acid (DNA) synthesis through its action as a purine analogue [9]. Il-2, through the activation and proliferation of T lymphocytes and the stimulation of natural killer (NK) cells, fostered a real interest in the treatment of cancers at the end of the 1990s, but it demonstrates a response rate of only 15% in metastatic renal cancers and metastatic malignant carcinoma. In high doses, IL-2 has demonstrated a clinical response of up to 50% in the with malignant melanoma with lymphopenia [9]. Type I and II interferons (IFNs) present antitumor action through signal transducer and activator of transcription (STAT) 1 and 2 activations and related mechanisms: the activation of mitogen-activated protein kinase (MAPK) pathway, of mammalian target of rapamycin (mTOR) pathway and of nucleotide Guanosine TriPhosphate (GTP)ases/cyclin-dependent kinases pathway. IFN-α stimulates tumor immunogenicity and the response of antitumor dendritic cells through the polarization, maturation and survival of dendritic cells. Therefore, IFNs have a proapoptotic, anti-proliferative, antiangiogenic and potent immunoregulatory effect [10,11].

Starting from the theory that tumor cells manage to "fool" the immune system through the "peripheral tolerance" mechanism—a process of regulating the activation of T lymphocytes by which autoimmunization is prevented, the treatment of advanced/metastatic malignant melanoma is being revolutionized by immune checkpoint inhibitors. An important role in this process is played by immune control pathways: cytotoxic T-lymphocyte–associated antigen 4 (CTLA-4) and programmed death receptor-1 (PD-1) [12,13]. The CTLA-4 pathway inhibits the autoreactive potential in the initial stage of activation of naïve T cells, at the level of lymph nodes, while the PD-1 pathway regulates this process, at the level of peripheral tissues, in the late phase of the immune process, on the previously activated T cell. PD-1 is expressed both at the level of T lymphocytes and at the level of B lymphocytes, monocytes, dendritic cells and tumor-infiltrating lymphocytes (TILs). By linking to its protein programmed cell death-ligands 1 and 2 (PD-L1 and PD-L2), PD-1 inhibits the activity of T cells [12–14]. Tumor cells from malignant melanoma overexpress PD-L1 to escape the inflammatory process. Thus, the treatment of advanced/metastatic malignant melanoma was revolutionized starting in the 2010s through: Ipilimumab (recombinant humanized monoclonal antibody that binds to CTLA-4), Pembrolizumab and Nivolumab (direct anti PD-1 monoclonal antibodies), and Atezolizumab (fully humanized, engineered monoclonal antibody of IgG1 isotype against PD-L1), including the association between anti-CTLA-4 and anti-PD-1 [12,13].

BRAF, named also as NS7; B-raf; BRAF1; RAFB1; B-RAF1; BRAF-1, is a B-Raf proto-oncogene, serine/threonine kinase, the gene that encodes a protein belonging to the RAF family of serine/threonine protein kinases, located on chromosome 7q34. It was noticed that 40–60% of the patients with malignant melanoma present BRAF(h) gene mutations— most frequently, the modification of a single amino acid at the level of codon 600 (BRAFV600E). The oncogenic effect of BRAF mutation occurs by activating the MAPK pathway/extracellular signal-regulated kinase (ERK), named MAPK/ERK kinase or MEK. MEK is activated by MEK phosphorylation and subsequently ERK phosphorylation [15–18]. This characteristic determines the sensitivity of these tumors to targeted therapies with the inhibition of the MAPK pathway. Thus, BRAF inhibitors (Dabrafenib, Vemurafenib, Encorafenib) demonstrated an objective antitumor response in approximately 50% of the patients with malignant melanoma, but with a short duration of approximately 6 months due to mechanisms of resistance to treatment [15,16,19]. The simultaneous inhibition of

BRAF kinase and MEK (trametinib, cobimetinib, binimetinib) determines long-term responses, an increased tumor response rate and a paradoxical decrease in skin toxicity, thus becoming the standard treatment in malignant melanoma with present BRAF mutations, with increases in the disease interval until progression of up to 1.8 months [20].

2. Materials and Methods

A 52-year-old patient from a rural area, with an occupational history of exposure to sun, presented to our Oncology Department 3 years ago. At that time, the patient had undergone a surgical treatment for a 2/2 cm pigmented ulcerated skin tumor formation located at the level of the left basal posterior chest, 0.5 cm superior to a post-resection scar from a similar skin lesion 12 months earlier (Figure 1).

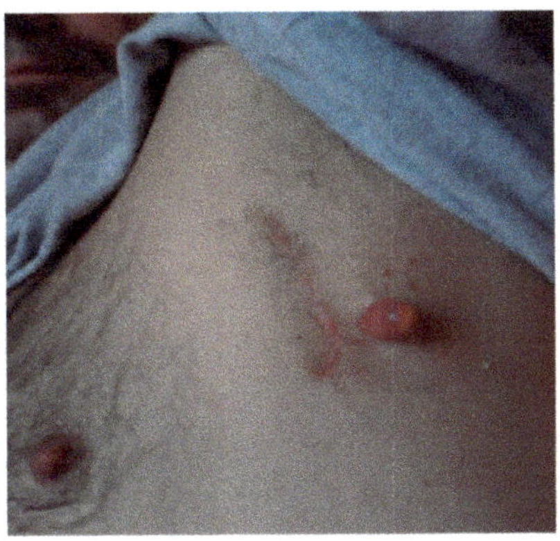

Figure 1. Melanoma relapse, at the time of the patient's admission to the Surgical Department of our team.

Histopathological and immunohistochemical tests revealed nodular pigmented malignant melanoma, Breslow 15 mm (local staging pT4b), Clark V, 8 mitoses/mm^2, in hot-spot, "non-brisk" lymphocytic inflammatory infiltrates, presenting angiotropism, without any lymphatic-vascular invasion, absent neurotropism, with V600E mutation detected in the BRAF gene (c.1799T > A). The pre-therapeutic assessment comprised a computerized tomography (CT) scan of the brain, thorax, abdomen and pelvis with contrast medium, which revealed a thoraco-abdominal aortic aneurysm and lymphoscintigraphy with surgical removal of the left axillary sentinel nodes. A complete left axillary excision was performed, followed by the histopathological and immunohistochemical analysis of the excised sentinel lymph nodes (3 lymph nodes with MM metastases, the largest metastatic mass of 18 mm, subcapsular and extramedullary location, presenting extra lymph node extension, p. V600E mutation in the present BRAF gene), pN2a stage. A positron emission tomography (PET) scan/CT assessment at approximately 1 month after the axillary lymph node removal pointed to a metabolically active upper left perijugular adenopathy with a diameter of 29/30 mm, moderately fixed lymph nodes for 18Fluor-fluorodeoxyglucose (18F-FDG) PET located bilaterally infrahilar and hilar pulmonary, possibly with a non-specific inflammatory substrate; non-capturing pulmonary micronode of 5 mm, medial segment, middle lobe. The subsequent therapeutic decision was established by the multidisciplinary board, taking into account the extremely difficult surgical approach for the metabolically active perijugular lesion and the coronavirus disease 2019 (COVID-19) pandemic at the onset with severe restrictive health regimen, deciding to initiate the systemic oncological treatment. The targeted treatment for the BRAF mutation was initiated with Dabrafenib in the standard dose (150 mg BID (taken two times a day)) and Trametinib 2 mg/day. The patient underwent this treatment for 6 months, with very good clinical tolerance, with a few febrile spikes at the beginning of the treatment that were controlled with antipyretic treatment. The stage reassessment after 6 months, by magnetic resonance imaging (MRI) of the cervical area with contrast medium and the CT of the thorax, abdomen and pelvis with contrast medium, described the progression of the disease at the level of the lymph nodes—a mass with a structure and non-homogeneous gadolinium of 30/22 mm in the axial plane and 44 mm in the craniocaudal diameter, located in the left jugular carotid, with a mass effect on the jugular vein—MRI appearance of adenopathy; multiple bilateral superior latero-cervical adenopathies of 10 mm, submandibular of 13 mm, bilateral deep latero-cervical of 11 mm, bilateral posterior laterocervical of 7 mm, lymph nodes of maximum 22/8 mm in Barety's space, infracentimeters in the aorto-pulmonary window, lateral-aortic, right hilar of 20/13 mm.

For the next 10 months, the patient underwent treatment with Nivolumab 240 mg every 14 days, with a temporary interruption of the treatment for 6 weeks due to an increase in serum transaminase values > 5xN, until the correction of the values under corticotherapy. The occurrence of skin depigmentation was also noticed in the last 3 months of nivolumab administration, predominantly at the level of the face and bilateral palms with a clinical aspect of "vitiligo" (Figure 2), and at the level of the distal lateral extremity of the postoperative scar, a 5 mm brown-blue lesion with a clinical aspect of malignant melanoma relapse.

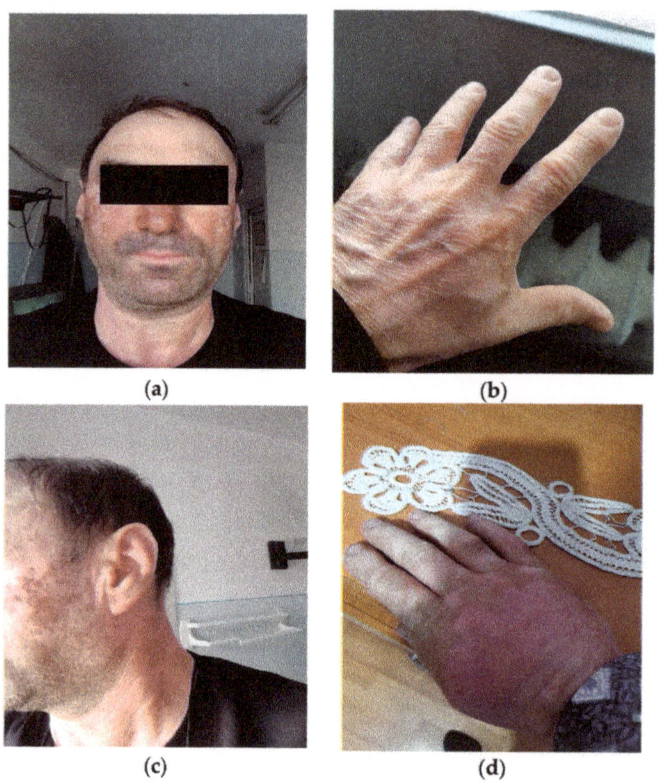

Figure 2. Vitiligo-like lesions in the areas exposed to the sun: (**a**) face and forehead; (**b**) dorsal side of the left hand; (**c**) face, forehead and left side of the neck; (**d**) dorsal side of the right hand.

After 10 months of treatment, the imaging evaluations, the CT scan of the cervical, thoracic and abdominal area with contrast medium, the ultrasound of soft tissues in the left lateral cervical area (Figure 3), and the angiography and the MRI of the cervical area with contrast medium revealed the increase in the tumor mass located in the left jugular carotid area, with no progression of the other lymph node lesions described in previous examinations (an image with adenopathy including necrotic areas, medial to the left jugular carotid vascular mass of 58.7/26/24 mm, with a structure similar to adenomegalies upon MRI examination). The patient was re-evaluated by the therapeutic indication board, and a surgical approach was recommended for the lesions in progression. The histopathological and immunohistochemical tests of the tumor lesion in the left cervical area revealed Sry-related HMg-Box gene 10 (S100 SOX10) intensely positive in the tumor proliferation, epithelial membrane antigen (EMA)—intensely positive at the level of the fibrous capsule, Ki67 < 5%—Schwannoma, and for the previously described skin formation—malignant melanoma relapse. The patient continued the treatment with Nivolumab, in the same therapeutic sequence for 4 more months, until the next PET/CT imaging evaluation when the metabolic and dimensional progression of a left lung adenopathy up to 22/11 mm and standardized uptake values (SUV) of 10.13 were revealed. The histopathological and immunohistochemical assessment of the lymph node lesion confirmed the progression of the disease under treatment through malignant melanoma metastasis with the presence of BRAF mutation.

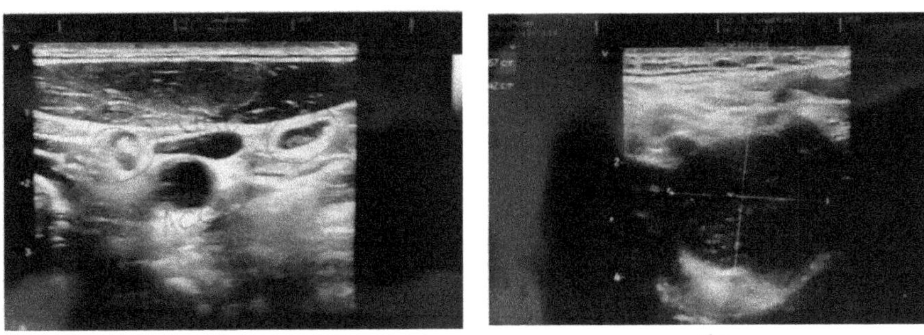

(a) (b)

Figure 3. (**a**) Two nodular images, with the hyperechoic center of 0.66 and, respectively, 0.53 cm, between external to common carotid artery (CCA) and internal jugular vein (IJV). (**b**) Hypoechoic nodular image, with Doppler signal in the hilum, the transverse diameter of 2.57/2.42 cm; longitudinal diameter of 4.54/2.14 cm, between internal carotid artery (ICA) and external carotid artery (ECA), in direct contact with the vessels.—April 2021.

After using the combination of BRAF + MEK inhibitors as the first line and treatment, with immune checkpoint inhibitors as a second line of treatment, i.e., the anti PD-1, treatment was limited to chemotherapy, anti CTLA-4 (ipilimumab), and radiotherapy (controversial)/re-challenge of the treatment with Dabrafenib + Trametinib (the BRAF + MEK inhibitor combination in Romania). In January 2022, the targeted treatment was reinitiated, keeping the standard doses of treatment—Dabrafenib 150 mg BID (taken two times a day), Trametinib 2 mg/day, but with extremely difficult clinical tolerance, manifested by daily chills, facial edema, including that of the lips, generalized pustular erythematous, pruriginous eruption, with areas of integumentary peeling and depigmentation—symptoms that are not controlled by corticoid and antihistaminic treatment. After a prolonged discontinuation of treatment for approximately 6 weeks, until the disappearance of the clinical complaints, the patient required admission to the Oncology Department for ongoing hospitalization for the titration of systemic treatment and clinical monitoring of the symptoms. The clinical oncological and dermatological evaluation upon the attempt to resume the treatment with Dabrafenib and Trametinib guided the diagnosis to allergic reactions to the treatment but could not rule out a cutaneous adverse reaction with severely increased potential.

Due to the progression of the disease in the intrathoracic ganglionic areas, especially at the level of the right pulmonary hilum, the treatment with Dacarbazine 500 mg/m^2 every

21 days was initiated from April 2022 until September 2022, when the patient exhibited partial motor epileptic seizures on the right side of the body, motor deficit of the right (crural > brachial) hemibody, dysphasia and headache. The patient's emergency assessment within the Neurosurgical Department revealed a left parasagittal frontal-parietal cerebral tumor with a cystic component, of 36/50/44 mm in dimension, subsequently completely resected. Histopathological and immunochemical tests made the diagnosis of amelanotic malignant melanoma metastasis. Treatment with external stereotactic irradiation of the target volumes was completed, with total dose (TD) of 25 Gy left frontal-parietal tumor bed, using intensity-modulated radiation therapy and volumetric modulated arc therapy (IMRT-VMAT) technique, respectively, with TD 30 Gy for tumor relapse of 7 mm, revealed upon cerebral MRI assessment with contrast medium. It is worth mentioning that the thoracic imaging evaluation described the complete response to the treatment with Dacarbazine at the level of the intrathoracic lymph node areas.

Three weeks after completing radiotherapy, the patient presented again as an emergency case to the Neurosurgical Department with severe neurologic symptoms for a significant cerebral edema with thrombosis in the sagittal sinus and continuation in the evolution of the cerebral disease through secondary leptomeningeal determinations. The patient ended his struggle with the disease on Christmas day, after 2 months of "best supportive care".

3. Discussion

We consider that the peculiarities of our case are as follows:

- The description, in all of the imaging investigations carried out initially, but also in the reassessment of the treatment, of a left latero-cervical adenopathy, stages II-III, with dimensional progression under treatment with BRAF/MEK inhibitors and Nivolumab—histopathologically and immunohistochemically invalidated during the surgical verification—schwannoma;
- The development of a "vitiligo"-like cutaneous lesion with evolution at a distance, during the treatment with Nivolumab;
- A 3rd degree severity skin reaction when BRAF/MEK inhibitors were reintroduced into the treatment—an allergic reaction to treatment vs. an adverse skin reaction;
- An imaging response of intrathoracic ganglia to the treatment with Dacarbazine, after the progression of these lesions under targeted treatment and immune checkpoint inhibitors;
- Throughout the evolution of the disease, a single cerebral secondary determination occurred, thus challenging the therapeutic decision.

Considering the clinical and imaging diagnosis of schwannomas at the level of the head and neck, the specialized literature only mentions one incidence of this pathology of 5% of all soft tissue tumors, 45% being diagnosed at the peripheral or autonomic nerve level of the head and neck, with symptoms associated with the nerve involved in the tumor transformation [21,22]. Most frequently, schwannoma is associated with type II neurofibromatosis, more rarely with type I, and it rarely becomes malignant [21]. It presents a slow increase in dimensions, and it is described either as a small mass, homogenous or more frequently well circumscribed tumors, heterogenous, iso/hypodense compared to muscular structures, with cystic formations inside and with the displacement of the adjacent structures upon the imaging evaluation [21,23]. The FDG-PET/CT scan revealed that benign schwannomas present a varied behavior, being able to present increased metabolic activity, acting like a malignant tumor [24]. The imaging descriptions, in our case, either ultrasound, computed tomography or magnetic resonance imaging, led to the diagnosis of an adenopathic image, indeed with a heterogeneous appearance of the tumor mass and with the displacement of adjacent structures, without their invasion and with an increase in dimensions of approximately 1.8 cm in 6 months, with increased metabolic activity at 18F-FDG-PET/CT, which determined the change of the systemic treatment, from a targeted treatment for the BRAF mutation to immune checkpoint inhibitors—Nivolumab, until the pathological examination.

The treatment with immune checkpoint inhibitors (anti-CTLA-4, anti-PD-1, anti-PD-L1) caused a prolonged T-lymphocyte activation [25]. This hyperactivation of CD8+ T-lymphocytes seems to be the mechanism that determines the occurrence of vitiligo-like skin lesions in patients with malignant melanoma under treatment with immune checkpoint inhibitors, called "vitiligo-like depigmentation" [25,26]. The incidence of occurrence is 10–28% among patients with malignant melanoma, with an average onset period of 9 months from the initiation of treatment, and more frequently described in patients under treatment with Pembrolizumab or Nivolumab [25,26]. Skin depigmentation was described in elderly patients and most frequently located in the areas exposed to the sun: the face, the neck region, the cleavage, and the palms, and it is described as a positive predictive factor for the response and survival rate. A decrease of 40–50% in the risk of disease progression was described [26]. We interpret the occurrence of vitiligo-like skin depigmentation in our patient 4 months after the initiation of treatment with Nivolumab as an adverse reaction to the treatment, with the expectation of a sustained and prolonged response to the administration of anti-PD-1, but the progression of the disease was initially noted after 6 months of treatment and confirmed after another 3 months of therapy.

After the progression of the disease at the level of the mediastinal lymph nodes, confirmed by imaging scans and lymph node biopsy, the option was to "re-challenge" the treatment with BRAF/MEK inhibitors [27,28]—once, due to the "false progression" established in the first therapeutic line (through the progression of a lesion that turned out to be schwannoma), the second time due to the scientific synthesis that reported a disease control rate of 57%, a complete response in 8% of the cases, partial response in 20% of the cases and stable disease in 28% of the cases [27,28]. If, upon the first administration of BRAF/MEK inhibitors, the clinical tolerance of the treatment was a very good one, with minimal adverse reactions, at the second administration, we faced skin reactions from the very beginning with extension of medium severity at the level of the jugular mucosa, difficult to manage due to the fact that the patient followed the treatment at home in another city, and he required the interruption of treatment and the initiation of treatment with corticoids and antihistaminic drugs at home, through recommendation by phone. Therefore, the severity of the skin reactions could only be quantified through the patient's case history and through the remaining skin lesions clinically evaluated upon presentation. The following are known as severe adverse skin reactions to the administration of BRAF/MEK inhibitors: Stevens–Johnson syndrome and toxic epidermal necrolysis, with a lethal potential in evolution [29]. The mean onset period of these skin manifestations was reported to be between 11 and 15 days from the beginning of the treatment [29]. The most potent agent in the occurrence of these adverse reactions proved to be the treatment with Vemurafenib, but they were also reported during the treatment with Dabrafenib [29]. During the evaluation by phone of the adverse reactions experienced by the patient at home, a potentially severe evolution was suspected toward the Stevens–Johnson syndrome, and the immediate interruption of the treatment was indicated along with the initiation of the corticoid treatment. The clinical and biological evaluation (specific markers for immediate or delayed allergic reactions) with the titration of the treatment under strict medical control delayed the therapeutic decision by approximately 4 weeks, and it was not possible to clearly differentiate between the two probabilities that could induce the described skin reactions. This had an impact on the subsequent therapeutic decisions, of delaying the treatment with Ipilimumab and initiating the treatment with Dacarbazine.

Around 40–50% of the patients with malignant melanoma develop cerebral metastases along the evolution of the disease, exhibiting high cerebral tropism after lung and breast cancers [30,31]. The occurrence of cerebral metastases represents a poor prognostic factor for the evolution of the disease, also influenced by the site of metastasis, the number of metastases and their size [30,31]. In the case of oligometastases, surgical treatment remains the first therapeutic option. However, radiotherapy and the new systemic treatments bring real benefits in disease control and survival rates. Stereotactic radiosurgery has added a benefit of 8–10 months to the mean survival period [32]. The combination of BRAF/MEK inhibitors determines greater responses than 58% in intracranial disease, but with short duration of response, and the treatment with anti-PD-1 demonstrated durable responses in 20% of the cases [30–32]. Ipilimumab and the combination of Ipilimumab with anti-PD-1 seem to have the most sustained benefit in the systemic treatment of brain metastases in patients with malignant melanoma so far [32]. However, the association of the surgical treatment with stereotactic radiosurgery, followed by the administration of systemic treatment, has been proven to provide the best benefit to survival [32]. Therefore, in the case of a young patient with very good performance status until the onset of the neurological symptoms, with proof of a single cerebral metastasis, the therapeutic approach was the neurosurgical treatment, followed by stereotactic radiotherapy. The complete imaging response of the disease to the systemic treatment with Dacarbazine did not show any benefit in the control of the intracranial disease, for which reason, after the local control of the cerebral disease, the patient's case was to be re-analyzed by the oncology indication board in order to establish the opportunity for the treatment with anti-CTLA-4 and the risk of immune-mediated adverse skin reactions with lethal potential.

The most frequent and severe complications of stereotactic radiotherapy are radionecrosis and hemorrhage. Radionecrosis occurs in 50% of the patients, with complications described in 30% of the patients, and exhibits benefits after the treatment with corticoids [32,33]. Unfortunately, our patient also presented extensive thrombosis of the sagittal sinus associated with the early radionecrosis after stereotactic radiotherapy, with unfavorable evolution under corticotherapy and low molecular weight heparin in the therapeutic dose, with death occurring as a result of these complications.

In an analysis of 463 patients with malignant melanoma who underwent chemotherapy after anti-PD-1 treatment, a number of 0.4% for complete responses were reported in a medial progression-free interval of 2.5 months [34]. The best response in terms of the objective response rate and the free interval until the progression of the disease was identified in the case of the treatment with Taxane. Studies on "rescue" chemotherapy after the administration of immune checkpoint inhibitors in other cancer sites have reported a possible benefit of this therapeutic sequence [34–36].

The innovative treatments of the past 20 years have brought clear benefits to the treatment of advanced/metastatic malignant melanoma:

- Highly increased clinical effectiveness;
- Combined therapies (e.g., BRAF/MEK inhibitors) that increase clinical tolerance and decrease toxicity;
- Active interventions in advanced disease in order to increase the disease-free interval with the approval of new therapies for adjuvant purposes (e.g., Nivolumab, Pembrolizumab, Dabrafenib/Trametinib);
- Therapeutic diversity with the possibility of disease attack, in case of progression under a therapeutic option, in patients with good performance status;
- Increased survival until disease progression, increased overall survival, lasting benefit of the treatment.

However, there are also new challenges:

- The development of new algorithms in the sequential administration of these therapeutic agents;
- The establishment of the best interval of the disease for the initiation of the specific treatment—adjuvant/"follow-up";
- The multidisciplinary approach to the disease—in establishing the sequence of the specific oncological treatments—surgery, systemic treatment, radiotherapy;
- The selection of patients for whom certain treatments bring maximum benefit;
- The early and correct diagnosis and the control of the adverse reactions of the new therapies by the multidisciplinary team.

4. Conclusions

Despite the current systemic therapeutic diversity, the choice of the therapeutic sequence (surgery/systemic treatment/radiotherapy) remains the first important decision-making step.

Any suspicious lesion (especially in the case of oligometastatic neoplasms) must be histopathologically examined.

In the context of the description, it is necessary to augment the interdisciplinary team with specialists in dermatology and allergy for the new therapeutic variants of skin reactions with severe evolutionary potential (e.g., drug-induced hypersensitivity syndrome, Stevens–Johnson syndrome and toxic epidermic necrolysis).

The complications of current treatments with severe potential must be evaluated, diagnosed and treated early to ensure the continuity of the oncological treatment.

After treatment with BRAF/MEK inhibitors or immune checkpoint inhibitors, systemic chemotherapy (e.g., Dacarbazine) can still have benefits in the treatment of metastatic melanoma.

Author Contributions: Conceptualization, A.-R.S., G.R.-N. and L.-F.R.; methodology, L.G., L.-F.A. and G.S.; software, G.R.-N. and G.S.; validation, A.-R.S., L.G., L.-F.A., R.-M.A., R.-A.R.-N., A.-N.C., I.S., G.R.-N., L.-F.R. and G.S.; formal analysis, I.S., R.-M.A. and R.-A.R.-N.; investigation, A.-R.S., L.G., L.-F.A., R.-M.A., R.-A.R.-N., A.-N.C., I.S. and G.R.-N.; resources, G.R.-N., L.-F.R. and G.S.; data curation, R.-A.R.-N., A.-N.C., L.-F.R. and G.S.; writing—original draft preparation, A.-R.S., G.R.-N. and L.-F.R.; writing—review and editing, G.R.-N. and G.S.; visualization, L.G., L.-F.A., R.-M.A., R.-A.R.-N., A.-N.C. and I.S.; supervision, A.-R.S., L.-F.R. and G.S.; project administration, A.-R.S., G.R.-N. and L.-F.R. All authors have read and agreed to the published version of the manuscript.

Funding: This research received no external funding.

Institutional Review Board Statement: The study was conducted in accordance with the Declaration of Helsinki, and approved by the Ethics Committee of "St. Luca" Chronic Disease Hospital, Bucharest: No. 3083/17.03.2022.

Informed Consent Statement: Informed consent was obtained from the subject involved in the study. Written informed consent was obtained from the patient to publish this paper.

Data Availability Statement: The data presented in this study are available on reasonable request from the corresponding authors.

Conflicts of Interest: The authors declare no conflict of interest.

References

1. Borik-Heil, L.; Endler, G.; Parson, W.; Zuckermann, A.; Schnaller, L.; Uyanik-Ünal, K.; Jaksch, P.; Böhmig, G.; Cejka, D.; Staufer, K.; et al. Cumulative UV Exposure or a Modified SCINEXA™-Skin Aging Score Do Not Play a Substantial Role in Predicting the Risk of Developing Keratinocyte Cancers after Solid Organ Transplantation—A Case Control Study. *Cancers* **2023**, *15*, 864. [CrossRef] [PubMed]
2. Juszko-Piekut, M.; Mozdzierz, A.; Koosza, Z.; Krolikowska-Jeruzalska, M.; Wawro-Bielecka, P.; Kowalska-Ziomek, G.; Olczyk, D.; Stojko, J. Incidence of melanoma and non-melanoma skin cancer in the inhabitants of the Upper Silesia, Poland. In *Highlights in Skin Cancer*; Vereecken, P., Ed.; IntechOpen: London, UK, 2013. Available online: https://www.intechopen.com/chapters/43189 (accessed on 15 January 2023).
3. Memon, A.; Bannister, P.; Rogers, I.; Sundin, J.; Al-Ayadhy, B.; James, P.; Mcnally, R. Changing epidemiology and age-specific incidence of cutaneous malignant melanoma in England: An analysis of the national cancer registration data by age, gender and anatomical site, 1981–2018. *Lancet Reg. Health Eur.* **2021**, *2*, 100024. [CrossRef] [PubMed]
4. Islami, F.; Ward, E.M.; Sung, H.; Cronin, K.A.; Tangka, F.K.L.; Sherman, R.L.; Zhao, J.; Anderson, R.N.; Henley, S.J.; Yabroff, K.R.; et al. Annual Report to the Nation on the Status of Cancer, Part 1: National Cancer Statistics. *J. Natl. Cancer Inst.* **2021**, *113*, 1648–1669. [CrossRef] [PubMed]
5. Morton, D.L.; Essner, R.; Kirkwood, J.M.; Wollman, R.C. History of melanoma. In *Holland-Frei Cancer Medicine*, 6th ed.; BC Decker: Hamilton, ON, USA, 2003. Available online: https://www.ncbi.nlm.nih.gov/books/NBK13962/ (accessed on 15 March 2023).
6. Rebecca, V.W.; Sondak, V.K.; Smalley, K.S. A brief history of melanoma: From mummies to mutations. *Melanoma Res.* **2012**, *22*, 114–122. [CrossRef] [PubMed]
7. Srivastava, N.; McDermott, D. Update on benefit of immunotherapy and targeted therapy in melanoma: The changing landscape. *Cancer Manag. Res.* **2014**, *6*, 279–289. [CrossRef] [PubMed]

8. Skudalski, L.; Waldman, R.; Kerr, P.E.; Grant-Kels, J.M. Melanoma: An update on systemic therapies. *J. Am. Acad. Dermatol.* **2022**, *86*, 515–524. [CrossRef]
9. Monroy-Garcia, A.; de Lourdes Mora-Garcia, M.; Hernandez- Montes, J. Cancer immunology and novel strategies for immunotherapy. In *Molecular Oncology: Principles and Recent Advances*; Bentham Science Publisher: Sharjah, UAE, 2012; Volume 1, pp. 130–147, ISBN 978-1-60805-016-1. [CrossRef]
10. Yu, R.; Zhu, B.; Chen, D. Type I interferon-mediated tumor immunity and its role in immunotherapy. *Cell. Mol. Life Sci.* **2022**, *79*, 191. [CrossRef]
11. Tarhini, A.A.; Gogas, H.; Kirkwood, J.M. IFN-α in the treatment of melanoma. *J. Immunol.* **2012**, *189*, 3789–3793. [CrossRef]
12. Buchbinder, E.I.; Desai, A. CTLA-4 and PD-1 Pathways: Similarities, Differences, and Implications of Their Inhibition. *Am. J. Clin. Oncol.* **2016**, *39*, 98–106. [CrossRef]
13. Fellner, C. Ipilimumab (yervoy) prolongs survival in advanced melanoma: Serious side effects and a hefty price tag may limit its use. *Pharm. Ther.* **2012**, *37*, 503–530.
14. Mireștean, C.C.; Iancu, R.I.; Iancu, D.T. Immunotherapy and Radiotherapy as an Antitumoral Long-Range Weapon—A Partnership with Unsolved Challenges: Dose, Fractionation, Volumes, Therapeutic Sequence. *Curr. Oncol.* **2022**, *29*, 7388–7395. [CrossRef] [PubMed]
15. Sullivan, R.J.; Flaherty, K.T. BRAF in Melanoma: Pathogenesis, Diagnosis, Inhibition, and Resistance. *J. Skin Cancer* **2011**, *2011*, 423239. [CrossRef] [PubMed]
16. Eroglu, Z.; Ribas, A. Combination therapy with BRAF and MEK inhibitors for melanoma: Latest evidence and place in therapy. *Ther. Adv. Med. Oncol.* **2016**, *8*, 48–56. [CrossRef] [PubMed]
17. Loras, A.; Gil-Barrachina, M.; Marqués-Torrejón, M.Á.; Perez-Pastor, G.; Martinez-Cadenas, C. UV-Induced Somatic Mutations Driving Clonal Evolution in Healthy Skin, Nevus, and Cutaneous Melanoma. *Life* **2022**, *12*, 1339. [CrossRef]
18. Ronchi, A.; Montella, M.; Zito Marino, F.; Caraglia, M.; Grimaldi, A.; Argenziano, G.; Moscarella, E.; Brancaccio, G.; Troiani, T.; Napolitano, S.; et al. Predictive Evaluation on Cytological Sample of Metastatic Melanoma: The Role of BRAF Immunocytochemistry in the Molecular Era. *Diagnostics* **2021**, *11*, 1110. [CrossRef]
19. Palamaris, K.; Moutafi, M.; Gakiopoulou, H.; Theocharis, S. Histone Deacetylase (HDAC) Inhibitors: A Promising Weapon to Tackle Therapy Resistance in Melanoma. *Int. J. Mol. Sci.* **2022**, *23*, 3660. [CrossRef]
20. Kim, K.B.; Kefford, R.; Pavlick, A.C.; Infante, J.R.; Ribas, A.; Sosman, J.A.; Fecher, L.A.; Millward, M.; McArthur, G.A.; Hwu, P.; et al. Phase II study of the MEK1/MEK2 inhibitor Trametinib in patients with metastatic BRAF-mutant cutaneous melanoma previously treated with or without a BRAF inhibitor. *J. Clin. Oncol.* **2013**, *31*, 482–489. [CrossRef]
21. Crist, J.; Hodge, J.R.; Frick, M.; Leung, F.P.; Hsu, E.; Gi, M.T.; Venkatesh, S.K. Magnetic Resonance Imaging Appearance of Schwannomas from Head to Toe: A Pictorial Review. *J. Clin. Imaging Sci.* **2017**, *7*, 38. [CrossRef]
22. Tamas, T.; Dinu, C.; Lenghel, L.M.; Boțan, E.; Tamas, A.; Stoia, S.; Leucuta, D.C.; Bran, S.; Onisor, F.; Băciuț, G.; et al. High-Frequency Ultrasound in Diagnosis and Treatment of Non-Melanoma Skin Cancer in the Head and Neck Region. *Diagnostics* **2023**, *13*, 1002. [CrossRef]
23. Skolnik, A.D.; Loevner, L.A.; Sampathu, D.M.; Newman, J.G.; Lee, J.Y.; Bagley, L.J.; Learned, K.O. Cranial Nerve Schwannomas: Diagnostic Imaging Approach. *Radiographics* **2016**, *36*, 1463–1477. [CrossRef]
24. Monson, K.; Dewey, B.; Ugorowski, M.; Broski, S. 18F-FDG PET/CT Imaging Features of Benign Schwannomas. *J. Nucl. Med.* **2019**, *60*, 2053.
25. Lommerts, J.E.; Bekkenk, M.W.; Luiten, R.M. Vitiligo induced by immune checkpoint inhibitors in melanoma patients: An expert opinion. *Expert Opin. Drug Saf.* **2021**, *20*, 883–888. [CrossRef] [PubMed]
26. Guida, M.; Strippoli, S.; Maule, M.; Quaglino, P.; Ramondetta, A.; Chiaron Sileni, V.; Antonini Cappellini, G.; Queirolo, P.; Ridolfi, L.; Del Vecchio, M.; et al. Immune checkpoint inhibitor associated vitiligo and its impact on survival in patients with metastatic melanoma: An Italian Melanoma Intergroup study. *ESMO Open* **2021**, *6*, 100064. [CrossRef] [PubMed]
27. Tietze, J.K.; Forschner, A.; Loquai, C.; Mitzel-Rink, H.; Zimmer, L.; Meiss, F.; Rafei-Shamsabadi, D.; Utikal, J.; Bergmann, M.; Meier, F.; et al. The efficacy of re-challenge with BRAF inhibitors after previous progression to BRAF inhibitors in melanoma: A retrospective multicenter study. *Oncotarget* **2018**, *9*, 34336–34346. [CrossRef] [PubMed]
28. Zaremba, A.; Eggermont, A.M.M.; Robert, C.; Dummer, R.; Ugurel, S.; Livingstone, E.; Ascierto, P.A.; Long, G.V.; Schadendorf, D.; Zimmer, L. The concepts of rechallenge and retreatment with immune checkpoint blockade in melanoma patients. *Eur. J. Cancer* **2021**, *155*, 268–280. [CrossRef] [PubMed]
29. Torres-Navarro, I.; de Unamuno-Bustos, B.; Botella-Estrada, R. Systematic review of BRAF/MEK inhibitors-induced Severe Cutaneous Adverse Reactions (SCARs). *J. Eur. Acad. Dermatol. Venereol.* **2021**, *35*, 607–614. [CrossRef] [PubMed]
30. Janavicius, M.; Lachej, N.; Anglickiene, G.; Vincerzevskiene, I.; Brasiuniene, B. Outcomes of Treatment for Melanoma Brain Metastases. *J. Skin Cancer* **2020**, *2020*, 7520924. [CrossRef]
31. Costa Svedman, F.; Das, I.; Tuominen, R.; Darai Ramqvist, E.; Höiom, V.; Egyhazi Brage, S. Proliferation and Immune Response Gene Signatures Associated with Clinical Outcome to Immunotherapy and Targeted Therapy in Metastatic Cutaneous Malignant Melanoma. *Cancers* **2022**, *14*, 3587. [CrossRef]
32. Tawbi, H.A.; Boutros, C.; Kok, D.; Robert, C.; McArthur, G. New Era in the Management of Melanoma Brain Metastases. *Am. Soc. Clin. Oncol. Educ. Book* **2018**, *38*, 741–750. [CrossRef]

33. Fetcko, K.; Lukas, R.V.; Watson, G.A.; Zhang, L.; Dey, M. Survival and complications of stereotactic radiosurgery: A systematic review of stereotactic radiosurgery for newly diagnosed and recurrent high-grade gliomas. *Medicine* **2017**, *96*, e8293. [CrossRef]
34. Goldinger, S.M.; Buder-Bakhaya, K.; Lo, S.N.; Forschner, A.; McKean, M.; Zimmer, L.; Khoo, C.; Dummer, R.; Eroglu, Z.; Buchbinder, E.I.; et al. Chemotherapy after immune checkpoint inhibitor failure in metastatic melanoma: A retrospective multicentre analysis. *Eur. J. Cancer* **2022**, *162*, 22–33. [CrossRef] [PubMed]
35. Tone, M.; Izumo, T.; Awano, N.; Kuse, N.; Inomata, M.; Jo, T.; Yoshimura, H.; Miyamoto, S.; Kunitoh, H. Treatment effect and safety profile of salvage chemotherapy following immune checkpoint inhibitors in lung cancer. *Lung. Cancer Manag.* **2019**, *4*, LMT12. [CrossRef] [PubMed]
36. Heinhuis, K.M.; Ros, W.; Kok, M.; Steeghs, N.; Beijnen, J.H.; Schellens, J.H.M. Enhancing antitumor response by combining immune checkpoint inhibitors with chemotherapy in solid tumors. *Ann. Oncol.* **2019**, *30*, 219–235. [CrossRef] [PubMed]

diagnostics

MDPI

Case Report

Cutaneous Melanoma and Glioblastoma Multiforme Association—Case Presentation and Literature Review

Olguța Anca Orzan [1,2], Călin Giurcăneanu [1,2], Bogdan Dima [2], Monica Beatrice Dima [2], Ana Ion [2], Beatrice Bălăceanu [2,*], Cornelia Nițipir [3,4], Irina Tudose [5], Cătălina Andreea Nicolae [5] and Alexandra Maria Dorobanțu [2,*]

1 Dermatology Department, "Carol Davila" University of Medicine and Pharmacy, 020021 Bucharest, Romania
2 Dermatology Department, "Elias" University Emergency Hospital, 011461 Bucharest, Romania
3 Oncology Department, "Carol Davila" University of Medicine and Pharmacy, 020021 Bucharest, Romania
4 Oncology Department, "Elias" University Emergency Hospital, 011461 Bucharest, Romania
5 Anatomic Pathology Laboratory, "Elias" University Emergency Hospital, 011461 Bucharest, Romania
* Correspondence: balaceanubeatrice@yahoo.com (B.B.); alexandramdorobantu@gmail.com (A.M.D.)

Abstract: The occurrence of both melanoma and glioma was first suggested by the observation of a familial association between these conditions, which was later confirmed by the description of the melanoma–astrocytoma syndrome, an extremely rare, inherited affliction in which people have an increased risk of developing melanoma and nervous system tumors. Taking into consideration the common embryologic precursor, the neuroectoderm, it was hypothesized that this syndrome is associated with a genetic disorder. While some families with germline *CDKN2A* mutations are prone to develop just melanomas, others develop both melanomas and astrocytomas or even other nervous-system neoplasms. Herein, we report the case of a 63-year-old male patient with no personal or family history of malignancy who had primary melanoma followed by glioblastoma. Our case report suggests that the occurrence of both melanoma and glioblastoma is most likely not coincidental but instead linked to genetic mutations of common embryologic precursors or signaling pathways.

Keywords: glioblastoma; melanoma; cancers; astrocytomas; nervous system tumors; melanoma–astrocytoma syndrome; *CDKN2A*

1. Introduction—Literature Review

Glioblastoma is a high-grade malignant glioma and is additionally one of the most harmful and lethal disorders of the central nervous system. Melanoma is classified as the most dangerous skin cancer with high mortality rates.

During embryogenesis, the skin is formed through a complex process involving cells derived from both ectodermal and endodermal germ layers. The former is further divided into surface ectoderm, which gives rise to epidermal keratinocytes, and neuroectoderm, which forms the neural tube through invagination (neurulation). Furthermore, the posterior portion of the neural tube separates into the neural crest which generates nervous cells, both neurons and glial cells. In addition, another cell type that derives from the neural crest is the melanocyte, which migrates towards the epidermis [1].

In 1993, Kaufman et al. described the melanoma–astrocytoma syndrome, reporting on eight members of a family spread over three generations that developed cutaneous malignant melanoma and/or cerebral astrocytoma [2]. Taking into consideration the common embryologic precursor, the neuroectoderm, it was hypothesized that this syndrome is associated with a genetic disorder [2].

One of the main candidates is *CDKN2A* (Cyclin-Dependent Kinase Inhibitor 2A), situated on chromosome 9p21, because it is involved in neuroblastoma cell growth, while also representing a high penetrance susceptibility gene for familial cutaneous melanoma and sporadic multiple cutaneous melanoma syndromes [3–5].

CDKN2A is a tumor suppressor gene with a unique characteristic of encoding two distinct proteins by alternative splicing [6]. The two transcripts share exons 2 and 3 but differ through the first exon: when exon 1α is used the resulting protein is p16^{INK4A}, whereas exon 1β encodes p14^{ARF}, by employing an alternative reading frame [6].

CDK4 encodes a catalytic subunit of a Ser/Thr protein kinase that takes part in the G1 phase progression of the cell cycle through phosphorylation of the Retinoblastoma protein (Rb). p16^{INK4A} binds through arginine to *CDK4* and inhibits its function, thus arresting the cell cycle in the G1 phase, preventing aberrant cell growth. Cen et al. outline the importance of p16^{INK4A}-*CDK4*-Rb pathway mutations in the development of glioblastoma multiforme, as well as in predicting the response to targeted therapies with cyclin-dependent kinase inhibitors [7].

On the other hand, p14^{ARF} stabilizes another tumor suppressor protein, p53, by protecting it from mediated degradation through E3 ubiquitin-protein ligase MDM2 (Murine Double Minute/Mouse Double Minute 2 homolog) [6]. Unlike p16^{INK4A}, p14^{ARF} can arrest the cell cycle in both G1 and G2 phases [6].

Petronzelli et al. cite exon 1β mutations affecting p14^{ARF} as the main trigger for the development of nervous system tumors, while larger germline deletions of *CDKN2A*, most likely exon 2 defects that affect both p16^{INK4A} and p14^{ARF}, are associated with melanoma-nervous system tumor syndromes [8]. Similar results have been published by Knappskog et al. who reported on a Norwegian melanoma-prone family (four out of fifteen members, two of them also having multiple cutaneous melanomas) which harbored a large *CDKN2A* deletion that led to truncated p14^{ARF} mRNA (lacking exon 2) and loss of p16^{INK4A} mRNA expression from the affected allele, thus altering the gene transcripts [9].

Loo et al., by sequencing exons and adjacent intronic regions, hypothesized that mutations affecting the noncoding, intronic regions of *CDKN2A* could also be responsible for altered p16^{INK4A} transcripts that are unable to properly bind to *CDK4* and inhibit it, thus explaining why some melanoma-prone families do not exhibit *CDKN2A* coding mutations [10].

It is well known that a positive family history is a significant risk factor for other family members to develop melanoma. Harland et al. identified in six English melanoma pedigrees a mutation situated deep in intron 2 of *CDKN2A*, more precisely at 105 bases from the 5′ end, towards exon 3, which results in aberrant splicing of mRNA [11]. The challenge in identifying mutations in noncoding regions is that they are extremely variable depending on the studied geographic population. Moreover, some mutations may not be included in the sequences amplified by RT-PCR due to the limitations of the primer selection.

CDKN2A gene silencing caused by mutations, deletions, or promoter hyper-methylation, has also been reported in a variety of other human cancer cell lines, such as: pancreatic adenocarcinoma, non-small cell lung cancer, Burkitt's lymphoma, gastric cancer, colorectal cancer, prostate cancer, breast cancer, and ovarian cancer [12].

Goldstein analyzed 189 melanoma-prone families, 42 of them with a positive history of pancreatic cancer, and identified 67 different germline *CDKN2A* mutations, exon 2 being the most frequently mutated in both groups [13]. Additionally, missense mutations were the most prevalent (56%) for both melanoma and pancreatic cancer families [13].

Unlike *CDKN2A*, *MC1R* is a low-moderate-risk melanoma susceptibility gene, and it encodes the G-protein-coupled receptor for α-melanocyte stimulating hormone (α-MSH) [4]. When α-MSH binds to *MC1R* in the presence of UVB-exposure, it triggers the activation of adenylate cyclase which leads to an increase in cAMP production [4]. In turn, an elevated level of cAMP causes a switch in melanin production from red/yellow pheomelanin pigments to protective brown/black eumelanin pigments [4]. In addition, *MC1R* induces cAMP-dependent expression of catalase and superoxide dismutase which protect the UVB-exposed cells from oxidative stress, thus reducing the incidence of DNA defects [14].

Harding et al. studied the diversity in *MC1R* variants on different continents and discovered that in non-African populations there are many *MC1R* aminoacid variants which account for the large diversity of pigmentation patterns [15]. On the other hand, in Africa, *MC1R* variants were evolutionary eliminated due to a strong functional constraint of conversion to eumelanin pigments [15].

MC1R variants are incapable of switching the pigmentation pattern to eumelanin production and this explains why up to 75% of skin type I people carry a variant *MC1R* allele, whereas in darker skin types the percentage of *MC1R* variants vastly decreases [15]. Three main *MC1R* variants have been found to be associated with red hair and fair skin: *R151C* (*Arg151Cys*), *R160W* (*Arg160Trp*) and *D294H* (*Asp294His*), and have been grouped under the term of "red hair color" variants [16]. Moreover, *MC1R* "red hair color" variants are also correlated with other pigmentation characteristics, such as: freckling, propensity to burn and failure to tan after sun exposure [17]. All these characteristics are known risk factors for the development of cutaneous malignant melanoma [17]. Moreover, Palmer et al. calculated that *MC1R* variants are associated with a 2.2-fold risk increase of melanoma and that the effect is additive in individuals carrying more than one *MC1R* variant [17]. In addition, this predisposition also persists in individuals with dark skin, thus confirming that *MC1R* variants are independent risk factors for melanoma, apart from their influence in pigmentation patterns [4].

Box et al. studied 15 Australian melanoma pedigrees harboring *CDKN2A* mutations and discovered that families carrying *MC1R* "red hair color" variants had a *CDKN2A* mutation penetrance of 84%, whereas families with a normal *MC1R* gene presented a *CDKN2A* mutation penetrance of 50% [18]. Therefore, apart from being an independent risk factor for melanoma, *MC1R* variants also increase the penetrance of other high-susceptibility mutations (*CDKN2A*). In addition, when a *MC1R* variant was present, the mean age at onset of melanoma was also considerably lower [18].

Cao et al. reported that UVB-induced inactivation of *PTEN* (through phosphorylation of Ser380 and Thr382/383) and activation of Akt (through phosphorylation of Ser473) were significantly increased in cells expressing *MC1R* "red hair color" mutant variants, compared to cells expressing wild-type non-mutated *MC1R* (*WT-MC1R*) [19]. In addition, *WT-MC1R* also reduces the proteosomal degradation of *PTEN* (phosphatase and tensin homologue) [19].

PTEN is a tumor suppressor gene that acts as a phosphatidylinositol-3,4,5-triphosphate (PIP_3) phosphatase, degrading PIP_3 to PIP_2 [20]. High levels of PIP_3 stimulate the phosphorylation and over-activation of the Akt-mTOR signaling pathway, inducing aberrant cell growth and inhibition of apoptosis [20]. As a result, in UVB-exposed cells, *WT-MC1R* protects *PTEN*'s inhibitory function over the PI3K-Akt-mTOR signaling pathway by keeping in check PIP_3 cell levels, thus preventing aberrant cell development, whereas *MC1R* "red hair color" mutant variants are more often associated with *PTEN* inactivation.

There are studies that identified a correlation between certain groups of gene defects and the nature and evolution of gliomas. The most noteworthy genetic hallmarks are: *IDH1/2* mutations, 1p/19q co-deletion, *MGMT* (O-Methylguanine-DNA Methyltransferase) promoter methylation status, *TP53* expression, Ki-67, *ATRX* (X-linked α-Thalassemia/mental retardation syndrome) mutation, *TERT* (telomerase reverse transcriptase) mutation, EGFR (epidermal growth factor receptor) expression, *PTEN* mutations, *CDKN2A* mutations, as well as *BRAF* mutations [21–24].

EGFR is a tyrosine kinase receptor that is involved in brain development, stimulating proliferation, migration and differentiation of all types of central nervous system cells. Secondary to ligand-receptor interaction, activated EGFR increases phosphorylation of the Ras pathway [24]. Ras is a G-protein that becomes active through the phosphorylation of GDP to GTP (Ras-GTP bound) and controls cell proliferation, survival and migration through the MAPK/ERK (mitogen-activated protein kinase) downstream signaling pathway. In addition, EGFR promotes the activation of PI3K (phosphatidylinositol-4,5-bisphophate 3-kinase) which phosphorylates the 3′ position of the inositol ring of PIP_2, thus generating high levels of PIP_3 [25]. In turn, PIP_3 triggers the phosphorylation of different targets from the Akt pathway, thus providing cell signals that promote survival, growth, and proliferation [25].

Therefore, the characteristic EGFR overexpression described in glioblastomas accounts for an excessive activation of Ras-MAPK linear and PI3K-Akt non-linear pathways which are responsible for aberrant cell growth and cell apoptosis inhibition [24,25]. In addition, *PTEN* gene silencing would lead to even more uncontrolled levels of PIP_3 due to the lack of degradation to PIP_2. As a result, *PTEN* inactivation may represent another genetic hallmark of the melanoma–astrocytoma syndrome.

2. Case Presentation

We present the case of a 63-year-old male patient who was admitted to our clinic with a 1.5 cm, asymptomatic, asymmetric, dark brown-black nodular tumor, located on the right posterior hemi-thorax. (Figure 1) According to the patient's history, it had developed in approximately three months, on the site of a former nevus. Dermoscopy revealed vascular polymorphism on an unevenly colored background (red, black, dark brown) with asymmetric borders, discretely covered by a bluish veil (Figure 2). The patient denied any personal or family history of malignancy. The clinical examination revealed two more dysplastic nevi on the anterior thorax, but no other noteworthy signs or symptoms suggestive of regional (palpable lymphadenopathy) or systemic metastatic disease (Figure 3).

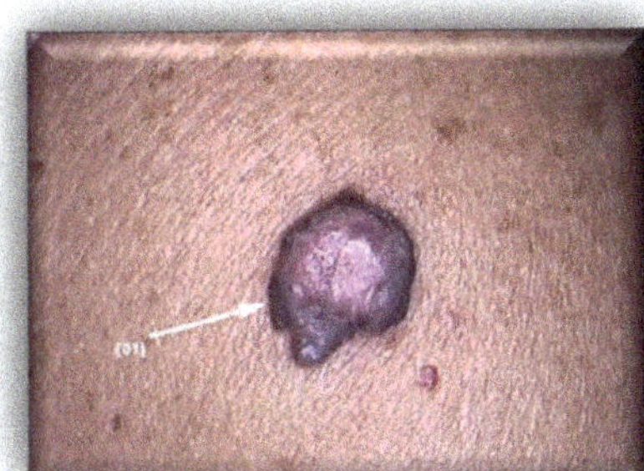

Figure 1. Clinical aspect: asymmetric, dark brown-black nodular tumor, located on the right posterior hemi-thorax.

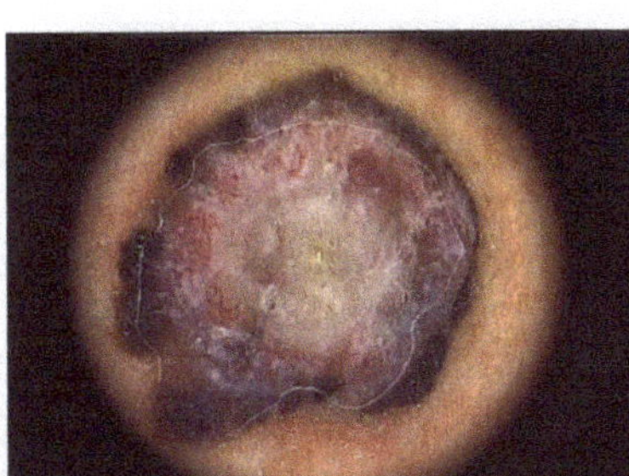

Figure 2. Dermoscopy findings: vascular polymorphism on an unevenly colored background (red, black, dark brown) with asymmetric borders, discretely covered by a bluish veil.

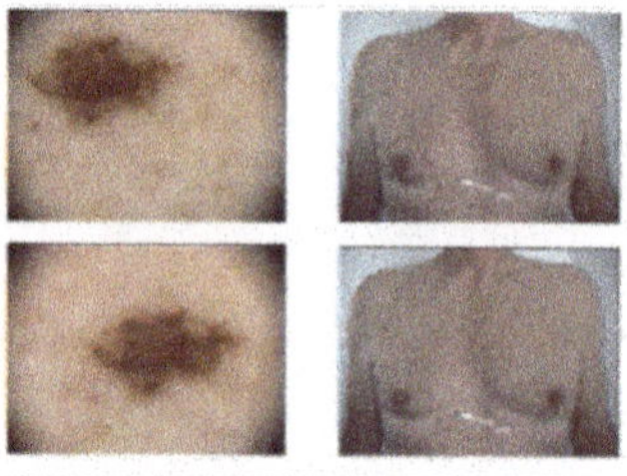

Figure 3. Clinical and dermoscopical images of two more dysplastic nevi on the anterior thorax.

We performed an excisional biopsy of the tumor and sent it for histopathological examination, which described nests of polymorphic melanocytes invading the reticular dermis (Clark level IV), with a Breslow index of 5 mm, intense mitotic activity (12 mitosis/mm^2), chronic inflammatory infiltrate surrounding the base of the tumor, absence of ulceration and lymphovascular invasion (Figure 4). In addition, immunohistochemical staining was also performed and it was positive for S100 protein and Melan-A, as well as Ki-67 which was positive in 40% of the tumor cells (Figure 5). Therefore, correlating these results, we established the diagnosis of T4a nodular malignant melanoma—clinical stage IIB according to AJCC 8th ed. melanoma staging [26]. Conformable to the American Cancer Society, the five-year survival rate for stage IIB melanoma is 70%, while the ten-year survival rate is around 57%.

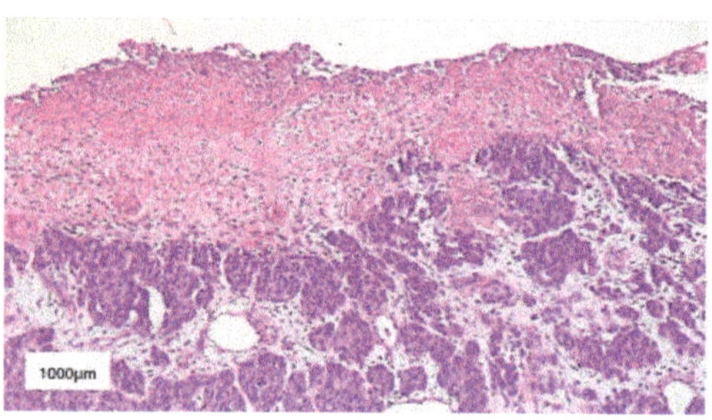

Figure 4. HE-stained (10×) histopathological section of Clark IV malignant melanoma. Scale bar: 1000 µm.

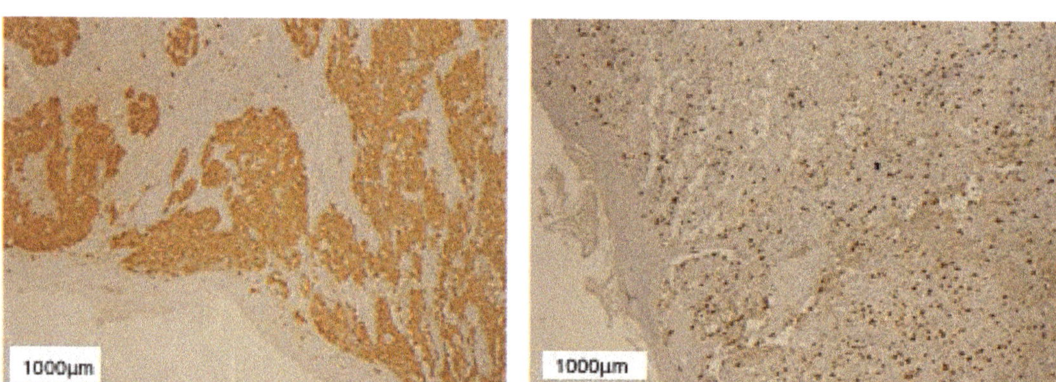

(a) (b)

Figure 5. *Cont.*

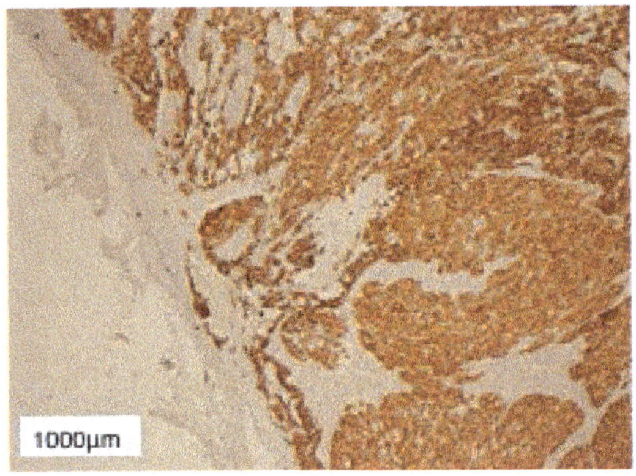

(c)

Figure 5. Microphotographs of malignant melanoma sections in our case. (**a**) Immunohistochemical stain (10×) positive for S100; (**b**) Immunohistochemical stain (10×) positive for Ki-67; (**c**) Immunohistochemical stain (10×) positive for Melan-A. Scale bars (**a**–**c**) images: 1000 μm.

In order to assess the presence of metastatic disease, we performed a complete ultrasound examination of the upper body main node stations, as well as thoracic, abdominal and pelvic computed tomography (CT) scans, and head magnetic resonance imaging (MRI). Because these imaging techniques did not reveal any abnormal modifications, we referred the patient for a sentinel lymph node biopsy, which was performed simultaneously with the wide local re-excision with 2 cm margins. The sentinel lymph node biopsy described no metastatic involvement of the axillary nodes (Figure 6). As a result, the patient was referred to the Oncology department with a pathological stage IIB (pT4aN0M0).

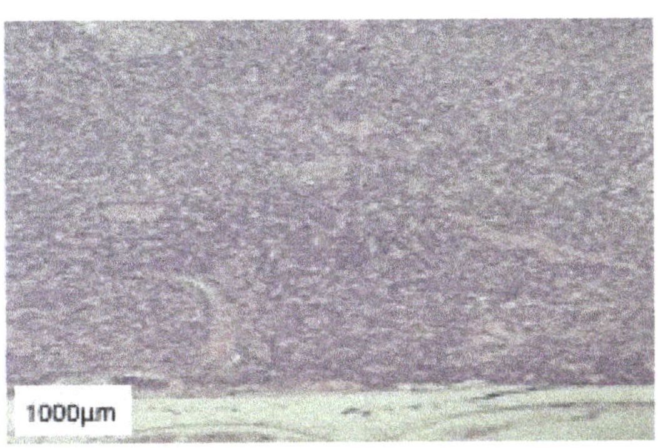

Figure 6. HE staining (20×) of sentinel lymph node biopsy which describes no metastatic involvement of the axillary nodes. Scale bar: 1000 μm.

Following the European interdisciplinary melanoma guideline, the Oncology department decided to initiate a low dose adjuvant therapy with interferon-α—3 million IU s.c., three days/week, for twelve months [27]. As far as follow-up was concerned, the patient was instructed to present for a dermatological skin examination every three months. In addition, the oncologist recommended a CT scan of the thorax, abdomen, and pelvis as well as a bone scintigraphy twice per year. Moreover, cerebral MRI and tumor markers dosage (CEA, CA, S100, PSA) every three months were included in the follow-up plan for the patient.

Considering the histopathological risk factors of developing metastasis, the patient was referred to test the *BRAF* gene mutation, which revealed no mutations of the V600 position (Figure 7).

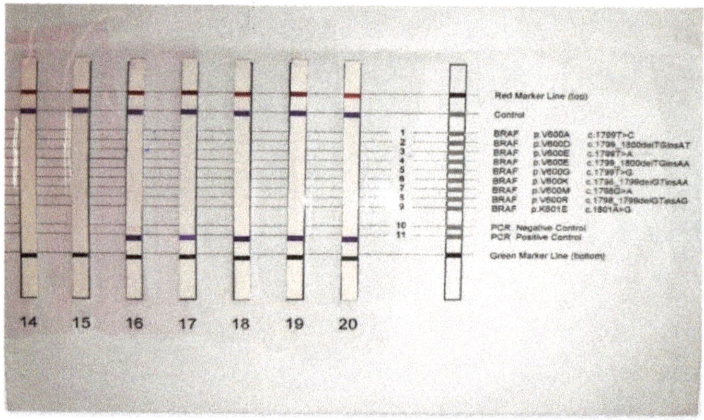

Figure 7. Negative BRAF gene mutation result. The genotyping was performed using the BRAF 600/601 StripAssay kit, which detects the mutations: V600E, V600A, V600D, V600E (c. 799T > A, c. 1799_1800TG > AA), V600G, V600K, V600M, V600R, K601E in the BRAF gene. The detection limit of the method is 1%.

Approximately 15 months after the primary excision, a follow-up cerebral MRI revealed a 30/20 mm left frontal lobe tumor, for which the patient was referred to the Neurosurgery department (Figure 8). The patient did not present any neurological symptoms and, due to his medical history, the first suspicion was that of a melanoma metastasis. As a result, Gamma knife therapy was recommended, but it failed to control the tumor, thus requiring direct excision.

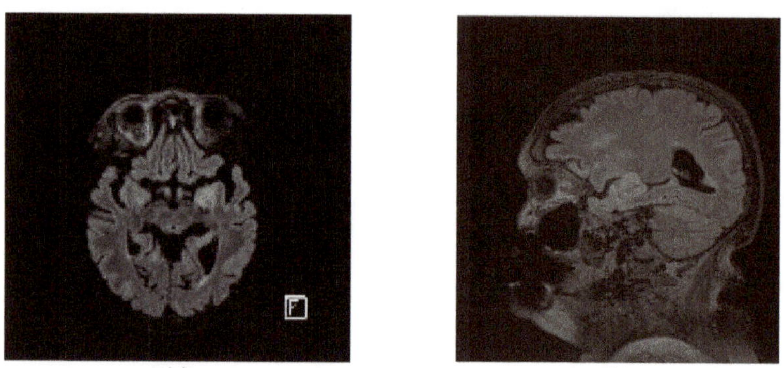

(a) **(b)**

Figure 8. (**a**) Axial MRI image of a large heterogeneous mass in the left frontal lobe. (**b**) Lateral MRI image of a large heterogeneous mass in the left frontal lobe.

The histopathological report described multinucleated giant cells as well as spindle and round cells with hyperchromatic nuclei and an increased nuclear to cytoplasmic ratio. In addition, numerous atypical mitoses, hyperplastic vascular structures, and necrotic foci were also present. These aspects, correlated with the rapid growth rate (less than three months), were suggestive of a diagnosis of glioblastoma multiforme (Figure 9).

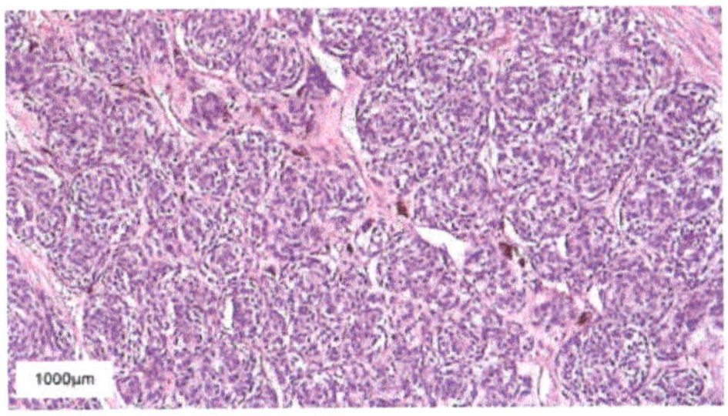

Figure 9. HE-stained (10×) histopathological section of glioblastoma multiforme. Scale bar: 1000 μm.

Immunohistochemical staining was positive for Vimentin, S100 protein, but negative for Melan-A, which excluded a melanocytic origin. In addition, glial fibrillary acidic protein (GFAP) and isocitrate dehydrogenase-1 *(IDH1)* were diffusely positive, while Ki-67 was positive in approximately 40% of the tumor cells. Immunohistochemistry confirmed the diagnosis of high-grade glioblastoma, but its primary or secondary nature was debatable.

On the one hand, *IDH1* staining targets the IDH1-R132H mutation characterized by the replacement of arginine with histidine at codon 132, which corresponds to the enzymatic active site [28]. This mutation is present in approximately 80–90% of secondary glioblastomas [29]. On the other hand, the patient's age (secondary glioblastomas usually affect younger people), Ki-67 positivity in 40% of the cells and the fact that the tumor developed rapidly, in less than three months, are arguments for a primary, more aggressive glioblastoma. The most probable explanation was that the last clear brain MRI missed an intermediate grade glioma (grade II or III), which transformed in the following three months in a secondary grade IV glioblastoma. However, there are also cited cases of primary glioblastoma with *IDH1* positive immunohistochemical staining [23].

Many authors questioned whether *IDH1* mutation is a main driver of oncogenesis or merely a secondary product of defective DNA repair attempts. *IDH1* oxidizes isocitrate to α-ketoglutarate by utilizing nicotinamide adenine dinucleotide phosphate (NADP+) as a cofactor which is transformed into NADPH during the reaction. As a result, *IDH1* protects the cell from oxidative damage, but it can also reduce α-ketoglutarate to isocitrate during periods of cellular hypoxia or mitochondrial dysfunction [29].

Dang et al. discovered that the IDH1-R132H mutation of the enzymatic active site leads to a gain-of-function, *IDH1* mutants promoting an NADPH-dependent reduction of α-ketoglutarate to 2-hydroxiglutarate [30]. Moreover, in the frequent cases of IDH1-R132H codominance with wild-type *IDH1*, the latter directly generates NADPH and α-ketoglutarate which are consumed by the mutant gene copy [30]. This explains why malignant gliomas expressing *IDH1* mutations are characterized by elevated levels of 2-hydroxiglutarate, similar to patients with 2-hydroxiglutarate dehydrogenase deficiencies. In these cases, as well, elevated levels of 2-hydroxiglutarate correlate with a higher risk of central nervous system malignancies [30].

Xu et al. termed 2-hydroxiglutarate an "oncometabolite" because it occupies the same active site as α-ketoglutarate, thus inhibiting α-ketoglutarate-dependent dioxygenases which normally control histone demethylation [31]. As a result, elevated levels of 2-hydroxiglutarate interfere with histone and DNA methylation levels which directly regulate transcriptional activity [31]. Moreover, depletion of α-ketoglutarate inhibits prolyl-hydroxylase domain enzymes (PHD) that normally degrade the hypoxia-inducible factor 1α (HIF-1α) [28]. Therefore, in an *IDH1* mutant tumor cell, high levels of HIF-1α promote angiogenesis and tumorigenesis [31].

Although secondary glioblastomas are less frequent than primary ones, establishing the correct nature of the tumor and its genetic signature has an important prognostic value and it enables the prescription of targeted therapies, if available.

In their meta-analysis, Crespo et al. gathered the incidences of specific genetic alterations that are associated with primary, as well as secondary, glioblastoma, thus contouring a more complete genetic background, based on a larger dataset [24]. The results revealed that primary glioblastoma is possibly triggered by the correlation of: EGFR amplification, *TERT* mutation, *CDKN2A*-p16^{INK4a} deletion, *PTEN* mutation, and loss of heterozygosity 10p/10q [24]. On the other hand, secondary glioblastoma is characterized by: promoter methylation of *MGMT, TIMP-3, CDKN2A*-p14^{ARF} and *CDKN2A*-p16^{INK4}, *IDH1* mutation, *p53* mutation, and loss of heterozygosity 22q/13q/19q [24].

In our case, because personalized targeted therapies were unavailable, and considering the economic factors, additional tests, such as determining the *MGMT* promoter methylation status or EGFR expression, were not performed.

The patient did not present any signs or symptoms of neurologic deficits post-surgery; therefore, his Karnofsky performance status (KPS) was calculated to 100 points, which allowed the oncologist to suggest a more aggressive treatment scheme. Following National Comprehensive Cancer Network (NCCN) guidelines, we administered a 60 Gy dose split in 2 Gy fractions over six weeks, five days/week. Once the patient completed the radiotherapy sessions, adjuvant chemotherapy was initiated with temozolomide 200 mg/m^2, on a 5/28 cycle (five days in a row, followed by a break of twenty-eight days) [32]. The follow-up brain MRIs did not reveal any tumor activity or recurrences.

3. Results

Due to the fact that the patient developed two distinct primary cancers, both having neuroectodermal origins, we started analyzing the possibility of facing a melanoma-astrocytoma syndrome [2]. Therefore, we investigated the common gene defects that can trigger this syndrome and considered *CDKN2A*, situated on chromosome 9p21, as the main candidate.

The NCCN guidelines for melanoma also recommend considering mutation testing for *p16/CDKN2A* in cases with three or more primary melanomas, or different associations of cancers, such as: melanoma, pancreatic cancer and/or astrocytoma diagnosed in the same patient or in blood relatives [32].

In our case, the patient agreed to *CDKN2A* gene sequencing from the primary melanoma, but it came back negative for any mutations in the coding region. However, we could not test for other mutations in the noncoding regions of *CDKN2A* that might affect the gene transcripts. The next step would be genetic testing for other associated gene defects, such as: *CDK4, TERT, MITF, BAP1, TP53,* and *POT1*.

4. Discussion

Several reports have been published describing the development of different malignancies with common embryologic precursors or signaling pathways, either in the same patient or in the same family, as well as the associated gene defects.

Frigerio et al. studied two young patients who developed multiple cutaneous melanomas eight to ten years after undergoing radiotherapy for childhood astrocytomas [33]. For the female patient, array comparative genomic hybridization revealed a large 9p21.3 hemizygotic deletion affecting the *CDKN2A* and *CDKN2B* (encoding p15 protein which inhibits *CDK4/CDK6*, alongside p16^{INK4A}) gene cluster which was also detected in her twin sister [33]. However, due to the fact that their parents did not carry the 9p21.3 deletion, the conclusion was that the gene defect occurred de novo as a result of paternal allele loss [33]. On the other hand, the male patient did not carry any *CDKN2A-CDKN2B, CDK4, MC1R* or *MITF* gene defects [33]. The authors concluded that, for young patients, radiotherapy for previous cancers is a more important risk factor for the development of cutaneous melanomas than the associated gene defects. For example, the twin sister who did not undergo radiotherapy did not develop cutaneous melanomas, even though she carried the 9p21.3 deletion, whereas the male patient presented congenital nevi associated melanoma 10 years after radiotherapy for astrocytoma, in the absence of associated gene defects [33].

Similar conclusions have been drawn by Braam et al. who studied 151,575 childhood cancer survivors and observed that 5.3% of all second malignant neoplasms were melanomas (212 out of 4010 patients that secondarily developed other cancers) and that the main associated risk factors were: radiotherapy, alkylating agents and anti-mitotic drugs [34].

Moreover, Pasmant et al. studied a large French family (approximately 100 members with eight individuals diagnosed with cutaneous melanoma and nine individuals affected by a variety of nervous system tumors, six of them presenting both entities) and discovered a large germline deletion, spanning over 403-kb, affecting the *CDKN2B-CDKN2A* gene cluster [35]. Furthermore, the authors additionally identified a large antisense noncoding RNA, termed *ANRIL*, whose gene overlapped the two exons of *CDKN2B* and had the 5′ end of its first exon located near the transcription start site of *p14/ARF* gene, which explained why *ANRIL* exhibited a stronger correlation for coordinated transcription with *p14/ARF* [35]. The authors hypothesized that in normal tissues noncoding RNAs may act as gene transcription regulators through different possible mechanisms, such as RNA interference, gene cosuppression, and DNA demethylation (i.e., *ANRIL* could regulate *p14/ARF*, as well as *p15/CDKN2B* and *p16/CDKN2A* expression), thus explaining why large deletions including the *ANRIL* gene could further affect tumor suppressor gene transcripts [35,36].

5. Conclusions

Though exceptionally, there are reports in the literature on the occurrence of both melanoma and glioblastoma, either in the same patient or in the same family. The genetic basis of these associations has neuroectodermal origins and was clarified by studies that detected depletion in genetic loci encoding essential tumor suppressor genes [37]. This association is now officially known as the familial melanoma–astrocytoma syndrome [37].

Even though secondary glioblastomas are less frequent than primary ones, establishing the correct nature of the tumor and its genetic signature has an important prognostic value and it enables the prescription of targeted therapies, if these are available. Studies show that survival rates from glioblastoma are poor, with a median survival of only three months in untreated patients [38]. Only a few patients survive two and a half years, and fewer than 5% survive five years following a glioblastoma diagnosis [39].

Despite current studies, an evident syndrome describing the association between melanoma and glioblastoma has yet to be determined. From our point of view, the presented case report describes a patient with a history of stage IIB malignant melanoma and primary glioblastoma. This unique case report illustrates both histologic and genetic diversity and raises awareness regarding the sporadic co-occurrence of these two medical entities. We believe that the co-occurrence of melanoma and glioblastoma is likely not coincidental. Therefore, more insight and research are needed to fully understand this unique association.

Author Contributions: Conceptualization, A.M.D. and O.A.O.; methodology, A.M.D. and O.A.O.; software, A.M.D., O.A.O., A.I., B.B., C.N., M.B.D., B.D., I.T., C.A.N.; validation, A.M.D., O.A.O., C.G., C.A.N. and I.T.; formal analysis, A.M.D., O.A.O., A.I., B.B., C.N., M.B.D., B.D., I.T., C.G., C.A.N.; investigation, A.M.D., O.A.O., M.B.D., B.D., C.N., C.A.N.; resources, A.M.D., O.A.O., A.I., B.B., C.N., M.B.D., B.D., I.T., C.A.N.; data curation, A.M.D., O.A.O., M.B.D., B.D., C.A.N. and A.I.; writing—original draft preparation, A.M.D., O.A.O., M.B.D. and B.D.; writing—review and editing, A.M.D., O.A.O., M.B.D., B.D. and B.B.; visualization, A.M.D.; supervision, A.M.D., O.A.O. and I.T.; project administration, A.M.D. and O.A.O.; All authors have read and agreed to the published version of the manuscript.

Funding: This research received no external funding.

Institutional Review Board Statement: Not applicable.

Informed Consent Statement: Not applicable.

Data Availability Statement: Not applicable.

Conflicts of Interest: The authors declare no conflict of interest.

References

1. Bolognia, J.L.; Schaffer, J.V.; Cerroni, L. *Dermatology*, 4th ed.; Elsevier: Amsterdam, The Netherlands, 2018; Chapter 113; pp. 1920–1933.
2. Kaufman, D.K.; Kimmel, D.W.; Parisi, J.E.; Michels, V.V. A familial syndrome with cutaneous malignant melanoma and cerebral astrocytoma. *Neurology* **1993**, *43*, 1728–1731. [CrossRef] [PubMed]
3. Takita, J.; Hayashi, Y.; Nakajima, T.; Adachi, J.; Tanaka, T.; Yamaguchi, N.; Ogawa, Y.; Hanada, R.; Yamamoto, K.; Yokota, J. The p16 (CDKN2A) gene is involved in the growth of neuroblastoma cells and its expression is associated with prognosis of neuroblastoma patients. *Oncogene* **1998**, *17*, 3137–3143. [CrossRef] [PubMed]
4. Hayward, N.K. Genetics of melanoma predisposition. *Oncogene* **2003**, *22*, 3053–3062. [CrossRef]
5. Soura, E.; Eliades, P.J.; Shannon, K.; Stratigos, A.J.; Tsao, H. Hereditary melanoma: Update on syndromes and management: Genetics of familial atypical multiple mole melanoma syndrome. *J. Am. Acad. Dermatol.* **2016**, *74*, 395–407. [CrossRef] [PubMed]
6. Laud, K.; Marian, C.; Avril, M.F.; Barrois, M.; Chompret, A.; Goldstein, A.M.; Tucker, M.A.; Clark, P.A.; Peters, G.; Chaudru, V.; et al. Comprehensive analysis of CDKN2A (p16INK4A/p14ARF) and CDKN2B genes in 53 melanoma index cases considered to be at heightened risk of melanoma. *J. Med. Genet.* **2006**, *43*, 39–47. [CrossRef]
7. Cen, L.; Carlson, B.L.; Schroeder, M.A.; Ostrem, J.L.; Kitange, G.J.; Mladek, A.C.; Fink, S.R.; Decker, P.A.; Wu, W.; Kim, J.S.; et al. p16-Cdk4-Rb axis controls sensitivity to a cyclin-dependent kinase inhibitor PD0332991 in glioblastoma xenograft cells. *Neuro Oncol.* **2012**, *14*, 870–881. [CrossRef]
8. Petronzelli, F.; Sollima, D.; Coppola, G.; Martini-Neri, M.E.; Neri, G.; Genuardi, M. CDKN2A germline splicing mutation affecting both p16(ink4) and p14(arf) RNA processing in a melanoma/neurofibroma kindred. *Genes Chromosomes Cancer* **2001**, *31*, 398–401. [CrossRef]
9. Knappskog, S.; Geisler, J.; Arnesen, T.; Lillehaug, J.R.; Lønning, P.E. A novel type of deletion in the CDKN2A gene identified in a melanoma-prone family. *Genes Chromosomes Cancer* **2007**, *45*, 1155–1163, Erratum in: *Genes Chromosomes Cancer.* **2007**, *46*, 716. [CrossRef]
10. Loo, J.C.; Liu, L.; Hao, A.; Gao, L.; Agatep, R.; Shennan, M.; Summers, A.; Goldstein, A.M.; Tucker, M.A.; Deters, C.; et al. Germline splicing mutations of CDKN2A predispose to melanoma. *Oncogene* **2003**, *22*, 6387–6394. [CrossRef]
11. Harland, M.; Mistry, S.; Bishop, D.T.; Bishop, J.A. A deep intronic mutation in CDKN2A is associated with disease in a subset of melanoma pedigrees. *Hum. Mol. Genet.* **2001**, *10*, 2679–2686. [CrossRef]
12. Zhao, R.; Choi, B.Y.; Lee, M.H.; Bode, A.M.; Dong, Z. Implications of Genetic and Epigenetic Alterations of CDKN2A (p16(INK4a)) in Cancer. *EBioMedicine* **2016**, *8*, 30–39. [CrossRef] [PubMed]
13. Goldstein, A.M. Familial melanoma, pancreatic cancer and germline CDKN2A mutations. *Hum. Mutat.* **2004**, *23*, 630. [CrossRef]
14. Castejón-Griñán, M.; Herraiz, C.; Olivares, C.; Jiménez-Cervantes, C.; García-Borrón, J.C. cAMP-independent non-pigmentary actions of variant melanocortin 1 receptor: AKT-mediated activation of protective responses to oxidative DNA damage. *Oncogene* **2018**, *37*, 3631–3646. [CrossRef] [PubMed]
15. Harding, R.M.; Healy, E.; Ray, A.J.; Ellis, N.S.; Flanagan, N.; Todd, C.; Dixon, C.; Sajantila, A.; Jackson, I.J.; Birch-Machin, M.A.; et al. Evidence for variable selective pressures at MC1R. *Am. J. Hum. Genet.* **2000**, *66*, 1351–1361. [CrossRef]
16. Box, N.F.; Wyeth, J.R.; O'Gorman, L.E.; Martin, N.G.; Sturm, R.A. Characterization of melanocyte stimulating hormone receptor variant alleles in twins with red hair. *Hum. Mol. Genet.* **1997**, *6*, 1891–1897. [CrossRef]
17. Palmer, J.S.; Duffy, D.L.; Box, N.F.; Aitken, J.F.; O'Gorman, L.E.; Green, A.C.; Hayward, N.K.; Martin, N.G.; Sturm, R.A. Melanocortin-1 receptor polymorphisms and risk of melanoma: Is the association explained solely by pigmentation phenotype? *Am. J. Hum. Genet.* **2000**, *66*, 176–186. [CrossRef]
18. Box, N.F.; Duffy, D.L.; Chen, W.; Stark, M.; Martin, N.G.; Sturm, R.A.; Hayward, N.K. MC1R genotype modifies risk of melanoma in families segregating CDKN2A mutations. *Am. J. Hum. Genet.* **2001**, *69*, 765–773. [CrossRef]
19. Cao, J.; Wan, L.; Hacker, E.; Dai, X.; Lenna, S.; Jimenez-Cervantes, C.; Wang, Y.; Leslie, N.R.; Xu, G.X.; Widlund, H.R.; et al. MC1R is a potent regulator of PTEN after UV exposure in melanocytes. *Mol. Cell.* **2013**, *51*, 409–422. [CrossRef]
20. Georgescu, M.M. PTEN Tumor Suppressor Network in PI3K-Akt Pathway Control. *Genes Cancer* **2010**, *1*, 1170–1177. [CrossRef]
21. Cai, J.; Zhang, C.; Zhang, W.; Wang, G.; Yao, K.; Wang, Z.; Li, G.; Qian, Z.; Li, Y.; Jiang, T.; et al. ATRX, IDH1-R132H and Ki-67 immunohistochemistry as a classification scheme for astrocytic tumors. *Oncoscience* **2016**, *3*, 258–265. [CrossRef]
22. Cai, J.; Zhu, P.; Zhang, C.; Li, Q.; Wang, Z.; Li, G.; Wang, G.; Yang, P.; Li, J.; Han, B.; et al. Detection of ATRX and IDH1-R132H immunohistochemistry in the progression of 211 paired gliomas. *Oncotarget* **2016**, *7*, 16384–16395. [CrossRef]
23. Lee, K.S.; Choe, G.; Nam, K.H.; Seo, A.N.; Yun, S.; Kim, K.J.; Cho, H.J.; Park, S.H. Immunohistochemical classification of primary and secondary glioblastomas. *Korean J. Pathol.* **2013**, *47*, 541–548. [CrossRef] [PubMed]
24. Crespo, I.; Vital, A.L.; Gonzalez-Tablas, M.; Patino Mdel, C.; Otero, A.; Lopes, M.C.; de Oliveira, C.; Domingues, P.; Orfao, A.; Tabernero, M.D. Molecular and Genomic Alterations in Glioblastoma Multiforme. *Am. J. Pathol.* **2015**, *185*, 1820–1833. [CrossRef]
25. Ogura, R.; Tsukamoto, Y.; Natsumeda, M.; Isogawa, M.; Aoki, H.; Kobayashi, T.; Yoshida, S.; Okamoto, K.; Takahashi, H.; Fujii, Y.; et al. Immunohistochemical profiles of IDH1, MGMT and P53: Practical significance for prognostication of patients with diffuse gliomas. *Neuropathology* **2015**, *35*, 324–335. [CrossRef] [PubMed]

26. Gershenwald, J.E.; Scolyer, R.A.; Hess, K.R.; Sondak, V.K.; Long, G.V.; Ross, M.I.; Lazar, A.J.; Faries, M.B.; Kirkwood, J.M.; McArthur, G.A.; et al. Melanoma staging: Evidence-based changes in the American Joint Committee on Cancer eighth edition cancer staging manual. *CA Cancer J. Clin.* **2017**, *67*, 472–492. [CrossRef] [PubMed]
27. Garbe, C.; Peris, K.; Hauschild, A.; Saiag, P.; Middleton, M.; Bastholt, L.; Grob, J.J.; Malvehy, J.; Newton-Bishop, J.; Stratigos, A.J.; et al. Diagnosis and treatment of melanoma. European consensus-based interdisciplinary guideline—Update 2016. *Eur. J. Cancer* **2016**, *63*, 201–217. [CrossRef]
28. Cohen, A.; Holmen, S.; Colman, H. IDH1 and IDH2 Mutations in Gliomas. *Curr. Neurol. Neurosci. Rep.* **2013**, *13*, 345. [CrossRef]
29. Horbinsky, C. What do we know about IDH1/2 mutations so far, and how do we use it? *Acta Neuropathol.* **2013**, *125*, 621–636. [CrossRef]
30. Dang, L.; White, D.W.; Gross, S.; Bennett, B.D.; Bittinger, M.A.; Driggers, E.M.; Fantin, V.R.; Jang, H.G.; Jin, S.; Keenan, M.C.; et al. Cancer-associated IDH1 mutations produce 2-hydroxyglutarate. *Nature* **2009**, *462*, 739–744. [CrossRef]
31. Xu, W.; Yang, H.; Liu, Y.; Yang, Y.; Wang, P.; Kim, S.H.; Ito, S.; Yang, C.; Wang, P.; Xiao, M.T.; et al. Oncometabolite 2-hydroxyglutarate is a competitive inhibitor of α-ketoglutarate-dependent dioxygenases. *Cancer Cell.* **2011**, *19*, 17–30. [CrossRef]
32. National Comprehensive Cancer Network. *Clinical Practice Guidelines in Oncology—Central Nervous System Cancers*; Version I.2018; NCCN.Org: Bethesda, MD, USA, 2018.
33. Frigerio, S.; Disciglio, V.; Manoukian, S.; Peissel, B.; Della Torre, G.; Maurichi, A.; Collini, P.; Pasini, B.; Gotti, G.; Ferrari, A.; et al. A large de novo 9p21.3 deletion in a girl affected by astrocytoma and multiple melanoma. *BMC Med. Genet.* **2014**, *15*, 59. [CrossRef]
34. Braam, K.I.; Overbeek, A.; Kaspers, G.J.; Ronckers, C.M.; Schouten-van Meeteren, A.Y.; Van Dulmen-Den Broeder, E.; Veening, M.A. Malignant melanoma as second malignant neoplasm in long-term childhood cancer survivors: A systematic review. *Pediatr. Blood Cancer* **2012**, *58*, 665–674. [CrossRef]
35. Pasmant, E.; Laurendeau, I.; Héron, D.; Vidaud, M.; Vidaud, D.; Bièche, I. Characterization of a germ-line deletion, including the entire INK4/ARF locus, in a melanoma-neural system tumor family: Identification of ANRIL, an antisense noncoding RNA whose expression coclusters with ARF. *Cancer Res.* **2007**, *67*, 3963–3969. [CrossRef]
36. Chan, A.K.; Han, S.J.; Choy, W.; Beleford, D.; Aghi, M.K.; Berger, M.S.; Shieh, J.T.; Bollen, A.W.; Perry, A.; Phillips, J.J.; et al. Familial melanoma-astrocytoma syndrome: Synchronous diffuse astrocytoma and pleomorphic xanthoastrocytoma in a patient with germline CDKN2A/B deletion and a significant family history. *Clin. Neuropathol.* **2017**, *36*, 213–221. [CrossRef] [PubMed]
37. Fiondella, L.; Cavallieri, F.; Napoli, M.; Froio, E.; Serra, S.; Borsari, S.; Giaccherini, L.; Ghadirpour, R.; Cozzi, S.; Pascarella, R.; et al. Glioblastoma and malignant melanoma: Serendipitous or anticipated association? *Neuropathology* **2021**, *41*, 489–491. [CrossRef] [PubMed]
38. Malmström, A.; Grønberg, B.H.; Marosi, C.; Stupp, R.; Frappaz, D.; Schultz, H.; Abacioglu, U.; Tavelin, B.; Lhermitte, B.; Hegi, M.E.; et al. Temozolomide versus standard 6-week radiotherapy versus hypofractionated radiotherapy in patients older than 60 years with glioblastoma: The Nordic randomised, phase 3 trial. *Lancet Oncol.* **2012**, *13*, 916–926. [CrossRef] [PubMed]
39. Tamimi, A.F.; Juweid, M. Epidemiology and Outcome of Glioblastoma. In *Glioblastoma [Internet]*; De Vleeschouwer, S., Ed.; Codon Publications: Brisbane, Australia, 2017; Chapter 8. Available online: https://www.ncbi.nlm.nih.gov/books/NBK470003/ (accessed on 12 February 2022). [CrossRef]

diagnostics

MDPI

Case Report

Clinical Remission in a 72-Year-Old Patient with a Massive Primary Cutaneous Peripheral T-Cell Lymphoma-NOS of the Eyelid, Following Combination Chemotherapy with Etoposide Plus COP

Sabina Iluta [1,2], Dragos-Alexandru Termure [3,4], Bobe Petrushev [5,6], Bogdan Fetica [7], Mindra-Eugenia Badea [4], Madalina Moldovan-Lazar [3], Manuela Lenghel [8], Csaba Csutak [8], Andrei Roman [8], Sergiu Pasca [1,5], Alina-Andreea Zimta [5], Ciprian Jitaru [2], Ciprian Tomuleasa [1,2,5,*] and Rares-Calin Roman [3]

1 Department of Hematology, Iuliu Hatieganu University of Medicine and Pharmacy, 400337 Cluj-Napoca, Romania; iluta.sabina@yahoo.com (S.I.); pasca.sergiu123@gmail.com (S.P.)
2 Department of Hematology, Ion Chiricuta Clinical Cancer Center, 400124 Cluj-Napoca, Romania; Ciprianjitaru.jitaru@gmail.com
3 Department of Oral and Maxillofacial Surgery, Iuliu Hatieganu University of Medicine and Pharmacy, 400001 Cluj-Napoca, Romania; dragos.tarmure@gmail.com (D.-A.T.); madilazar@yahoo.com (M.M.-L.); rares.roman@umfcluj.ro (R.-C.R.)
4 Department of Preventive Dentistry, Iuliu Hatieganu University of Medicine and Pharmacy, 400001 Cluj-Napoca, Romania; rbadea2012@gmail.com
5 Medfuture Research Center for Advanced Medicine, Iuliu Hatieganu University of Medicine and Pharmacy, 400337 Cluj-Napoca, Romania; bobe.petrushev@gmail.com (B.P.); zimta.alina.andreea@gmail.com (A.-A.Z.)
6 Department of Pathology, Octavian Fodor Regional Institute of Gastroenterology and Hepatology, 400139 Cluj-Napoca, Romania
7 Department of Pathology, Ion Chiricuta Clinical Cancer Center, 400124 Cluj-Napoca, Romania; feticab@yahoo.com
8 Department of Radiology, Iuliu Hatieganu University of Medicine and Pharmacy, 400124 Cluj-Napoca, Romania; lenghel.manuela@gmail.com (M.L.); csutakcsaba@yahoo.com (C.C.); andrei.roman678@gmail.com (A.R.)
* Correspondence: ciprian.tomuleasa@umfcluj.ro

Received: 25 July 2020; Accepted: 16 August 2020; Published: 24 August 2020

Abstract: Peripheral T-cell lymphoma not otherwise specified (PTCL-NOS) is the rarest subtype of primary cutaneous lymphoma, accounting for approximately 2% of cutaneous lymphomas. The rarity of primary cutaneous PTCL-NOS means that there is a paucity of data regarding clinical and histopathological features and its clinical course. This malignancy is an aggressive and life-threatening hematological malignancy that often presents mimicking other less severe plaque-like skin conditions. Due to the nonspecific nature of these lesions, CD4-positive cutaneous T-cell lymphoma (CTCL) is often misdiagnosed as either mycosis fungoides or Sezary syndrome. We describe a patient who presented with a large tumoral mass in the right frontal area, with involvement of the right upper eyelid and the ocular globe, causing loss of vision greatly impacting the quality of life. Biopsy revealed primary cutaneous PTCL-NOS, treated successfully with cyclophosphamide, doxorubicin, vincristine, and prednisone (CHOP) plus etoposide combination chemotherapy. As elderly patients are indicated to receive attenuated doses of chemotherapy, CHOP-based regimens represent viable options.

Keywords: primary cutaneous T-cell lymphoma; eyelid; chemotherapy; elderly patient

1. Introduction

Peripheral T-cell lymphomas not otherwise specified (NOS) are a heterogeneous group of lymphomas that do not have well-defined criteria for being able to fit into a particular type of lymphoma [1–4]. The incidence of T-cell lymphomas in our center is small, only 8.2%. Of these, the most common is the angioimmunoblastic subtype. Primary cutaneous lymphomas are a heterogeneous group of non-Hodgkin lymphomas presenting in the skin with no evidence of lymph nodes, bone marrow, or viscera involvement at the time of diagnosis.

Cutaneous lymphomas exhibit various clinical, histological, immunophenotypic, and genetic features. Moreover, they differ in prognosis and treatment from systemic lymphomas with similar histological features [5–8]. Up to 75–80% of cutaneous lymphomas originate from T-cells, and only 20–25% from B-cells [9–11]. Many subtypes of cutaneous T-cell lymphomas were described, and the most common ones are mycosis fungoides and Sézary syndrome. Primary cutaneous peripheral T-cell lymphoma-NOS (PTCL-NOS) is the rarest form of cutaneous T-cell lymphomas, accounting for approximately 2% of all cutaneous lymphomas [12–14]. PTCL-NOS is a diagnosis of exclusion, based on the presence of typical histopathological features of lymphoma, an aberrant T-cell immunophenotype, often with a loss of CD5 and CD7, and a clonal T-cell receptor (TCR) gene rearrangement, in the appropriate clinical context [15,16].

Primary cutaneous PTCL-NOS are clinically aggressive and heterogeneous, varying from localized to generalized plaques or nodules commonly associated with constitutional symptoms [17]. Patients often present with advanced stage, lesions are often misdiagnosed, which both worsen the prognosis: five-year overall survival is about 20% [18,19]. Due to the rarity of cases, PTCL-NOS algorithms for diagnosis, workup, and treatment are lacking, especially in the elderly population. Thus, in the present manuscript we describe the successful clinical management of a PTCL-NOS of massive dimensions of the eye in a 72-year-old lady, with cardiac comorbidities.

2. Patient Information

A 72-year-old Romanian woman came into the Hematology service of Ion Chiricuta Clinical Cancer Center (IOCN) from Cluj-Napoca, Romania. The primary concerns as expressed by the patient were as follows: two constantly growing tumors, the loss of vision in the right eye, and the overall negative impact on her quality of life. The main symptom was itching in the tumor area, accompanied by occasional mild pain. Bleeding was noticed by the patient after low-intensity trauma to the lesion, usually caused by clothing. No spontaneous bleeding was reported. The patient was in good general health, with general B signs that included weight loss (about 15 kg in the last month) and excessive sweating. The patient had grade II hypertension but denied personal or family history of malignancy. On her initial visit, physical examination revealed a 6 cm ill-defined tumor-like lesion, red-violaceous with visible telangiectasia that presented with an ulcerated central area and crusting (Figure 1a–c). On palpation, the lesion was immobile, indurated and nontender. Upon clinical examination, we also noticed the tumefaction and skin erythema of the lower-central frontal area and glabellar area together with a 1 cm round-shaped crust. The patient did not present with palpable lymph nodes or hepato-splenomegaly. Because of the infected appearance of the lesion and the presence of odor, microbiological samples were obtained from the deepest ulcerated area and were taken to the laboratory. Microbiological analysis confirmed the clinical suspicion of infection and identified methicillin-resistant *Staphylococcus aureus* (MRSA) and *Enterobacter cloacae* complex. The patient was put on teicoplanin after the antibiogram confirmed the susceptibility of both bacterial pathogens to this antibiotic. Biologically, there was mild leukocytosis with neutrophils, and increased lactate dehydrogenase (LDH). No atypically cells were found on the blood smear. cytomegalovirus (CMV), Epstein-Barr virus (EBV), human T-cell leukemia virus (HTLV) viral investigations were performed and all came back negative.

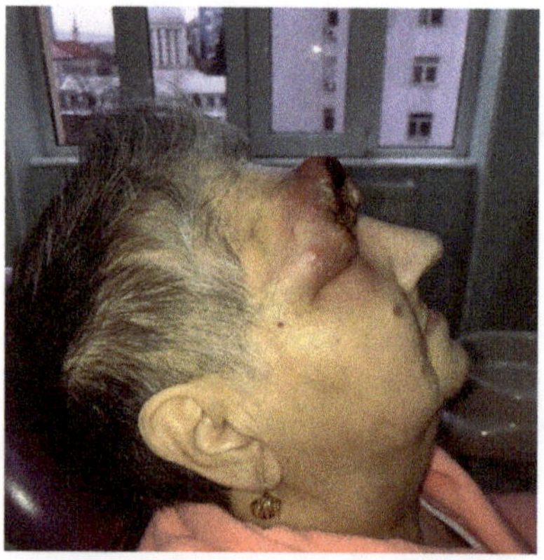

(a)

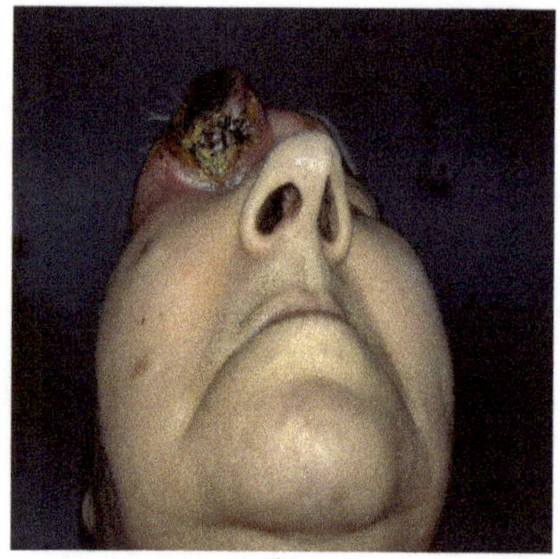

(b)

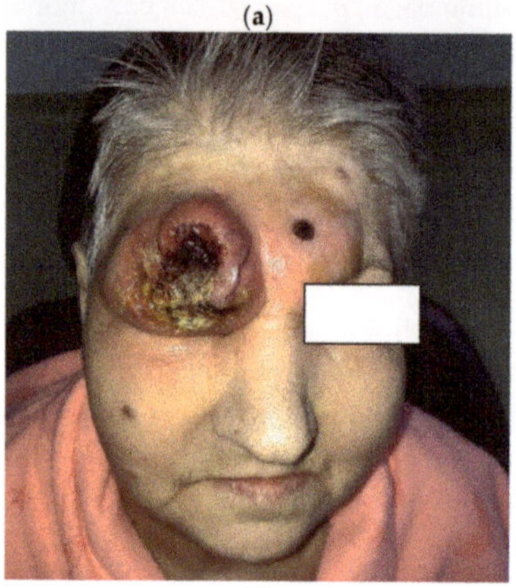

(c)

Figure 1. (**a**–**c**) The initial visit, which revealed a 6 cm ill-defined tumor-like lesion, red-violaceous with visible telangiectasia that presented with an ulcerated central area and crusting.

3. Diagnostics

The patient reported the appearance of the first tumor two years before her first visit into our hospital. She initially presented to the doctor's office with an infiltrated, slightly itchy, erythematous-purple plaque in the frontal level. Skin biopsy was performed in Regional County Emergency Hospital. Histopathological examination revealed polymorphous infiltrate colored mainly with the CD3, a T-cell marker. There were also numerous CD20+ B cells, which formed one or two reactive-looking lymphoid structures. The T infiltrate stained with CD4 in most cells. The Bcl-6-negativity and CD10 excluded a centrofollicular skin lymphoma. The absence of nodular CD20-positive collections did not favor the diagnosis of cutaneous lymphoma of the marginal area. The conclusion was that the clinical aspect excluded cutaneous lymphoma with small-to-medium pleomorphic T-cells, and it was in favor of a lupus erythematosus tumor. Treatment with Plaquenil (200 mg, 1 tb/day) and control over 3 months was indicated. The patient followed the treatment, but the general condition did not improve. The eyelid tumor increased, obstructing the view. The patient was then referred to a dermatology department for further investigations. The biopsy taken at the eyelid tumor level was performed again, and the biopsy was examined in a private practice pathological department. Histopathological examination revealed polymorphous inflammatory infiltrates, with predominantly small- and medium-sized lymphocytes, several granulocytes, histiocytes, and diffusely distributed plasmocytes in small nodules throughout the examined material. Immunohistochemistry analysis detected CD3-positive, CD4-positive, CD8-low positive, CD20-very low positive, CD68-positive, and CD30-negative cells. The histological aspect of the tissue sent for analysis suggested either Spiegler–Fendt sarcoid, Jessner's lymphocytic infiltrate, or lupus erythematosus tumidus. The patient was started on hydroxychloroquine on the basis of the diagnosis, but the therapeutic choice proved to be ineffective in the following 6 months.

The symptoms did not improve, and thus the patient was referred to our department. On the basis of primary clinical evaluation, our primary diagnostic was that of mycosis fungoides (MF)/Sézary syndrome (SS). The conclusion of a final diagnosis was even more difficult because our pathologists did not have access to previous histopathological preparations. At this stage, the possibility of diagnosing MF—tumor stage having the immunophenotype of cells by their positivity for CD3, CD4, and CD7 was discussed, thus posing even more difficulty in formulating a final diagnostic. Following this, we decided to take a new biopsy from the tumor formation. The eyelid tumor was re-biopsied, and the biopsy preparation was analyzed by our team of pathologists. The hematoxylin–eosin-stained samples highlighted that the skin presented with hypodermis and ancillary structures; the tissue had a faded morphology and it was infiltrated by a malignant tumor proliferation areas, with diffuse architecture that infiltrated the dermis and subcutaneous adipose tissue, with isolated and discrete phenomena of epidermotropism, angiocentricity, and focal infiltration of skin annexes (Figure 2a,b). It was composed of atypical cells with lymphoid appearance, having a small-to-medium size with reduced cell-to-cell cohesiveness and low-to-moderate pleomorphism. The boundaries of tumor cells were hard to define and showed pale eosinophilic cytoplasm and amphophilia. The nuclei were round or oval, some of them even slightly angled. Most of the nuclei had vesicular chromatin with small nucleoli that were difficult to see, but also relatively evenly dispersed fine-grained chromatin (Figure 2c,d). Numerous mitotic areas with an increased mitotic index were also noted.

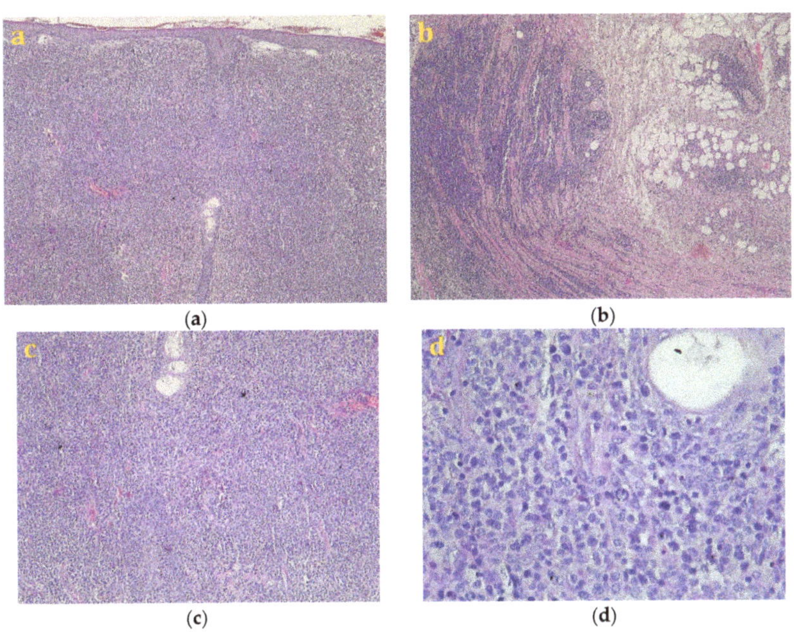

(**a**) (**b**) (**c**) (**d**)

Figure 2. Pathology diagnosis of a primary cutaneous peripheral T-cell lymphoma not otherwise specified—Hematoxylin–Eosin staining. ((**a**), Magnification 40×, (**b**), magnification 100×, (**c**), Magnification 200×, (**d**), Magnification 400×).

Following the immunohistochemical study, we obtained the following conclusive results. The tumor cells were intensely positive for CD3 (Figure 3b) and CD7 (Figure 3g), a relatively large number of tumor cells were positive for CD2 (Figure 3d) and CD4 (Figure 3e), and only a small population of T lymphocytes or some atypical larger cells were positive for CD8 (Figure 3f), with a $CD4^+/CD8^+$ ratio favoring CD4+ cells. Tumor cells were negative for CD30 (Figure 3i), CD5 (Figure 3c), and CD 56 (Figure 3h). A small population of atypical lymphocytes were positive in granzyme B after immunohistochemical staining. The cell proliferation index quantified by immunolabeling for Ki-67 was estimated and quantified at 80–85% (Figure 3j). Following the morphological examination correlated with special stains, the diagnosis of primary cutaneous T-cell lymphoma NOS was established. Whole-body CT was performed and revealed tumor formation in the soft epicranial parts, right frontal area, and supraorbital area, with dimensions of 22/50/45 mm. The lesion affected the upper eyelid and the paranasal region, with no signs of ocular or intracranial invasion. In the frontal on the median and parasagital line on the left, epicranial area, adjacent to the described lesion, another structure with 35/15 mm contrast socket was highlighted. The dynamic contrast-enhanced magnetic resonance imaging (MRI) revealed a large, inhomogeneous lesion, with infiltrative, invasive contour. The origin of the tumor was most probably from the periorbital superficial tissues, with invasion in the orbit and direct contact to the anteromedial aspect of the ocular globe without the presence of a cleavage plane between the two structures (Figures 4a,b and 5a,b). Figure 4a shows that the T1 FSE (fast spin echo) axis and sagittal T1 FSE contrast-enhanced with fat-saturation sequence was not enhanced (a), whereas Figure 4b shows inhomogeneous enhancement of the lesion, with superficial tissue origin and invasion in the right orbit. Figure 5a,b depict the axis T2 PROPELLER sequence showing the necrotic areas in the primary and secondary lesion, satellite lesion in the superficial tissues of the frontal region (marked by arrows in the picture). A restricted diffusion on diffusion weighted imaging (DWI) sequences was observed, with an apparent diffusion coefficient (ADC) value of 0.873×10^{-3} mm^2/s on the ADC map (Figure 6a). Dynamic contrast-enhanced MRI (DCE-MRI) imaging showed a type III (C) curve highly suggestive for malignancy (Figure 6b), and a high constant transfer (ktrans) value was observed, pointing to a possible good response to radiochemotherapy (Figure 7). In the proximity of the left frontal area, two nodular superficial lesions were identified, suspected of secondary nodules.

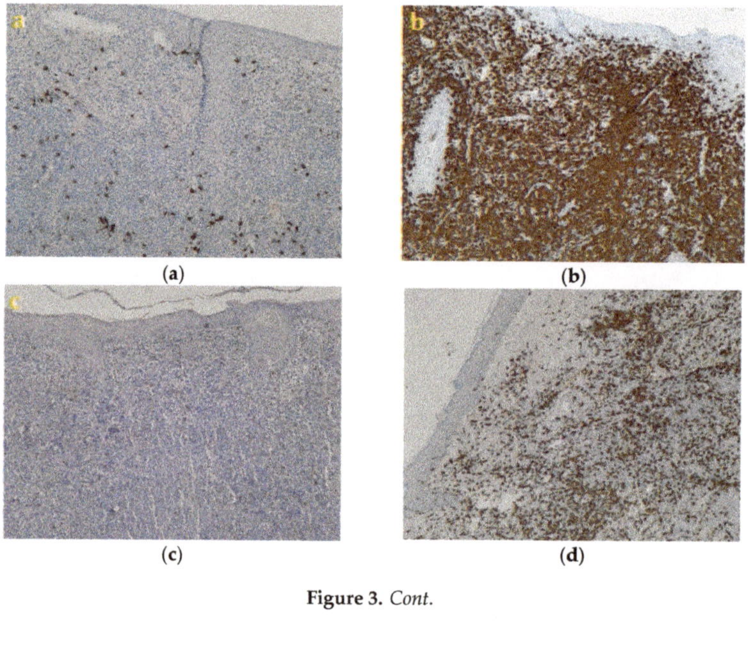

(a) (b) (c) (d)

Figure 3. *Cont.*

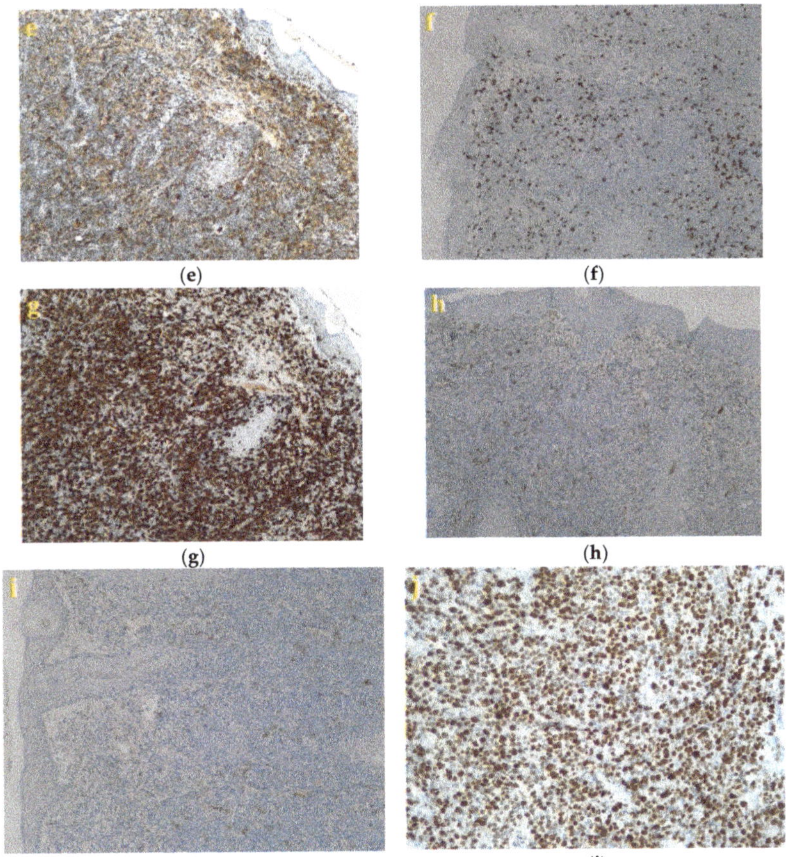

(**e**) (**f**)
(**g**) (**h**)
(**i**) (**j**)

Figure 3. Immunohistochemistry staining for the following markers: CD20 (**a**, Magnification 200×), CD3 (**b**, Magnification 200×), CD5 (**c**, Magnification 200×), CD2 (**d**, Magnification 200×), CD4 (**e**, Magnification 200×), CD8 (**f**, Magnification 200×), CD7 (**g**, Magnification 200×), CD56 (**h**, Magnification 200×), CD30 (**i**, Magnification 200×), Ki-67 (**j**, Magnification 200×).

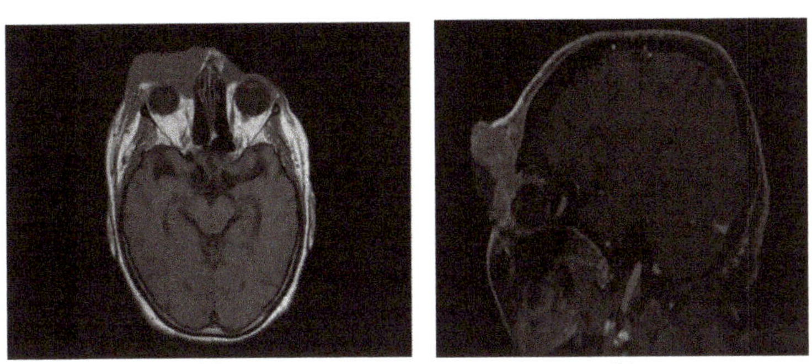

(**a**) (**b**)

Figure 4. Ax T1 FSE non-enhanced (**a**) and sagittal T1 FSE contrast-enhanced with fat-saturation sequence (**b**). Inhomogeneous enhancement of the lesion, with superficial tissue origin and invasion in the right orbit.

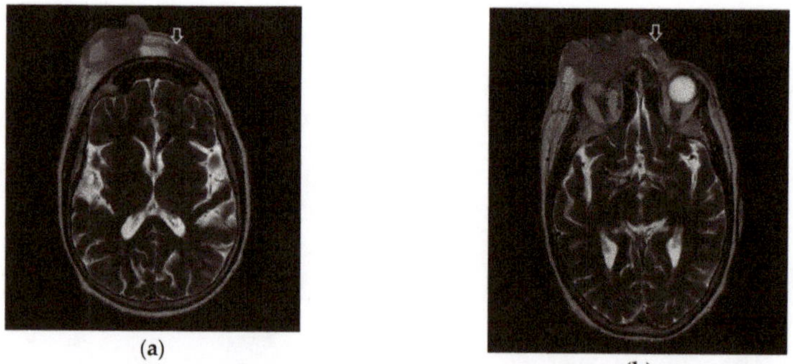

(a) (b)

Figure 5. (**a**,**b**) The ax T2 PROPELLER sequence showing the necrotic areas in the primary lesion and secondary, satellite lesion in the superficial tissues of the frontal region (arrows).

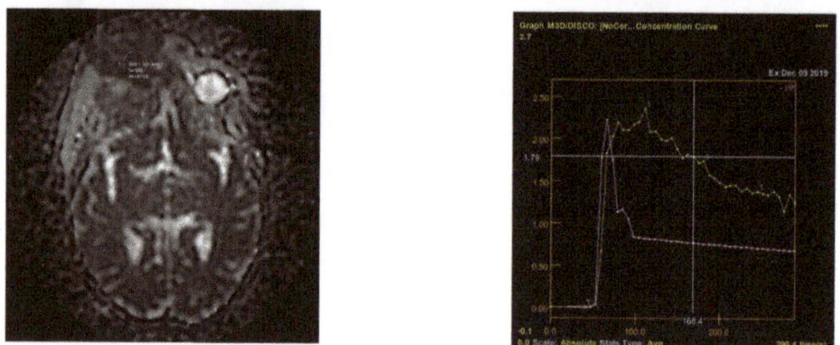

(a) (b)

Figure 6. (**a**) Restricted diffusion on DWI sequences was observed, with an ADC value of 0.873×10^{-3} mm^2/s on the ADC map; (**b**) Dynamic contrast-enhanced MRI (DCE-MRI) imaging showed a type III (C) curve highly suggestive for malignancy.

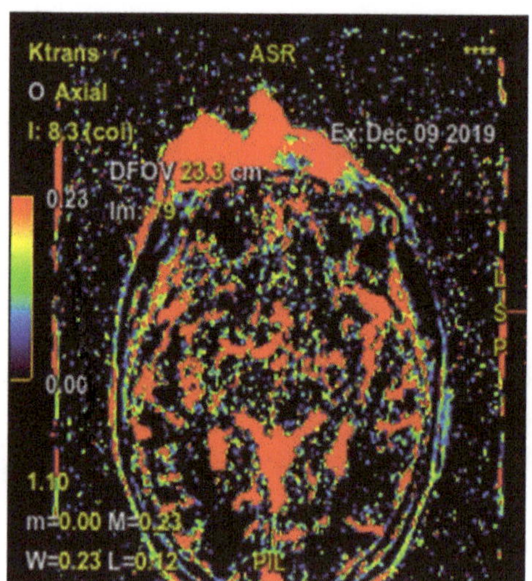

Figure 7. High constant transfer (ktrans) value was observed, pointing to a possible good response for radiochemotherapy.

4. Therapy

First, in the regional hospital, where the patient was first attended, the initial therapeutic choice was a six-months regiment of hydroxychloroquine. This showed no result, since the diagnostic was incorrect, due to limited histopathological and imaging analysis of the case.

When she presented to our department, we performed a second biopsy of the tumor and gave the accurate diagnostic. Considering the age of the patient and the associated cardiac pathology, we decided to initiate combination chemotherapy using the COP plus etoposide regimen, without anthracycline. During initial evolution, we also identified that she had bacterial superinfection with MRSA in the tumor. Under appropriate antibiotic treatment, the evolution of MRSA infection was favorable. After six chemotherapy cycles, the evolution of the tumor was spectacular, with complete oncological remission of the tumoral mass. The patient regained her vision in the right eye (Figure 8a–c).

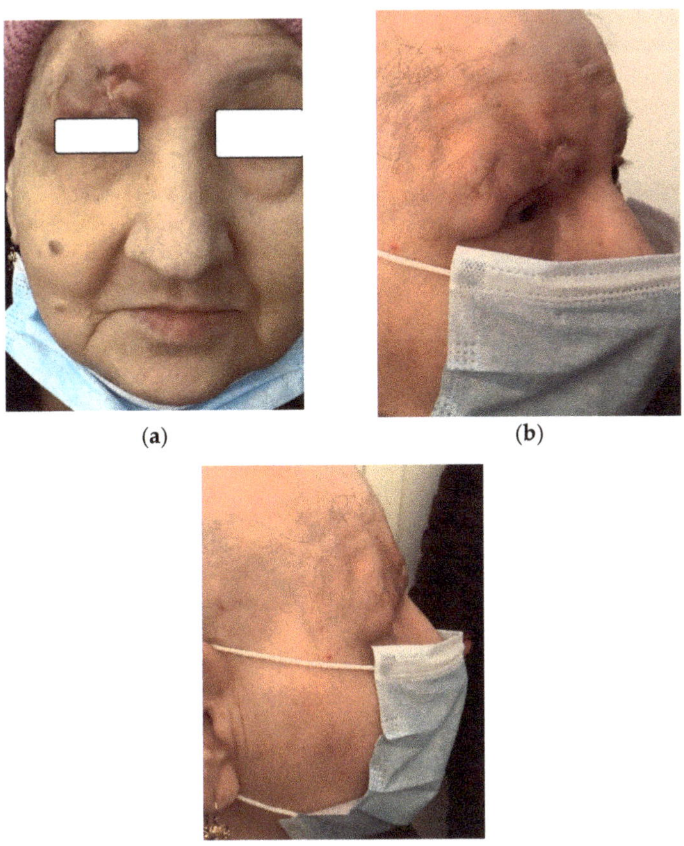

(a) (b) (c)

Figure 8. (a–c) Clinical remission following chemotherapy.

5. Follow-Up

The case report presented the immediate response to therapy in an impressive tumor. Unfortunately, the follow-up was not possible as the patient died of acute respiratory distress syndrome, within a month after completion of chemotherapy. This event was most probably related to her cardiovascular pathology since the lymphoma tumor was in remission. The exact cause of death was not fully established, since the pathology report was done in a local hospital.

6. Discussions

Over the past years, the incidence of non-Hodgkin lymphomas has steadily risen, mainly the B-cell lymphomas as they represent the great majority of non-Hodgkin's lymphomas (NHL). T-cell and NK-cell lymphomas are generally underrepresented due to their overall low incidence, these two diseases also have a wide variability in different geographical regions and racial populations [20,21]. T- and NK-cell lymphomas are relatively rare, clinically, these subtypes are aggressive and quite heterogenous and they originate from post-thymic T-lymphocytes and NK-cells. Of all NHL, 10–15% are represented by T- and NK-cell lymphomas [22,23]. The peak incidence of T-cell lymphomas is in adults found in their sixties. These pathologies usually develop from mature T-cells and they rarely affect children. Men have a higher incidence of this disease, in comparison with women. The occurrence of adult T-cell lymphomas may be linked to previous viral infections with human T-lymphotropic virus type 1 (HTLV-1) and Epstein–Barr virus (EBV). Some medical conditions may be linked to special forms of T-cell lymphomas, like enteropathy-associated T-cell lymphoma (EATL). This form of T-cell lymphoma is associated with celiac disease and represents a rare gastrointestinal form of NHL that originates from intraepithelial T-lymphocytes [24–26]. Considering that, in general, T-cell lymphomas are a rarely occurring type of malignancies with a high number of subtypes, the lessons learned from each rare case are valuable. Our case report shows the importance of performing accurate and detailed histopathological and immunohistochemical characterization in case of suspecting a case of T-cell lymphoma. Our patient, after receiving a correct diagnostic of peripheral T-cell lymphoma-NOS with CD3+/CD7+/CD2+/CD7+/CD30−/CD5−/CD56− and high CD4+/CD8+ ratio, which also had MRSA infection, showed complete remission in a short period of time, after six chemotherapy cycles of COP plus etoposide regimen associated with antibiotics for MRSA.

Peripheral T-cell lymphoma not otherwise specified (PTCL-NOS) is overall the most common lymphoma type of the highly heterogenous group of T-cell lymphomas worldwide. It represents nearly 35% of all T-cell lymphomas in Europe and North America. The PTCL-NOS is a World Health Organization (WHO)-defined diagnostic category of lymphomas [27,28]. It received its name because it cannot be classified into a specific type of T-cell lymphoma. The PTCL-NOS shows association with EBV and exhibits EBV-positive neoplastic T-cells. Most commonly, older adults are affected with a slight male predominance. Generalized lymphadenopathy and B-symptoms are frequently presented, and in advanced stages, bone marrow, spleen, liver, and other tissues may be involved. Extranodal presentation is uncommon but may occur. The most frequent extranodal site is represented by the skin and gastrointestinal tract. With a five-year survival of 20–30%, the prognosis is relatively poor [4,29–32].

T-cell lymphoproliferations remain a real problem and diagnostic challenge for pathologists. Primary cutaneous T-cell lymphoma NOS is largely a diagnosis of exclusion. Among the variants and subtypes of primary cutaneous T-cell lymphoma, the following are listed: lymphoepithelioid (Lennert's lymphoma): numerous histiocytes clustered along with T-cell; T-cells, commonly scattered, larger, more atypical, and pleomorphic cells, including occasional Reed–Sternberg-like cells (usually EBV+); in these, the T-cell subtype, often CD8+. The T-zone lymphoma is no longer considered a variant of PTCL-NOS. Instead, it is considered to have a nonspecific morphologic pattern, which can be seen in PTCL-NOS cases with helper T-cell phenotype that are reclassified as angioimmunoblastic T-cell lymphoma (AITL) or other nodal lymphomas of T follicular helper cell origin. The other excluded variant is the follicular variant that is also no longer considered a variant of PTCL-NOS and was reclassified by WHO as AITL and other nodal lymphomas of T follicular helper cell origin. In peripheral T-cell lymphoma associated with B-cell proliferation, the B-cells are small, alongside mature plasma cells and plasmacytoid large B lymphocytes or plasmablastic B-cells. These are often EBV+.

The differential diagnosis of primary cutaneous T-cell lymphoma NOS includes the following entities: angioimmunoblastic T-cell lymphoma (AITL) which has similar morphology to NOS; but it may be more prominent in arborizing high endothelial venules in AITL. The infiltrate may be more polymorphous than in PTCL-NOS. B-cells EBV+ are present in most cases of AITL. The primary cutaneous presentation being extreme; anaplastic large-cell lymphoma, ALK-positive and ALK-negative (ALCL), has cells with hallmark morphology, sinus involvement, and anaplastic cytologic features supporting ALCL. The ALCL is strong and has uniform CD30+ cells in a membranous and paranuclear pattern. CD30 marker can also be focally positive and represented in a lower proportion in PTCL-NOS. ALK-negative subtype shows no evidence of ALK locus abnormalities; whereas the ALK-positive subtype presents with ALK locus abnormalities, more precisely(2;5)(p23;q35) NPM1/ALK. Histologically and immunophenotypically, PTCL and adult T-cell leukemia/lymphoma (ATLL) can be indistinguishable. However, adult T-cell leukemia/lymphoma is caused by infection by human T-cell leukemia virus type 1 (HTLV-1). HTLV-1 is integrated into the genome of neoplastic cells. Patients with ATLL often present with hypercalcemia. The leukemic phase is commonly encountered in ATLL patients. Multilobate and flower-shaped cells are in blood smear. In subcutaneous panniculitis-like T-cell lymphoma, the patients present with subcutaneous nodules, usually without other signs of disease. Atypical lymphoid infiltrates in subcutaneous tissue that involves fat lobules, typically spares septa, overlying dermis, and epidermis, and the tumor cells are CD3+ and CD8+. In typical cases of the nasal type of extranodal NK-/T-Cell lymphoma, the patients present only with extranodal disease. Rare cases have been described at the upper aerodigestive tract level and in extranasal sites: skin, soft tissue, lower gastrointestinal, and testes. This entity is constituted by polymorphous lymphoid infiltrate of variable morphology and/or angioinvasion, often associated with extensive fibrinoid necrosis. The tumor cells are CD3+/−, CD2+, CD56+, and EBV+/−. Skin involvement of enteropathy-associated T-cell lymphoma and hepatosplenic T-cell lymphoma is unusual, and, in our case, they were clinically excluded.

When it comes to interpreting immunohistochemistry (IHC) biomarkers staining for the diagnostic of PTCL-NOS, it is important to include as many as possible since these can only be interpreted as a cluster. For instance, CD30 can either be absent or slightly positive, while CD56 can rarely be positive. CD68 expression with high positivity of CD8 is typical for indolent acral CD8+ T-cell lymphoma that is usually a benign tumor. CD20 can be either less than 1% positive or entirely negative. CD3+/CD4+/CD8− shows a less aggressive type of PTCL-NOS, as compared with CD8 strongly positive PTCL-NOS. An aggressive type of PTCL-NOS is also characterized by positivity for granzyme B [33–36].

A more detailed description of each immunohistochemical marker that we used in our study and how it helped us differentiate between different types of cutaneous lymphomas is found in Table 1.

Table 1. Track-record of the pathology diagnosis

Round	Marker	Status	Meaning
2nd Biopsy	CD3	Positive	A marker typical for T-cell, used to validate T-cell origin of the tumor
	CD4	Positive	Specific with peripheral T-cell lymphoma, not otherwise specified (PTL-NOS), which is also commonly associated with loss of other T-cell markers
	CD8	Positive	Excluding Reed–Sternberg-like cells (usually EBV+) diagnostic and subcutaneous panniculitis-like T-cell, high positivity of CD8 shows a more aggressive type of PTCL Lymphoma, which has a higher population of CD8+ cells
	CD20	A very low proportion of cells	CD20+ should be less than 1% of stained cells
	CD68	Positive	Specific with PTL-NOS, however little information is known about this marker
	CD30	Negative	Usually negative in PTCL-NOS

Table 1. *Cont.*

Round	Marker	Status	Meaning
3rd biopsy	CD3	Positive	A marker typical for T-cell, used to validate T-cell origin of the tumor
	CD7	Positive	Typical T-cell marker, PTCL-NOS commonly express it
	CD2	High number of positive cells	Typical T-cell marker, PTCL-NOS commonly express it
	CD4	High number of positive cells	High positivity of CD4 shows a less aggressive type of PTCL lymphoma
	CD8	Small population of positive cells	excluding Reed–Sternberg-like cells (usually EBV+) diagnostic and subcutaneous panniculitis-like T-cell, high positivity of CD8 shows a more aggressive type of PTCL lymphoma, which has a higher population of CD8+ cells
	CD30	Negative	Exclusion of anaplastic large-cell lymphoma ALK-positive and ALK-negative (ALCL), which is strongly CD30+
	CD5	Negative	Loss of CD5 is specific for PTCL-NOS
	CD56	Negative	Exclusion of extranodal NK-/T-Cell lymphoma, which is CD56+
	Granzyme B	Small population of positive cells	Specific for cytotoxic T-cell lymphomas, a more aggressive phenotype
	Ki-67	Proliferation index of 80–85%	Verify the proliferation activity of cells

7. Conclusions

As elderly patients are indicated to receive attenuated doses of chemotherapy, CHOP-based regimens represent viable options [4,5]. Still, considering that our patient had cardiac comorbidities, we chose to remove anthracyclines from the combination treatment and replaced it with etoposide. The particularity of this case is the impressive huge dimensions of the tumoral mass of the eye. Moreover, the impressive feature of this case is that the patient even regained her vision, proving thus that etoposide plus COP is a viable alternative for such a clinical scenario: elderly patients with invasion of internal organs.

Author Contributions: S.I. and D.-A.T. wrote the manuscript. B.P. and B.F. were responsible for the pathology diagnosis. M.-E.B., M.M.-L., M.L., C.C. and A.R. were responsible for the imaging and diagnostic assessment. S.P., A.-A.Z. and C.J. were responsible for the clinical management of the patient. C.T. and R.-C.R. supervised the work. All authors have read and agreed to the published version of the manuscript.

Funding: Dragos Termure was funded from an internal grant of the School of Doctoral Studies—Iuliu Hatieganu University of Medicine and Pharmacy Cluj Napoca (PCD 2019–2022). This work was supported by a grant from the Romanian Ministry of Research and Innovation, CCCDI-UEFISCDI, Project No. PN-III-P4-ID-PCCF-2016-0112 within PNCDI III, as well as by awarded for Young Research Teams 2020–2022 (Grant No. PN-III-P1-1.1-TE-2019-0271), and by an international collaborative grant of the European Economic Space between Romania and Iceland 2020–2022 (Grant No. 19-COP-0031).

Acknowledgments: Sabina Iluta and Dragos Termure contributed equally and are both considered first-author. The patient has provided informed consent for publication of the case, in accordance with all legal and ethical standings.

Conflicts of Interest: The authors declare no conflict of interest

References

1. Cucuianu, A.; Petrushev, B.; Crisan, D.; Buiga, R.; Rosu, A.-M.; Dima, D.; Tomuleasa, C. Chronic granulomatous dermatosis as a presenting sign of cutaneous T-cell lymphomas. *Eur. J. Dermatol.* **2015**, *25*, 210. [CrossRef] [PubMed]
2. Desmirean, M.; Deak, D.; Rus, I.; Dima, D.; Iluta, S.; Preda, A.; Moldovan, T.; Roman, A.; Tomuleasa, C.; Petrushev, B. Paraneoplastic hypereosinophilia in a patient with peripheral T cell lymphoma, not otherwise specified. *Med. Pharm. Rep.* **2019**, *92*, 421–426. [CrossRef] [PubMed]

3. Gafencu, G.A.; Selicean, S.E.; Petrushev, B.; Cucuianu, A.; Dima, D.; Frinc, I.; Irimie, A.; Pileczki, V.; Berindan-Neagoe, I.; Berce, C.; et al. Clinicopathological analysis of a case series of peripheral T-cell lymphomas, not otherwise specified, of lymphoepithelioid variant (Lennert's lymphoma). A Central European single-center study. *Hum. Pathol.* **2016**, *53*, 192–194. [CrossRef] [PubMed]
4. Ma, H.; Cheng, B.; Falchi, L.; Marchi, E.; Sawas, A.; Bhagat, G.; O'Connor, O.A. Survival benefit in patients with peripheral T-cell lymphomas after treatments with novel therapies and clinical trials. *Hematol. Oncol.* **2019**, *38*, 51–58. [CrossRef] [PubMed]
5. Zain, J.M. Aggressive T-cell lymphomas: 2019 updates on diagnosis, risk stratification, and management. *Am. J. Hematol.* **2019**, *94*, 929–946. [CrossRef]
6. Janíková, A.; Chloupkova, R.; Campr, V.; Klener, P.; Hamouzova, J.; Belada, D.; Prochazka, V.; Pytlik, R.; Pirnos, J.; Duras, J.; et al. First-line therapy for T cell lymphomas: A retrospective population-based analysis of 906 T cell lymphoma patients. *Ann. Hematol.* **2019**, *98*, 1961–1972. [CrossRef]
7. Siaghani, P.J.; Wong, J.T.; Chan, J.; Weisenburger, D.D.; Song, J.Y. Epidemiology and Pathology of T-and NK-Cell Lymphomas. *Cancer Treat. Res.* **2018**, *176*, 1–29. [CrossRef]
8. Griesel, C.; Desmirean, M.; Esterhuizen, T.; Pașca, S.; Petrushev, B.; Selicean, C.; Roman, A.; Fetica, B.; Teodorescu, P.; Swanepoel, C.; et al. Differential Diagnosis of Malignant Lymphadenopathy Using Flow Cytometry on Fine Needle Aspirate: Report on 269 Cases. *J. Clin. Med.* **2020**, *9*, 283. [CrossRef]
9. Willemze, R.; Cerroni, L.; Kempf, W.; Berti, E.; Facchetti, F.; Swerdlow, S.H.; Jaffe, E.S. The 2018 update of the WHO-EORTC classification for primary cutaneous lymphomas. *Blood* **2019**, *133*, 1703–1714. [CrossRef]
10. Willemze, R.; Meijer, C.J. Classification of cutaneous T-cell lymphoma: From Alibert to WHO-EORTC. *J. Cutan. Pathol.* **2006**, *33*, 18–26. [CrossRef]
11. Walia, R.; Yeung, C. An Update on Molecular Biology of Cutaneous T Cell Lymphoma. *Front. Oncol.* **2020**, *9*, 1558. [CrossRef] [PubMed]
12. Ferreira, C.R.; Zhao, S.; Sanches, J.A.; Miyashiro, D.; Cury-Martins, J.; Azevedo, R.S.; Zerbini, M.C.N.; Natkunam, Y.; Gratzinger, D. Clinicopathologic and microenvironmental analysis of primary cutaneous CD30-positive lymphoproliferative disorders: A 26 year experience from an academic medical center in Brazil. *Diagn. Pathol.* **2019**, *14*, 115–118. [CrossRef] [PubMed]
13. Martires, K.J.; Ra, S.; Abdulla, F.; Cassarino, D.S. Characterization of primary cutaneous CD8+/CD30+ lymphoproliferative disorders. *Am. J. Dermatopathol.* **2015**, *37*, 822–833. [CrossRef] [PubMed]
14. Kempf, W. A new era for cutaneous CD30-positive T-cell lymphoproliferative disorders. *Semin. Diagn. Pathol.* **2017**, *34*, 22–35. [CrossRef]
15. Szpor, J.; Dyduch, G.; Gałazka, K.; Bahyrycz, J.; Stój, A.; Tomaszewska, R. Primary cutaneous CD30+ lymphoproliferative disorder–a 10-year follow-up. A case report and differential diagnosis. *Pol. J. Pathol.* **2009**, *60*, 43–48. [PubMed]
16. Kempf, W.; Kerl, K.; Mitteldorf, C. Cutaneous CD30-positive T-cell lymphoproliferative disorders—Clinical and histopathologic features, differential diagnosis, and treatment. *Semin. Cutan. Med. Surg.* **2018**, *37*, 24–29. [CrossRef]
17. Rezania, D.; Sokol, L.; Cualing, H.D. Classification and treatment of rare and aggressive types of peripheral T-cell/natural killer-cell lymphomas of the skin. *Cancer Control* **2007**, *14*, 112–123. [CrossRef]
18. Amber, K.T.; Bloom, R.; Nouri, K. Second Primary Malignancies in CTCL Patients from 1992 to 2011: A SEER-Based, Population-Based Study Evaluating Time from CTCL Diagnosis, Age, Sex, Stage, and CD30+ Subtype. *Am. J. Clin. Dermatol.* **2015**, *17*, 71–77. [CrossRef]
19. Fujii, K.; Hamada, T.; Shimauchi, T.; Asai, J.; Fujisawa, Y.; Ihn, H.; Katoh, N. Cutaneous lymphoma in Japan, 2012–2017: A nationwide study. *J. Dermatol. Sci.* **2020**, *97*, 187–193. [CrossRef]
20. Zhang, J.-C.; Wang, Y.; Wang, X.-F.; Zhang, F.-X. Type I enteropathy-associated T-cell lymphoma in the colon of a 29-year-old patient and a brief literature review. *OncoTargets Ther.* **2016**, *9*, 863–868. [CrossRef]
21. Phan, A.; Veldman, R.; Lechowicz, M.J. T-cell Lymphoma Epidemiology: The Known and Unknown. *Curr. Hematol. Malign. Rep.* **2016**, *11*, 492–503. [CrossRef] [PubMed]
22. Satou, A.; Asano, N.; Kato, S.; Katsuya, H.; Ishitsuka, K.; Elsayed, A.A.; Nakamura, S. FoxP3-positive T-cell Lymphoma Arising in non-HTLV1 carrier: Clinicopathological analysis of 11 cases of PTCL-NOS and 2 cases of Mycosis Fungoides. *Histopathology* **2015**, *68*, 1099–1108. [CrossRef] [PubMed]

23. Jawed, S.I.; Myskowski, P.L.; Horwitz, S.; Moskowitz, A.; Querfeld, C. Primary cutaneous T-cell lymphoma (mycosis fungoides and Sézary syndrome): Part I. Diagnosis: Clinical and histopathologic features and new molecular and biologic markers. *J. Am. Acad. Dermatol.* **2014**, *70*, 205-e1. [CrossRef] [PubMed]
24. Yumeen, S.; Girardi, M. Insights into the Molecular and Cellular Underpinnings of Cutaneous T Cell Lymphoma. *Yale J. Biol. Med.* **2020**, *93*, 111–121. [PubMed]
25. Wilcox, R.A. Cutaneous T-cell lymphoma: 2017 update on diagnosis, risk-stratification, and management. *Am. J. Hematol.* **2017**, *92*, 1085–1102. [CrossRef]
26. Wilcox, R.A. Cutaneous T-cell lymphoma: 2014 Update on diagnosis, risk-stratification, and management. *Am. J. Hematol.* **2014**, *89*, 837–851. [CrossRef]
27. Jiang, M.; Bennani, N.N.; Feldman, A.L. Lymphoma classification update: T-cell lymphomas, Hodgkin lymphomas, and histiocytic/dendritic cell neoplasms. *Expert Rev. Hematol.* **2017**, *10*, 239–249. [CrossRef]
28. Satou, A.; Bennani, N.N.; Feldman, A.L. Update on the classification of T-cell lymphomas, Hodgkin lymphomas, and histiocytic/dendritic cell neoplasms. *Expert Rev. Hematol.* **2019**, *12*, 833–843. [CrossRef]
29. Marchi, E.; O'Connor, O.A. The rapidly changing landscape in mature T-cell lymphoma (MTCL) biology and management. *CA A Cancer J. Clin.* **2019**, *70*, 47–70. [CrossRef]
30. Reddy, N.M.; Evens, A.M. Chemotherapeutic Advancements in Peripheral T-Cell Lymphoma. *Semin. Hematol.* **2014**, *51*, 17–24. [CrossRef]
31. Choi, E.-J.; Hong, J.Y.; Yoon, D.H.; Kang, J.; Park, C.-S.; Huh, J.; Chae, E.J.; Lee, Y.; Ryu, J.-S.; Suh, C. Treatment outcomes of dose-attenuated CHOP chemotherapy in elderly patients with peripheral T cell lymphoma. *Blood Res.* **2017**, *52*, 270–275. [CrossRef] [PubMed]
32. Fox, C.P.; Civallero, M.; Ko, Y.-H.; Manni, M.; Skrypets, T.; Pileri, S.; Kim, S.J.; Cabrera, M.E.; Shustov, A.R.; Chiattone, C.S.; et al. Survival outcomes of patients with extranodal natural-killer T-cell lymphoma: A prospective cohort study from the international T-cell Project. *Lancet Haematol.* **2020**, *7*, e284–e294. [CrossRef]
33. Tanase, A.D.; Colita, A.; Craciun, O.G.; Lipan, L.; Varady, Z.; Stefan, L.; Ranete, A.; Pasca, S.; Bumbea, H.; Andreescu, M.; et al. Allogeneic Stem Cell Transplantation for Adult T-Cell Leukemia/Lymphoma—Romanian Experience. *J. Clin. Med.* **2020**, *9*, 2417. [CrossRef] [PubMed]
34. Colita, A.; Colita, A.; Bumbea, H.; Croitoru, A.; Orban, C.; Lipan, L.E.; Craciun, O.-G.; Soare, D.; Ghimici, C.; Manolache, R.; et al. LEAM vs. BEAM vs. CLV Conditioning Regimen for Autologous Stem Cell Transplantation in Malignant Lymphomas. Retrospective Comparison of Toxicity and Efficacy on 222 Patients in the First 100 Days after Transplant, On Behalf of the Romanian Society for Bone Marrow Transplantation. *Front. Oncol.* **2019**, *9*. [CrossRef]
35. Grewal, R.-K.; Chetty, M.; Abayomi, E.-A.; Tomuleasa, C.; Fromm, J.R. Use of flow cytometry in the phenotypic diagnosis of hodgkin's lymphoma. *Cytom. Part B Clin. Cytom.* **2018**, *96*, 116–127. [CrossRef]
36. Grewal, R.; Irimie, A.; Naidoo, N.; Mohamed, N.; Petrushev, B.; Chetty, M.; Tomuleasa, C.; Abayomi, E.-A. Hodgkin's lymphoma and its association with EBV and HIV infection. *Crit. Rev. Clin. Lab. Sci.* **2018**, *55*, 102–114. [CrossRef]

diagnostics

MDPI

Case Report

Somatostatinoma and Neurofibromatosis Type 1-A Case Report and Review of the Literature

Sorina Martin [1,2,*], Simona Fica [1,2], Ovidiu Parfeni [2], Liliana Popa [3,4], Teodora Manuc [5], Oana Rizea [6,7], Ioana Lupescu [6,7], Mirela Gherghe [6,8,*], Gabriel Becheanu [9,10] and Adina Croitoru [11,12]

1 Endocrinology Department, Carol Davila University of Medicine and Pharmacy, 020021 Bucharest, Romania; simonafica55@gmail.com
2 Endocrinology Department, Elias Hospital, 011461 Bucharest, Romania; parfeniovidiu@gmail.com
3 Dermatology Department, Carol Davila University of Medicine and Pharmacy, 020021 Bucharest, Romania; lilidiaconu@yahoo.com
4 Dermatology Department, Elias Hospital, 011461 Bucharest, Romania
5 Gastroenterology Department, Fundeni Clinical Institute, 022328 Bucharest, Romania; teodora.manuc@gmail.com
6 Radiology Department, Carol Davila University of Medicine and Pharmacy, 020021 Bucharest, Romania; rizea.oana@gmail.com (O.R.); ioana.lupescu@umfcd.ro (I.L.)
7 Radiology Department, Fundeni Clinical Institute, 022328 Bucharest, Romania
8 Nuclear Medicine Department, Alexandru Trestioreanu Oncological Institute, 022328 Bucharest, Romania
9 Pathology Department, Carol Davila University of Medicine and Pharmacy, 020021 Bucharest, Romania; gbecheanu@yahoo.com
10 Laboratory of Histopathology and Immunohistochemistry, Victor Babes National Institute of Pathology, 050096 Bucharest, Romania
11 Oncology Department, Titu Maiorescu University of Medicine, 040051 Bucharest, Romania; adina.croitoru09@yahoo.com
12 Oncology Department, Fundeni Clinical Institute, 022328 Bucharest, Romania
* Correspondence: sorina.martin@gmail.com (S.M.); mirela_gherghe@yahoo.com (M.G.); Tel.: +0040-722-71-4897 (S.M.); +0040-72-054-4679 (M.G.)

Received: 18 July 2020; Accepted: 17 August 2020; Published: 21 August 2020

Abstract: Somatostatinomas are rare neuroendocrine tumors (NET) that arise in the gastrointestinal (GI) tract. Because of their insidious growth, they are usually asymptomatic until late stages, presenting as malignant disease. We report the case of a 50-year-old woman who presented with epigastric abdominal pain, diarrhea and significant weight loss in the last two years. On clinical examination the patient met the criteria for neurofibromatosis type 1 (NF1). Abdominal CT and MRI revealed an infiltrative duodenal mass, with pancreatic invasion, locoregional enlarged lymph nodes and disseminated hepatic nodules. Microscopy and immunohistochemistry uncovered a neuroendocrine tumor, staining positive for chromogranin A (CgA), synaptophysin and somatostatin, with a Ki67 = 1%. Somatostatin receptors (SSTRs) type 2 were negative and SSTRs type 5 were positive in less than 50% of tumoral cells. Our patient was classified as a T3N1M1 stage IV metastatic duodenal grade 1 somatostatinoma and treatment with somatostatin analogues and chemotherapy with capecitabine and temozolomide was started, with so far abdominal imaging follow-up showing stable disease. When a patient is diagnosed with a rare NET, such as a somatostatinoma, it is of utmost importance to determine if it is a sporadic tumor or just a feature of a genetic disorder.

Keywords: neuroendocrine tumor; somatostatinoma; neurofibromatosis type 1; somatostatin analogues; chemotherapy; capecitabine; temozolomide

1. Introduction

Somatostatinomas are neuroendocrine tumors of D-cell origin that contain and sometimes secrete somatostatine [1]. The estimated incidence of this tumor is about one in 40 million [2] and it represents about 4% of gastrointestinal neuroendocrine tumors (GI-NETs) [3]. Approximately 55% of somatostatinomas are in the pancreas and the reminder arise in the ampulla and periampullary region of the duodenum and rarely in the jejunum [4]. Usually, somatostatinomas present as a metastatic disease, 75% of them being malignant tumors [5]. Somatostatinomas can occur in association with multiple endocrine neoplasia (MEN)-1, neurofibromatosis type 1 (NF1), tuberous sclerosis and von-Hipple Lindau syndrome (VHL), and a new syndrome, via hypoxia-inducible factor 2 somatic mutation, which includes somatostatinoma, paraganglioma/pheochromocytoma and polycythemia, found only in females, had been also described [6].

NF1 is the most common hereditary multitumor syndrome with an incidence at birth of approximately 1:3000 and it is caused by mutations in the NF1 gene. NF1 is localized on chromosome band 17q11.2 and it acts as a tumor suppressor gene. Biallelic NF1 inactivation results in complete loss of functional neurofibromin activity. Neurofibromin acts as a negative regulator of the RAS-MAPK pathway, and it has a GTPase activating domain that interacts with RAS and converts active RAS-GTP to its inactive form. Apart from NF1 promotor hypermethylation, other somatic inactivation events occur in tumors associated with NF1, such as intragenic mutations and loss of heterozygosity [7]. Persons with NF1 develop both benign and malignant tumors at increased frequency throughout life, neurofibromas being the most common type of benign tumor. Optic pathway gliomas are the predominant type of intracranial neoplasms and can lead to endocrine disorders, mostly in pediatric population [8]. The frequency of intrabdominal manifestations of NF1 ranges from 5 to 25%, and they appear during midlife, usually later than cutaneous manifestations. NF1 associates multiple gastrointestinal stromal tumors and neuroendocrine tumors of the adrenal, duodenal ampulla, pancreas and other sites [7]. Approximately 10% of the patients with NF1 develop somatostatinomas, which are typically located in the duodenum, are rarely associated with somatostatinoma syndrome and are less likely to metastasize as compared with sporadic somatostatinomas [9]. Patients presenting with this tumor are usually over 50, with a roughly equal gender distribution [2].

Somatostatin is a tetradecapeptide that inhibits numerous endocrine hormones, including insulin; pancreatic polypeptide; cholecystokinin; gastrin; glucagon; gastric inhibitory polypeptide; secretin; motilin; and exocrine functions such as gastrointestinal transit time, intestinal motility and absorption of nutrients from the small intestine. The classic somatostatinoma syndrome is described as an association of weight loss; abdominal pain; and less often diabetes mellitus, cholelithiasis and diarrhea/ steatorrhea. This syndrome occurs more often in tumors that are localized in the pancreas than those in the duodenum [9]. Most tumors appear to be well-differentiated islet cell or carcinoid-type tumors on light microscopy and they have positive somatostatin and chromogranin A (CgA) immunohistochemistry. Sporadic and NF1 related duodenal somatostatinomas frequently present with psammoma bodies [10].

Somatostatinomas are usually found during exploratory laparotomy, computed tomography (CT), magnetic resonance imaging (MRI), ultrasound or gastrointestinal endoscopy studies that are performed due to various symptoms and signs associated with somatostatinoma syndrome. The diagnosis can also be established by the presence of a fasting plasma somatostatin level exceeding 30 pg/ml [9,11]. Imaging can localize the tumor and stage the extent of disease. The evaluation of patients with somatostationoma begins with a helical, multiphasic, contrast-enhanced CT. MRI is performed when CT shows indeterminate lesions that need further characterization. Endoscopic ultrasound (EUS) can also be performed in patients with inconclusive CT or MRI. Somatostatin receptor scintigraphy or Ga-68 DOTATATE positron emission tomography (PET)/CT represent high sensitivity imaging evaluation methods and are performed if the finding of extra-abdominal metastases would change the treatment [12,13].

Usually, at diagnosis, patients present with metastatic disease, but survival rate is high, especially in those with metastatic duodenal somatostatinoma [6]. In cases of nonmetastatic somatostatinoma the surgical removal of the tumor represents the only chance of cure. Surgery can also be used in metastatic or locoregional extensive disease, for debulking, to provide symptomatic relief and prolong survival rate [14,15]. Metastatic liver disease can be treated by radiofrequency ablation, cryoablation, hepatic artery embolization or chemoembolization with doxorubicin, cisplatin and mitomycin C and selective intraarterial irradiation with yttrium labeled microspheres, alone or in combination with surgery [15,16]. Somatostatin analogues (SSAs) inhibit somatostatin secretion, providing relief of symptoms such as diarrhea, diabetes and weight loss and they are used for patients with metastatic or inoperable somatostatinoma [17–19]. It had been shown that lanreotide is associated with significantly prolonged progression free survival (PFS) compared to placebo [20]. Molecularly targeted therapy (everolimus, sunitinib) is used for patients with progressive advanced somatostatinomas. Conventional chemotherapy (CHT) combined with SSAs has been used as an initial treatment for highly symptomatic disease due to tumor bulk or for those who have rapidly enlarging metastases, but the experience with systemic chemotherapy in patients with somatostatinomas is limited [9]. Radiotherapy using SSAs coupled with Lutetium-177 represent a promising approach in the treatment of NETs and was associated with best tumor response, especially in patients with a high uptake on Octreoscan, limited hepatic tumor mass and a high Karnofsky performance score [20].

There are not enough data on somatostatinomas to give accurate estimates of survival, but the prognosis with any therapy is poor in patients with metastatic disease. The follow-up extends to a maximum of 10 years post-resection, and it should consist of history and physical examination, a check of the fasting plasma somatostatin level and imaging studies every 6 to 12 months or more often in the first year [9].

2. Case Presentation

We report the case of a 50 years old woman, without any significant personal or family history, who presented with epigastric abdominal pain, diarrhea and 30 kg weight loss in the last 2 years. The patient signed a written informed consent and the ethics committee of Elias Hospital approved the publication of this case report (no 5316/2020). The clinical examination revealed elevated blood pressure (160/100 mmHg) and the presence of a few café au lait spots located on the upper limbs; large, poorly circumscribed areas of hyperpigmentation on the trunk; bilateral axillary freckling (Figures 1 and 2); and multiple asymptomatic, soft, flesh-colored, sessile or dome-shaped cutaneous tumors, 1–2 cm in diameter, nonadherent to the underlying structures, distributed on the trunk and limbs, highly suggestive of neurofibromas (Figure 3). Several painless, violaceous, subcutaneous pseudoatrophic macules, with a rubbery consistency, ranging in size from 5 mm to 2 cm, typically found in NF1, were also present on the abdominal wall (Figure 4).

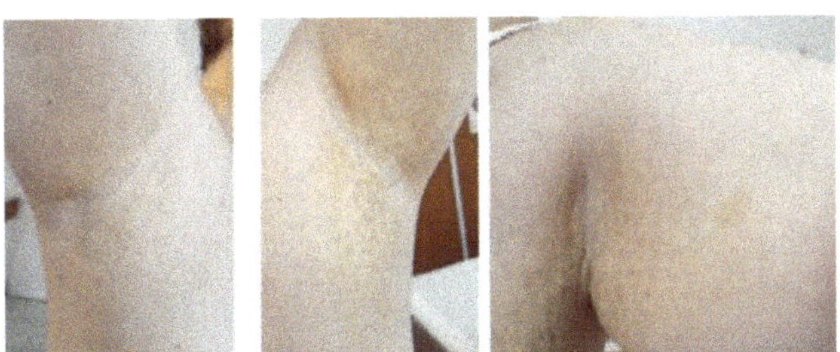

Figure 1. Bilateral axillary freckling and and café au lait spot on the left arm (scale 1:13).

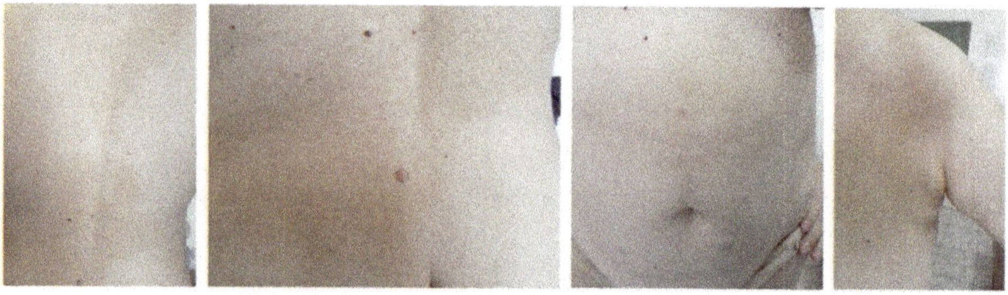

Figure 2. Large, poorly circumscribed areas of cutaneous hyperpigmentation and freckling on the trunk (scale 1:13/1:6/1:13/1:13).

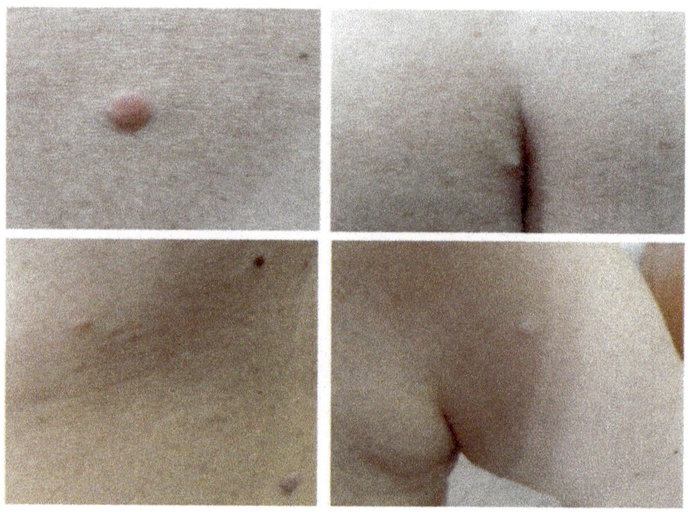

Figure 3. Multiple cutaneous neurofibromas distributed on the trunk and limbs (scale 1:7/1:13/1:11/1:11).

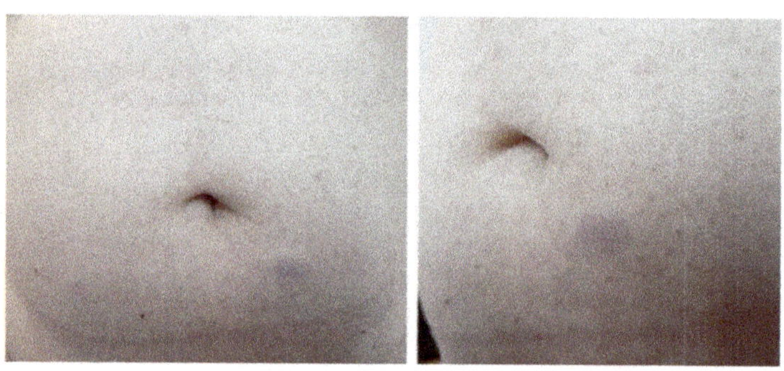

Figure 4. Asymptomatic, violaceous subcutaneous tumors in the abdominal area-pseudoatrophic macules typically found in NF1 (scale 1:11/1:8).

The mucous membranes did not show any pathologic changes and Lisch nodules were absent. Based upon the presence of three out of seven characteristic clinical features, our patient met diagnostic criteria developed by the United States National Institutes of Health (NIH) Consensus Conference in 1987 and updated in 1997 and was clinically diagnosed with neurofibromatosis type 1 (NF1) [21].

Biological evaluation was unremarkable except for diabetes mellitus (HbA1c% = 7%, fasting glycemia = 136mg/dl) and vitamin D deficiency (25 hydroxy-vitamin D = 10.53 ng/mL).

Upper digestive endoscopy revealed a gastric mucosa with prepyloric, circumferential linear and round-oval ulcerations, the largest of 6 mm, with the base covered with fibrin debris and an infiltrated area of 7 mm that could be seen between the first and second portion of the duodenum, on the medial wall, with rigidity and exaggerated friability upon biopsy. The normal papilla was visualized distally to the infiltrated area. A multiphasic contrast-enhanced CT diagnosed a 32 × 37 × 47 mm infiltrative duodenal mass, with heterogenous contrast enhancement, which narrowed the duodenal lumen and compressed the pancreatic head, and multiple low-density hepatic nodules strongly suggestive of metastatic disease. On MRI, the tumor displayed irregular borders with high contrast subtraction and heterogenous texture; it was located near the ampulla, in the second segment of the duodenum, with pancreatic invasion, associated locoregional enlarged lymph nodes and multiple disseminated hepatic nodules, with maximum diameters of 12 × 10 mm, suggestive of miliary metastatic disease (Figures 5 and 6).

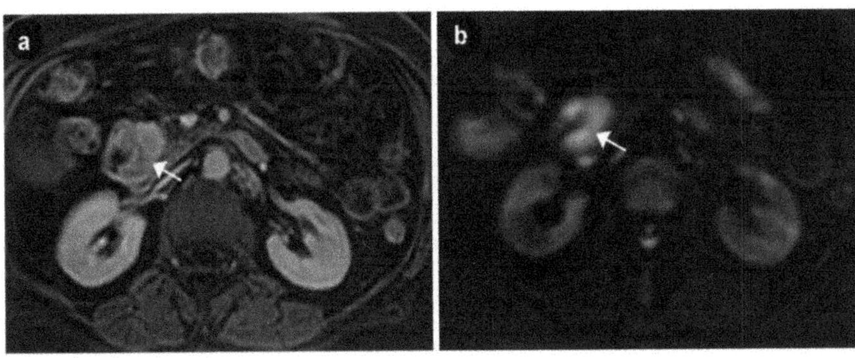

Figure 5. Axial MRI exam—duodenal tumor. (**a**) Gadolinium enhanced T1w image showing an hyperintense lesion in the arterial phase suggestive of the neuroendocrine etiology (arrow). (**b**) Diffusion Weighted Imaging (DWI) showing an hyperintense lesion with intense diffusion restriction (arrow).

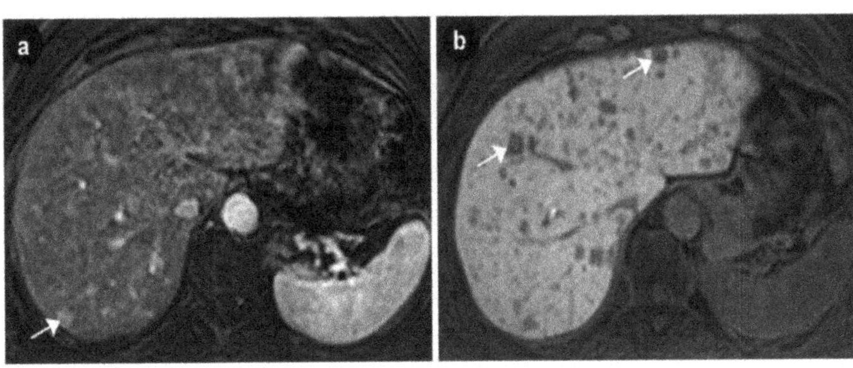

Figure 6. Axial MRI exam—multiple hepatic nodular lesions suggesting miliary metastic disease. (**a**) Gadolinium enhanced T1w showing hypervascular lesions (arrow). (**b**) T1w image, hepato-biliary phase after gadolinium enhancemet, showing hypointense lesions suggestive of the tumoral etiology (arrows).

Pathological exam of the duodenal biopsy described in our case a tumor with glandular and cribriform structures in the *lamina propria*, with an infiltrative aspect, focal foveolar metaplasia of the villous epithelia (Figure 7), cribriform structures with cuboidal or low columnar epithelial cells, with eosinophilic cytoplasm and round, central, monotonous nuclei, without visible nucleoli (Figure 8) and the tumor had a glandular (acinar) growth pattern with calcospherites (psammoma bodies). Immunohistochemistry described a tumor positive for chromogranin A, synaptophysin and somatostatin with a Ki67 of 1% (Figure 9). Somatostatin receptors (SSTR) type 2 were negative and SSTR type 5 were positive in less than 50% of tumoral cells.

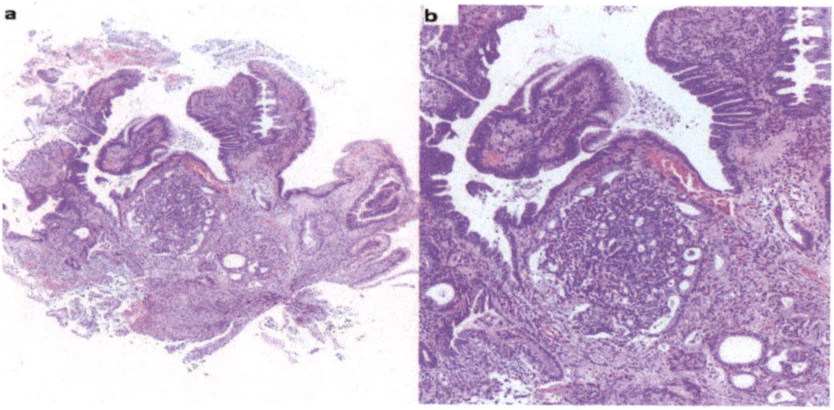

Figure 7. Histological findings. (**a**) HE, 40X. Duodenal biopsy with glandular and cribriform structures in the lamina propria, with the infiltrative aspect. (**b**) HE, 100X. Detail of the previous picture with cribriform structures and focal foveolar metaplasia of the villous epithelia.

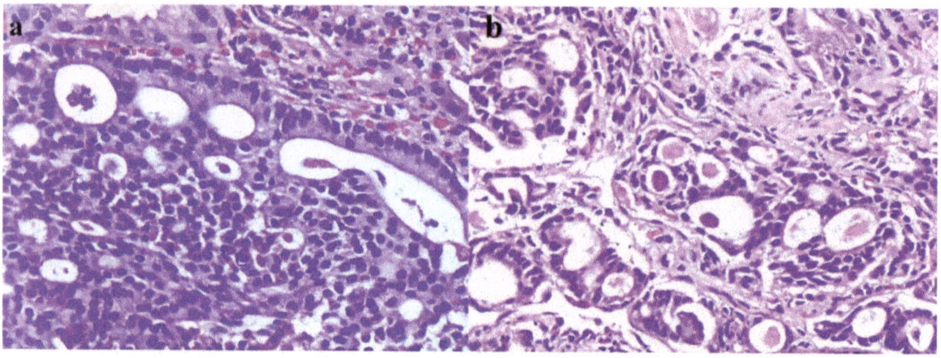

Figure 8. Histological findings. (**a**) HE, 200X. Cribriform structure with cuboidal or low columnar epithelial cells, with eosinophilic cytoplasm and round, central, monotonous nuclei, without visible nucleoli. Some necrotic cells and calcospherites are visible in the glandular lumina. (**b**) Glandular (acinar) growth pattern of the tumor with calcospherites (psammoma bodies).

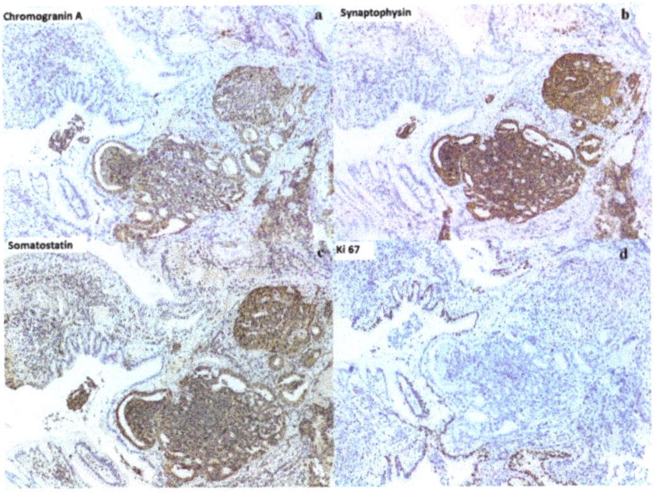

Figure 9. Imnuohistochemistry findings 40X. (**a**) Chromogranin A (Cell Marque, monoclonal, dilution 1:100): positive reaction of the tumoral cells, cytoplasmic, with internal positive control. (**b**) Synaptophysin (Immunopath, monoclonal, dilution 1:500): positive reaction of the tumoral cells, cytoplasmic, with internal positive control. (**c**) Somatostatin (Abcam, monoclonal, dilution 1:200): positive reaction of the tumoral cells, cytoplasmic, with internal positive control. (**d**) Ki67 (Leika, monoclonal, dilution 1:100): 1% in tumoral cells, with internal positive control in cryptic epithelial cells.

Our patient was evaluated using the radiopharmaceutical labelled with 99 mTc, available at the moment of diagnosis; 740 MBq 99mTc-EDDA/HYNIC-Tyr3-Octreotide was injected intravenously, with an acquisition protocol of early (1 h after injection) and late (4 h after injection) whole body scans with a two-headed, large field-of-view gamma camera equipped with low energy high resolution (LEHR) collimator and SPECT-CT acquisition 2 h after injection, on the region of interest (abdomen). On somatostatin receptor scintigraphy (SRS), SPECT-CT acquisition, we noticed the duodenal tumor compressing the pancreatic head and the particular aspect of very small disseminated nodules in both hepatic lobes, but without any uptake of the specific radiopharmaceutical; the aspect correlates with the immunohistochemistry results (Figures 10 and 11).

In this context we performed an endocrinological screening for MEN1 and pheochromocytoma which were negative. Plasma chromogranin A and 24 h urinary 5-hydroxyindoleacetic acid (5-HIAA) were in normal ranges. Unfortunately, plasma levels of somatostatin could not be measured at diagnosis.

According to the American Joint Committee on Cancer TNM staging system for neuroendocrine tumors of the duodenum and ampulla of Vater [22], our patient was classified as a T3, N1, M1 stage IV metastatic duodenal, grade 1, SSTR negative somatostatinoma. The patient was discussed in our multidisciplinary team meeting. NCCN guidelines version 1.2019 recommend for clinically significant tumor burden: lanreotide or octreotide and/or everolimus, peptide receptor radionuclide therapy with radiolabeled somatostatin analogues (PRRT), hepatic-directed therapy or interferon alfa-2b or cytotoxic chemotherapy if no other options feasible [23]. The MRI appearance of miliary liver metastases, with significant tumor burden, led us to decide on the association of SSAs with chemotherapy, as in our country we did not have other available reimbursed therapeutic options. We began the treatment with lanreotide Autogel 120 mg every 28 days and capecitabine 750 mg/m2/BID on days 1–14, and temozolomide 200 mg/m2/ID on days 10–14 every 28 days (CAPTEM). Follow-up at 3 (Figure 12), 6, 9 and 12 months with abdominal MRI and thorax CT showed stable disease. In May 2020, follow-up CT scan at 16 months from diagnosis showed a 14 × 8 mm segment 6 liver lesion, suggestive of a new metastasis, which needed to be monitored (Figure 13).

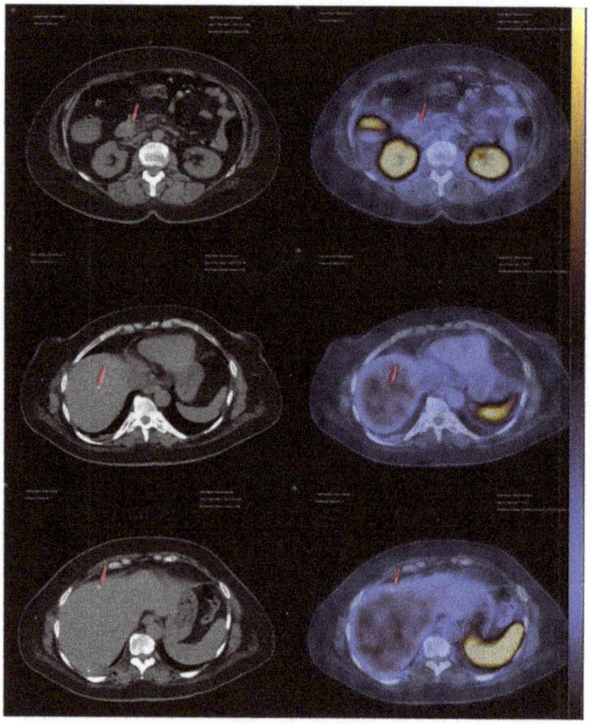

Figure 10. Axial unenhanced CT and fused SPECT-CT images showing duodenal tumor and multiple small hepatic metastasis with no uptake of 99mTc-EDDA/HYNIC-Tyr3-Octreotide (red arrows).

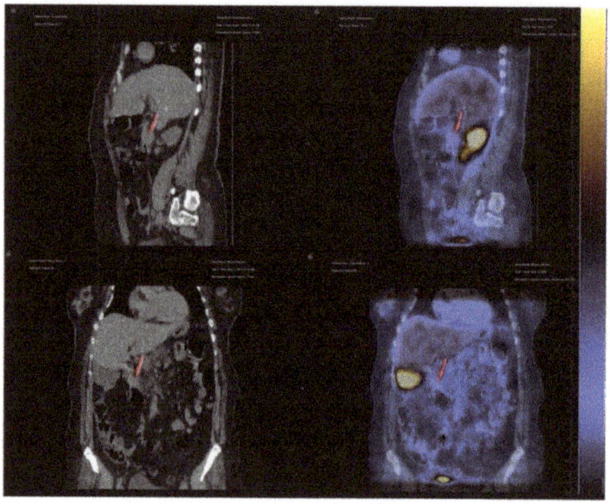

Figure 11. Sagital and coronal CT and SPECT-CT acquisition; duodenal tumor without 99mTc-EDDA/HYNIC-Tyr3-Octreotide uptake, suggesting no expression of SSTR2 or SSTR5 (red arrows).

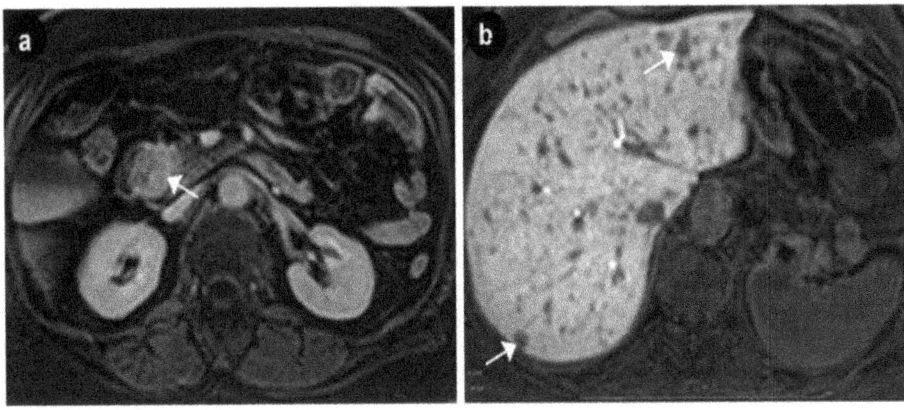

Figure 12. Follow up MRI at 3 months. (**a**) Slight regression of the duodenal tumor on T1w image after gadolinium enhancement (arrow). (**b**) Minimal regression of the metastatic hepatic lesions on the hepato-biliary phase (arrows).

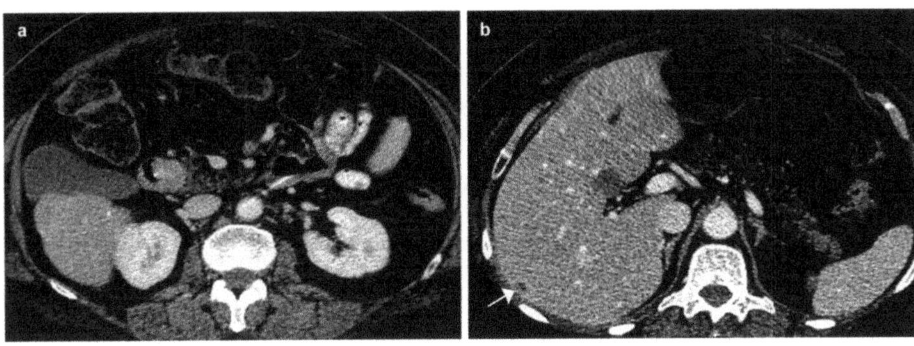

Figure 13. Follow up CT scan at 16 months from diagnosis (**a**) CECT - arterial phase- stable duodenal tumor. (**b**) CECT-portal phase-nodular lesion located in the sixth segment of the liver with an enhancing peripheral rim and a hypovascular center suggestive of a new metastasis (arrow).

There was a very good tolerability of both SSAs and chemotherapy. The patient presented grade 2 thrombocytopenia at cycles 3 and 4 and grade 2 palmar-plantar erythrodysesthesia at cycles 3, 4 and 5. Diarheea disappearead after the first two months of therapy.

3. Discussion

NF1 is an autosomal dominant disorder with complete penetrance and variable expression, associated with an increased risk of developing benign and malignant tumors of the skin, central and peripheral nervous system, adrenals and gastrointestinal tract, including pancreatic tumors [7,24]. The NF1 gene mutations determine a reduction or a complete loss of function of the cytoplasmatic protein, neurofibromin, which regulates cellular proliferation by inactivating p21 RAS and the MAP kinase [7,25]. The diagnosis of NF1 is based upon the presence of characteristic clinical features. According to NIH criteria, at least two of the following clinical features must be present to make the diagnosis of NF1: six or more café au lait macules (0.5 cm in children or >1.5 cm in adults), two or more cutaneous/subcutaneous neurofibromas or one plexiform neurofibroma, axillary or groin freckling, optic pathway glioma, two or more Lisch nodules (iris hamartomas seen on slit lamp examination) and bony dysplasia (sphenoid wing dysplasia, bowing of long bone ± pseudoarthrosis), first degree relative with NF1 [21]. As in our case, approximately 50% of NF1 patients have no family history of disease because the NF1 gene has high mutation rates in humans [26]. The multidisciplinary team management of this particular case also included genetic counseling. Genetic testing can be considered for a molecular diagnosis and can help direct screening of family members. Often, genetic testing is not required, but can it be a useful tool in confirming the diagnoses of children who do not meet diagnostic criteria or only exhibit café-au-lait macules and axillary freckling. Although of great academic interest, given the unquestionable clinical diagnosis, the fact that our patient did not have any first degree relatives that could benefit from the testing and that a positive NF1 mutation result is not a predictor of the severity or complications of the disorder, with some specific exceptions, we decided not to perform a genetic testing. The disorder is associated with an 8–15-year reduction in average life expectancy in both genders, primarily due to malignant neoplasms and cardiovascular disorders. Malignant peripheral nerve sheath tumor, breast cancer, cutaneous neurofibromas and significant psychiatric and neurologic diagnoses are common problems in patients with NF1 [27]. Patients with NF1 are also prone to developing gastrointestinal and retroperitoneal lesions, such as true neurogenic neoplasms, interstitial cell of Cajal lesions, NETs and adrenal tumors (pheochromocytoma) [7]. In 1982, Cantor et al. first documented the association between duodenal somatostatinoma and NF1 [28]. The association was afterwards described as a common gastrointestinal manifestation in patients with NF1 [11,29]. A case report and review of the literature, published in 1995, described 29 cases of duodenal somatostatinoma unassociated with NF1 and 27 cases of duodenal somatostatinoma associated with NF1, out of which four were also found to have adrenal pheochromocytomas [30]. An analysis of 162 patients with somatostatinomas (81 pancreatic somatostatinomas vs. 81 duodenal somatostatinomas) found a statistically significant difference between these two groups, regarding the incidence of NF1 (20% vs. 43.2%) [4]. In a more recent analysis, Garbrecht et al. confirmed the occurrence of duodenal somatostatinoma in NF1 patients, but with a lower frequency (14%), and found no pancreatic somatostatinoma associated with NF1 [10]. A 2010 review of literature showed that 47% of periampullary tumors in NF1 patients were of neuroendocrine origin with 40% being reported as somatostatinoma, 6% as carcinoid and 1% as malignant endocrine tumor [31].

Somatostatinoma is a rare NET [9] that originates in the enteric endocrine cells [32]. More than half of them are located in the pancreas and the rest in the ampulla or periampullary region of the duodenum or rarely in the jejunum. The tumors can be classified as functioning or non-functioning, depending on the secretion of somatostatin and often can present as metastatic disease upon diagnosis [9]. Duodenal somatostatinomas frequently display obstructive jaundice, abdominal pain, cholelithiasis, vomiting and abdominal bleeding [11] and can be accompanied by a somatostatin syndrome (steatorrhea, diabetes mellitus, cholelithiasis, weight loss), usually only if they are larger than 4 cm in size [33]. In our case, although the tumor was larger than 4 cm, located near the duodenal ampulla, with pancreatic invasion, the patient did not develop obstructive jaundice or cholelithiasis, but, instead, she presented diarrhea, weight loss and secondary diabetes mellitus. Although high plasma somatostatin levels (>30 pg/mL) may be a valuable tool to establish the biological diagnosis of somatostatinoma [9] and our patient associated somatostatin syndrome, we were not able to measure the somatostatin levels before the treatment initiation. Cg A is a neuroendocrine protein located in the secretory vesicles of neurons and endocrine cells, plays an important role in vesicle formation, is the precursor for many functional peptides (vasostatin, pancreastatin, catestatin, parastatin, etc.) and can be measured in order to provide information about diagnosis, prognosis and follow-up in NETs [34,35]. However, Cg A concentration may be normal in some low proliferative index NETs, as was the case for our patient. Pheochromocytomas are associated with VHL disease, MEN and NF1. In patients with NF1 they can occur with a frequency of 0.1–5.7%, higher than in the general population [36]. Moreover, there is also a hypoxia-inducible factor 2 somatic mutation, detected in ampullary somatostatinomas as part of a syndrome of paraganglioma and somatostatinoma, associated with polycythemia and found only in females [6]. Our patient had normal plasma normetanephrines and metanephrines, but further regular screening for other malignancies associated with NF1 is mandatory [27].

Duodenal neuroendocrine tumors account for about 20% of gastrointestinal tract neuroendocrine tumors and may show G-cell differentiation (gastrinomas) or D-cell differentiation expressing somatostatin, termed somatostatinomas, if they are functional. Acinar and psammomatous somatostatinomas are commonly found in the periampullary region in patients with NF1 and represent one third of somatostatin expressing tumors. Sometimes patients with NF1 may present other mesenchymal lesions: neurofibromas and gastrointestinal stromal tumors. Biopsy was performed and light microscopy revealed the presence of psammoma bodies, which had been previously identified in 68% of duodenal somatostatinomas [11]. Immunohistochemistry showed positive staining for Cg A, synaptophysin and somatostatin, Ki67 % = 1%, SSTR5 positive in less than 50% of tumor cells and negative SSTR2. Somatostatin receptor-based imaging can provide useful information on overall tumoral burden and location, and positive imaging can also confirm the presence of SSTRs [37]. SRS is a molecular imaging procedure for NET diagnosis and staging; it is more sensitive and specific at the biological than anatomical level, in contrast to conventional imaging, which it complements; the detection rate was reported to be between 80–100% in different studies [38–40]. There are two radiopharmaceuticals available on the market 111In-pentetreotide ([111In-DTPA0]-octreotide) with affinity for SSTR 2 and SSTR 5 and 99 mTc-EDDA/HYNIC-Tyr3-Octreotide, with high affinity for SSTR 2, but with a lesser extent for SSTR 3 and SSTR 5. 68Ga-DOTATATE is superior to other imaging modalities [6], but it was not available; therefore, we performed a 99 mTc-tektrotyde scintigraphy with no tumoral uptake for radiolabeled SSA.

Conventional and functional imaging revealed in our case, on diagnosis, a stage IV duodenal somatostatinoma (T3, N1, M1) with pancreatic invasion, locoregional lymph nodes enlargement and multiple disseminated hepatic nodules described as miliary metastases. Although it seems unreasonable to include SSAs in the therapeutic approach of a mostly SSTR negative somatostatinoma, in patients with symptomatic disease, clinically significant tumor burden or progressive disease, there is evidence in favor of SSA therapy (octreotide LAR or lanreotide) that can improve the related symptoms by reducing somatostatin secretion and potentially control tumor growth [20,41]. The correct mechanism by which SSAs inhibit tumoral somatostatin secretion is poorly understood. A possible explanation could be that octreotide has a higher affinity for the somatostatin receptor than for natural somatostatin and the enzymatic degradation of synthetic SSAs is on a slower rate than that of natural somatostatin [19]. High expression of SSTR 2 in NETs was associated with improved PFS and overall survival (OS) in patients treated with SSAs. Yet, there was no clear correlation between high vs. low expression of SSTRs 1–5 and uptake on somatostatin scintigraphy [42]. In one study, three out of ten patients with SSTR 2 negative and SSTR 5 positive NETs were responsive to SSA therapy [43]. Novel strategies such as SSTR 2 receptor gene transfer might prove useful as new therapies for unresectable tumors [44]. Diarrhea and weight loss, presenting signs in our patient and components of the somatostatinoma syndrome, ameliorated in the first couple of months as a result of SSA therapy. Intriguing enough, some studies have shown that patients with metastatic or inoperable NETs and no radiolabeled analogue uptake on Octreoscan or 68Ga-DOTA-peptides PET/CT, may also respond well to SSA therapy [43]. Although SSAs are highly effective at improving the hormone hypersecretion-associated symptoms, there is limited experience and little evidence that SSA therapy inhibits tumor growth, especially in somatostatinomas. Considering the results of randomized controlled studies for metastatic midgut NETs (octreotide) and gastroenteropancreatic NETs (lanreotide) without hormone-mediated symptoms, a reduction in tumor growth is expected. Therefore, due to the miliary aspect of liver metastases that could lead to acute liver failure, we decided to add SSAs to chemotherapy, as first-line therapy.

Cytotoxic chemotherapy with 5-fluorouracil, capecitabine, dacarbazine, oxaliplatin, streptozocin and temozolomide can be used, in association with SSAs, in patients with progressive metastatic disease [26]. ESMO guidelines recommend CHT in advanced pancreatic NETs (pNETs) and in neuroendocrine neoplasms (NEN) G3 of any site [45]. A systematic review showed that patients treated with CHT for locally advanced or metastatic well-differentiated G1/G2 GI NETs experienced poor results (OR rate range 5.8–17.2%) [46]. The use of the CAPTEM regimen is associated with objective responses and promising survival outcomes in metastatic, well-differentiated, intermediate and high-grade NETs (two-year OS = 42%, median PFS = 10 months, median PFS = 17 months in patients who received CAPTEM in first line) [47]. More studies reported similar results regarding CAPTEM regimen in NEN, with significantly higher PFS, median OS, objective response rate and disease control rate in pNENs compared with non-pNENs [48,49]. Pancreatic NENs have a lower O (6)-methylguanine-DNA methyltransferase (MGMT) expression than non-pNENs, and this might be the rationale for a superior treatment response to CAPTEM regimen. Yet, the use of MGMT expression or promoter methylation as a sole predictor for response to CAPTEM is not recommended and more biomarkers are needed for clinical decision making [49,50].

CAPTEM regimen is rarely associated with serious toxicities (grade 3 or 4 stratified by Common Terminology Criteria for Adverse Events (CTCAE) v5.0 Grade) and has low discontinuation rates, even in patients who follow the treatment for more than a year [48]. The ENETs guidelines currently recommend this regimen for patients with intermediate and high-grade NETs as a second line therapy [51]. In a retrospective study, patients with metastatic intermediate and high-grade NETs (pancreatic and nonpancreatic included) treated with CAPTEM had median PFS of 15.9 months as compared to a median PFS of 3.1 months when subsequent lines of therapy were used [52]. Similarly, in a series of 30 patients with metastatic well or moderately differentiated pNETs who received CAPTEM in the first-line setting, the objective radiographic response rate was 70% with a median PFS of 18 months, a 2-year OS rate of 92% and a 12% rate of grade 3 and 4 toxicities [53]. This regimen clearly proves its superiority compared to other cytotoxic regimens (triplet combination of streptozocin, doxorubicin and 5-fluorouracil) with a response rate of 39%, a 2-year OS rate of 74% and a 23% rate of grade 3 and 4 toxicities [54].

RADIANT-4, a phase 3 study, found that treatment with everolimus is associated with significant improvement in PFS vs. placebo, in patients with advanced, progressive, well-differentiated, non-functional neuroendocrine tumors of lung or gastrointestinal origin [55]. The appropriate sequencing or combination of various drugs remains unclear and is mostly dependent on patient individual factors, such as comorbidities or side effects.

Ablative methods, including radiofrequency ablation and cryoablation, and hepatic artery chemoembolization with doxorubicin, cisplatin and mitomycin C, can result in tumor regression and control of the symptoms. Additionally, surgery is recommended for symptoms palliation [6]. PRRT is recommended for patients with inoperable or metastasized NETs. Lutetium-177 coupled with SSAs is a very promising approach in treatment of NETs [56]. The combined therapy with 177Lu-DOTATE plus octreotide LAR showed better tumor response rate vs. therapy with octreotide LAR alone [57]. Furthermore, combining PRRT with capecitabine and temozolomide (radiosensitizing chemotherapy) may be feasible with minimal incremental toxicity [58]. There are some inclusion criteria for PRRT, such as inoperable/metastatic, well differentiated (G1/G2) NET (well-differentiated G3 NET may be considered), sufficient tumor uptake on the diagnostic somatostatin scintigraphy, sufficient bone marrow reserves, creatinine clearance >50 mL/min, Karnofsky performance status >50, expected survival >3 months and signed informed consent [56]. In our case, due to no somatostatin analogue uptake on the scintigraphy, PRRT could not be considered.

4. Conclusions

In conclusion, the lesson to be learned from this case resides in the fact that somatostatinomas are rare NETs, seldom associated with somatostatinoma syndrome, which can sometimes occur in the settings of a genetic syndrome such as NF1. Despite the progress achieved in the management of metastatic, progressive, unresectable NETs, in general, and for somatostatinoma, in particular, prognosis is poor, and the optimal treatment of choice still remains a challenge for the multidisciplinary team involved.

Funding: The APC was funded by Carol Davila University of Medicine and Pharmacy, Bucharest based on the law 3132/30.01.2020.

Conflicts of Interest: The authors declare no conflict of interest. The funders had no role in the design of the study; in the collection, analyses, or interpretation of data; in the writing of the manuscript, or in the decision to publish the results.

References

1. Friesen, S.R. Update on the diagnosis and treatment of rare neuroendocrine tumors. *Surg. Clin. North. Am.* **1987**, *67*, 379–393. [CrossRef]
2. Harris, G.J.; Tio, F.; Cruz, A.B., Jr. Somatostatinoma: A case report and review of the literature. *J. Surg. Oncol.* **1987**, *36*, 8–16. [CrossRef] [PubMed]
3. Klöppel, G.; Anlauf, M. Epidemiology, tumour biology and histopathological classification of neuroendocrine tumours of the gastrointestinal tract. *Best Pract. Res. Clin. Gastroenterol.* **2005**, *19*, 507–517. [CrossRef] [PubMed]
4. Soga, J.; Yakuwa, Y. Somatostatinoma/inhibitory syndrome: A statistical evaluation of 173 reported cases as compared to other pancreatic endocrinomas. *J. Exp. Clin. Cancer Res.* **1999**, *18*, 13–22.
5. Doherty, G.M. Rare endocrine tumours of the GI tract. Best practice & research. *Clin. Gastroenterol.* **2005**, *19*, 807–817. [CrossRef]
6. Vinik, A.; Pacak, K.; Feliberti, E.; Perry, R.R. Somatostatinoma. In *Endotext [Internet]*; Feingold, K.R., Anawalt, B., Boyce, A., Chrousos, G., Dungan, K., Grossman, A., Hershman, J.M., Kaltsas, G., Koch, C., Kopp, P., et al., Eds.; MDText.com, Inc.: South Dartmouth, MA, USA, 2017.
7. Agaimy, A.; Vassos, N.; Croner, R.S. Gastrointestinal manifestations of neurofibromatosis type 1 (Recklinghausen's disease): Clinicopathological spectrum with pathogenetic considerations. *Int. J. Clin. Exp. Pathol.* **2012**, *5*, 852–862.
8. Santoro, C.; Perrotta, S.; Picariello, S.; Sicilipoti, M.; Cirillo, M.; Quaglietta, L.; Cinalli, G.; Cioffi, D.; Di lorgi, N.; Maghnie, M.; et al. Pretreatment endocrine disorders due to optic pathway gliomas in pediatric neurofibromatosis type 1: Multicenter study. *J. Clin. Endocrinol. Metab.* **2020**, *105*, dgaa138. [CrossRef]
9. Bergsland, E. Somatostatinoma: Clinical Manisfestations, Diagnosis, and Management. In Post; Nathan, D.M., Whitcomb, D.C., Grover, S., Eds.; UptoDate. Available online: https://www.uptodate.com/contents/somatostatinoma-clinical-manifestations-diagnosis-and-management (accessed on 19 September 2019).
10. Garbrecht, N.; Anlauf, M.; Schmitt, A.; Henopp, T.; Sipos, B.; Raffel, A.; Eisenberger, C.F.; Knoefel, W.T.; Pavel, M.; Fottner, C.; et al. Somatostatin-producing neuroendocrine tumors of the duodenum and pancreas: Incidence, types, biological behavior, association with inherited syndromes, and functional activity. *Endocrine-Relat. Cancer* **2008**, *15*, 229–241. [CrossRef]
11. Tanaka, S.; Yamasaki, S.; Matsushita, H.; Ozawa, Y.; Kurosaki, A.; Takeuchi, K.; Hoshihara, Y.; Doi, T.; Watanabe, G.; Kawaminami, K. Duodenal somatostatinoma: A case report and review of 31 cases with special reference to the relationship between tumor size and metastasis. *Pathol. Int.* **2000**, *50*, 146–152. [CrossRef]
12. Deppen, S.A.; Liu, E.; Blume, J.D.; Clanton, J.; Shi, C.; Jones-Jackson, L.B.; Lakhani, V.; Baum, R.P.; Berlin, J.; Smith, G.T.; et al. Safety and efficacy of 68Ga-DOTATATE PET/CT for diagnosis, staging, and treatment management of neuroendocrine tumors. *J. Nucl. Med.* **2016**, *57*, 708–714. [CrossRef]
13. Sadowski, S.M.; Neychev, V.; Millo, C.; Shih, J.; Nilubol, N.; Herscovitch, P.; Pacak, K.; Marx, S.J.; Kebebew, E. Prospective study of 68Ga-DOTATATE positron emission tomography/computed tomography for detecting gastro-entero-pancreatic neuroendocrine tumors and unknown primary sites. *J. Clin. Oncol.* **2016**, *34*, 588–596. [CrossRef] [PubMed]
14. Fendrich, V.; Langer, P.; Celik, I.; Bartsch, D.K.; Zielke, A.; Ramaswamy, A.; Rothmund, M. An aggressive surgical approach leads to long-term survival in patients with pancreatic endocrine tumors. *Ann. Surg.* **2006**, *244*, 845–853. [CrossRef] [PubMed]
15. Brentjens, R.; Saltz, L. Islet cell tumors of the pancreas: The medical oncologist's perspective. *Surg. Clin. North. Am.* **2001**, *81*, 527–542. [CrossRef]
16. King, J.; Quinn, R.; Glenn, D.M.; Janssen, J.; Tong, D.; Liaw, W.; Morris, D.L. Radioembolization with selective internal radiation microspheres for neuroendocrine liver metastases. *Cancer* **2008**, *113*, 921–929. [CrossRef] [PubMed]

17. Vinik, A.; Moattari, A.R. Use of somatostatin analog in management of carcinoid syndrome. *Dig. Dis. Sci.* **1989**, *34* (Suppl. S3), 14S–27S. [CrossRef]
18. Oberg, K.; Eriksson, B.; Lundqvist, M. Neuroendocrine tumours of the upper gastrointestinal tract and pancreas. *Acta Chir. Scand. Suppl.* **1988**, *541*, 76–85.
19. Angeletti, S.; Corleto, V.D.; Schillaci, O.; Marignani, M.; Annibale, B.; Moretti, A.; Silecchia, G.; Scopinaro, F.; Basso, N.; Bordi, C.; et al. Use of the somatostatin analogue octreotide to localise and manage somatostatin-producing tumours. *Gut* **1998**, *42*, 792–794. [CrossRef]
20. Caplin, M.E.; Pavel, M.; Ćwikła, J.B.; Phan, A.T.; Raderer, M.; Sedláčková, E.; Cadiot, G.; Wolin, E.M.; Capdevila, J.; Wall, L.; et al. Lanreotide in metastatic enteropancreatic neuroendocrine tumors. *N. Engl. J. Med.* **2014**, *371*, 224–233. [CrossRef]
21. Neurofibromatosis. Conference statement. National Institutes of Health Consensus Development Conference. *Arch. Neurol.* **1988**, *45*, 575–578.
22. Edge, S.B.; Compton, C.C.; Fritz, A.G. *The American Joint Committee on Cancer: The 7th Edition of the AJCC Cancer Staging Manual and the Future of TNM*; Greene, F.L., Trotti, A., Eds.; Springer: New York, NY, USA, 2010; ISBN 978-0-387-88442-4.
23. National Comprehensive Cancer Network. *Neuroendocrine and Adrenal Tumors (Version 1.2019)*; National Comprehensive Cancer Network: Plymouth Meeting, PA, USA, 2019.
24. Kumar, T.; Gupta, B.; Das, P.; Jain, D.; Jain, H.A.; Madhusudhan, K.S.; Dash, N.R.; Gupta, S.D. Combined presence of multiple gastrointestinal stromal tumors along with duodenal submucosal somatostatinoma in a patient with neurofibromatosis type 1. *Indian J. Pathol. Microbiol.* **2016**, *59*, 359–361. [CrossRef]
25. Lodish, M.B.; Stratakis, C.A. Endocrine tumours in neurofibromatosis type 1, tuberous sclerosis and related syndromes. *Best Pract. Res. Clin. Endocrinol. Metab.* **2010**, *24*, 439–449. [CrossRef] [PubMed]
26. Ferner, R.E. Neurofibromatosis 1 and neurofibromatosis 2: A twenty first century perspective. *Lancet Neurol.* **2007**, *6*, 340–351. [CrossRef]
27. Stewart, D.R.; Korf, B.R.; Nathanson, K.L.; Stevenson, D.A.; Yohay, K. Care of adults with neurofibromatosis type 1: A clinical practice resource of the American College of Medical Genetics and Genomics (ACMG). *Genet. Med.* **2018**, *20*, 671–682. [CrossRef] [PubMed]
28. Cantor, A.M.; Rigby, C.C.; Beck, P.R.; Mangion, D. Neurofibromatosis, phaeochromocytoma, and somatostatinoma. *BMJ* **1982**, *285*, 1618–1619. [CrossRef] [PubMed]
29. Cappelli, C.; Agosti, B.; Braga, M.; Cumetti, D.; Gandossi, E.; Rizzoni, D.; Agabiti Rosei, E. Von Recklinghausen's neurofibromatosis associated with duodenal somatostatinoma. A case report and review of the literature. *Minerva Endocrinol.* **2004**, *29*, 19–24.
30. Mao, C.; Shah, A.; Hanson, D.J.; Howard, J.M. Von Recklinghausen's disease associated with duodenal somatostatinoma: Contrast of duodenal versus pancreatic somatostatinomas. *J. Surg. Oncol.* **1995**, *59*, 67–73. [CrossRef]
31. Relles, D.; Baek, J.; Witkiewicz, A.; Yeo, C.J. Periampullary and duodenal neoplasms in neurofibromatosis type 1: Two cases and an updated 20-year review of the literature yielding 76 cases. *J. Gastrointest. Surg.* **2010**, *14*, 1052–1061. [CrossRef]
32. Adams, M.S.; Bronner-Fraser, M. Review: The role of neural crest cells in the endocrine system. *Endocr. Pathol.* **2009**, *20*, 92–100. [CrossRef]
33. Kim, J.A.; Choi, W.H.; Kim, C.N.; Moon, Y.S.; Chang, S.H.; Lee, H.R. Duodenal somatostatinoma: A case report and review. [published correction appears in Korean J. Intern. Med. 26 June 2011, 253]. *Korean J. Intern. Med.* **2011**, *26*, 103–107. [CrossRef]
34. Al-Risi, E.S.; Al-Essry, F.S.; Mula-Abed, W.S. Chromogranin A as a biochemical marker for neuroendocrine tumors: A single center experience at Royal Hospital, Oman. *Oman Med. J.* **2017**, *32*, 365–370. [CrossRef]
35. Gut, P.; Czarnywojtek, A.; Fischbach, J.; Bączyk, M.; Ziemnicka, K.; Wrotkowska, E.; Gryczyńska, M.; Ruchała, M. Chromogranin A—Unspecific neuroendocrine marker. Clinical utility and potential diagnostic pitfalls. *Arch. Med. Sci.* **2016**, *12*, 1–9. [CrossRef] [PubMed]
36. Dluhy, R.G. Pheochromocytoma—Death of an axiom. *N. Engl. J. Med.* **2002**, *346*, 1486–1488. [CrossRef] [PubMed]
37. Srirajaskanthan, R.; Kayani, I.; Quigley, A.M.; Soh, J.; Caplin, M.E.; Bomanji, J. The role of 68Ga-DOTATATE PET in patients with neuroendocrine tumors and negative or equivocal findings on 111In-DTPA-octreotide scintigraphy. *J. Nucl. Med.* **2010**, *51*, 875–882. [CrossRef] [PubMed]

38. Krenning, E.P.; Kwekkeboom, D.J.; Bakker, W.H.; Breeman, W.A.; Kooij, P.P.; Oei, H.Y.; van Hagen, M.; Postema, P.T.; de Jong, M.; Reubi, J.C. Somatostatin receptor scintigraphy with [111In-DTPA-D-Phe1]- and [123I-Tyr3]-octreotide: The Rotterdam experience with more than 1000 patients. *Eur. J. Nucl. Med.* **1993**, *20*, 716–731. [CrossRef]
39. Asnacios, A.; Courbon, F.; Rochaix, P.; Bauvin, E.; Cances-Lauwers, V.; Susini, C.; Schulz, S.; Boneu, A.; Guimbaud, R.; Buscail, L. Indium-111-pentetreotide scintigraphy and somatostatin receptor subtype 2 expression: New prognostic factors for malignant well-differentiated endocrine tumors. *J. Clin. Oncol.* **2008**, *26*, 963–970. [CrossRef]
40. Wong, K.K.; Cahill, J.M.; Frey, K.A.; Avram, A.M. Incremental value of 111-in pentetreotide SPECT/CT fusion imaging of neuroendocrine tumors. *Acad. Radiol.* **2010**, *17*, 291–297. [CrossRef]
41. Rinke, A.; Wittenberg, M.; Schade-Brittinger, C.; Aminossadati, B.; Ronicke, E.; Gress, T.M.; Müller, H.H.; Arnold, R.; PROMID Study Group. Placebo-controlled, double-blind, prospective, randomized study on the effect of octreotide LAR in the control of tumor growth in patients with metastatic neuroendocrine midgut tumors (PROMID): Results of long-term survival. *Neuroendocrinology* **2017**, *104*, 26–32. [CrossRef]
42. Qian, Z.R.; Li, T.; Ter-Minassian, M.; Yang, J.; Chan, J.A.; Brais, L.K.; Masugi, Y.; Thiaglingam, A.; Brooks, N.; Nishihara, R.; et al. Association between somatostatin receptor expression and clinical outcomes in neuroendocrine tumors. *Pancreas* **2016**, *45*, 1386–1393. [CrossRef]
43. Vezzosi, D.; Bennet, A.; Rochaix, P.; Courbon, F.; Selves, J.; Pradere, B.; Buscail, L.; Susini, C.; Caron, P. Octreotide in insulinoma patients: Efficacy on hypoglycemia, relationships with Octreoscan scintigraphy and immunostaining with anti-sst2A and anti-sst5 antibodies. *Eur. J. Endocrinol.* **2005**, *152*, 757–767. [CrossRef]
44. Baldelli, R.; Barnabei, A.; Rizza, L.; Isidori, A.M.; Rota, F.; Di Giacinto, P.; Paoloni, A.; Torino, F.; Corsello, S.M.; Lenzi, A.; et al. Somatostatin analogs therapy in gastroenteropancreatic neuroendocrine tumors: Current aspects and new perspectives. *Front. Endocrinol.* **2014**, *5*, 7. [CrossRef]
45. Pavel, M.; Öberg, K.; Falconi, M.; Krenning, E.P.; Sundin, A.; Perren, A.; Berruti, A.; ESMO Guidelines Committee. Gastroenteropancreatic neuroendocrine neoplasms: ESMO clinical practice guidelines for diagnosis, treatment and follow-up. *Ann. Oncol.* **2020**, *31*, 844–860. [CrossRef] [PubMed]
46. Lamarca, A.; Elliott, E.; Barriuso, J.; Backen, A.; McNamara, M.G.; Hubner, R.; Valle, J.W. Chemotherapy for advanced non-pancreatic well-differentiated neuroendocrine tumours of the gastrointestinal tract, a systematic review and meta-analysis: A lost cause? *Cancer Treat. Rev.* **2016**, *44*, 26–41. [CrossRef] [PubMed]
47. Sahu, A.; Jefford, M.; Lai-Kwon, J.; Thai, A.; Hicks, R.J.; Michael, M. CAPTEM in Metastatic well-differentiated intermediate to high grade neuroendocrine tumors: A single centre experience. *J. Oncol.* **2019**, *2019*, Article. [CrossRef] [PubMed]
48. Thomas, K.; Voros, B.A.; Meadows-Taylor, M.; Smeltzer, M.P.; Griffin, R.; Boudreaux, J.P.; Thiagarajan, R.; Woltering, E.A.; Ramirez, R.A. Outcomes of Capecitabine and Temozolomide (CAPTEM) in advanced Neuroendocrine Neoplasms (NENs). *Cancers* **2020**, *12*, 206. [CrossRef]
49. Owen, D.H.; Alexander, A.J.; Konda, B.; Wei, L.; Hemminger, J.A.; Schmidt, C.R.; Abdel-Misih, S.; Dillhoff, M.E.; Sipos, J.A.; Kirschner, L.S.; et al. Combination therapy with capecitabine and temozolomide in patients with low and high grade neuroendocrine tumors, with an exploratory analysis of O6-methylguanine DNA methyltransferase as a biomarker for response. *Oncotarget* **2017**, *8*, 104046–104056. [CrossRef]
50. Walter, T.; van Brakel, B.; Vercherat, C.; Hervieu, V.; Forestier, J.; Chayvialle, J.A.; Molin, Y.; Lombard-Bohas, C.; Joly, M.O.; Scoazec, J.Y. O6-Methylguanine-DNA methyltransferase status in neuroendocrine tumours: Prognostic relevance and association with response to alkylating agents. *Br. J. Cancer* **2015**, *112*, 523–531. [CrossRef]
51. Garcia-Carbonero, R.; Sorbye, H.; Baudin, E.; Raymond, E.; Wiedmann, B.; Niederle, B.; Sedlackova, E.; Toumpanakis, C.; Anlauf, M.; Cwifla, J.B.; et al. ENETS consensus guidelines for high-grade gastroenteropancreatic neuroendocrine tumors and neuroendocrine carcinomas. *Neuroendocrinology* **2016**, *103*, 186–194. [CrossRef]
52. Peixoto, R.D.; Noonan, K.L.; Pavlovich, P.; Kennecke, H.F.; Lim, H.J. Outcomes of patients treated with capecitabine and temozolamide for advanced pancreatic neuroendocrine tumors (PNETs) and non-PNETs. *J. Gastrointest. Oncol.* **2014**, *5*, 247–252. [CrossRef]
53. Strosberg, J.R.; Fine, R.L.; Choi, J.; Nasir, A.; Coppola, D.; Chen, D.T.; Helm, J.; Kvols, L. First-line chemotherapy with capecitabine and temozolomide in patients with metastatic pancreatic endocrine carcinomas. *C Cancer* **2011**, *117*, 268–275. [CrossRef]

54. Kouvaraki, M.A.; Ajani, J.A.; Hoff, P.; Wolff, R.; Evans, D.B.; Lozano, R.; Yao, J.C. Fluorouracil, doxorubicin, and streptozocin in the treatment of patients with locally advanced and metastatic pancreatic endocrine carcinomas. [published correction appears in J. Clin. Oncol. 1 January 2005, 23, 248]. *J. Clin. Oncol.* **2004**, *22*, 4762–4771. [CrossRef]
55. Fazio, N.; Buzzoni, R.; Delle Fave, G.; Tesselaar, M.E.; Wolin, E.; Van Cutsem, E.; Tomassetti, P.; Strosberg, J.; Voi, M.; Bubuteishvili-Pacaud, L.; et al. Everolimus in advanced, progressive, well-differentiated, non-functional neuroendocrine tumors: RADIANT-4 lung subgroup analysis. *Cancer Sci.* **2018**, *109*, 174–181. [CrossRef] [PubMed]
56. Hicks, R.J.; Kwekkeboom, D.J.; Krenning, E.; Bodei, L.; Grozinsky-Glasberg, S.; Arnold, R.; Borbath, I.; Cwikla, J.; Toumpanakis, C.; Kaltsas, G.; et al. ENETS consensus guidelines for the standards of care in neuroendocrine neoplasia: Peptide receptor radionuclide therapy with radiolabeled somatostatin analogues. *Neuroendocrinology* **2017**, *105*, 295–309. [CrossRef] [PubMed]
57. Strosberg, J.; El-Haddad, G.; Wolin, E.; Hendifar, A.; Yao, J.; Chasen, B.; Mittra, E.; Kunz, P.L.; Kulke, M.H.; Jacene, H.; et al. NETTER-1 trial investigators. Phase 3 trial of 177Lu-dotatate for midgut neuroendocrine tumors. *N. Engl. J. Med.* **2017**, *376*, 125–135. [CrossRef] [PubMed]
58. Claringbold, P.G.; Price, R.A.; Turner, J.H. Phase I-II study of radiopeptide 177Lu-octreotate in combination with capecitabine and temozolomide in advanced low-grade neuroendocrine tumors. *Cancer Biother. Radiopharm.* **2012**, *27*, 561–569. [CrossRef] [PubMed]

MDPI AG
Grosspeteranlage 5
4052 Basel
Switzerland
Tel.: +41 61 683 77 34

Diagnostics Editorial Office
E-mail: diagnostics@mdpi.com
www.mdpi.com/journal/diagnostics

MDPI

www.ingramcontent.com/pod-product-compliance
Lightning Source LLC
Chambersburg PA
CBHW061615210926
51850CB00009B/2329